D1690625

L'édition de ce Mémoire a été réalisée sous la direction de Philippe BOUCHET Directeur de la publication de 1986 à 1992.

L'équipe de rédaction et l'auteur sont donc heureux de lui dédier ce volume en remerciement de son dévouement et de sa compétence pendant ces six années.

The editing of this memoir was completed under the direction of Philippe BOUCHET, Editor-in-Chief from 1986 to 1992.

The editing team and the author wish to dedicate this volume to him in thanks for his devotion and competence during these six years.

The genus Chicoreus *and related genera
(Gastropoda: Muricidae) in the Indo-West Pacific*

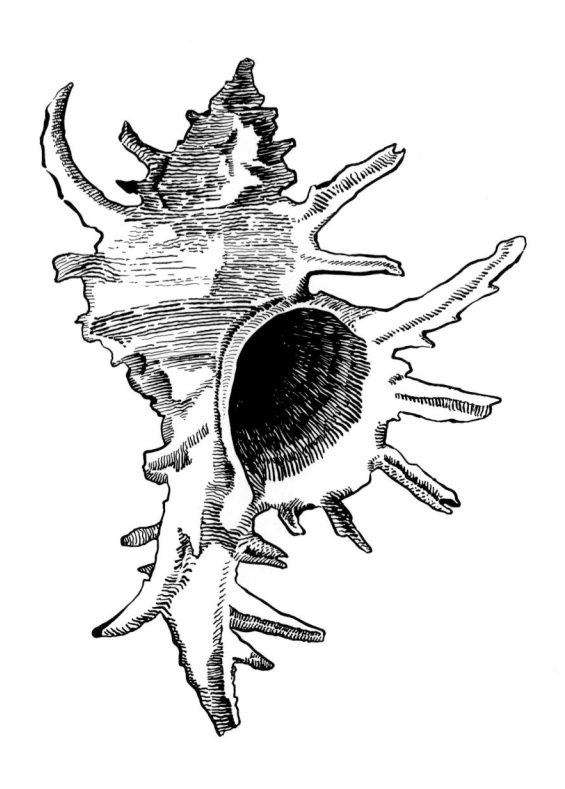

ISBN : 2-85653-194-6
ISSN : 0078-9747
© Éditions du Muséum national d'Histoire naturelle, Paris, 1992

MÉMOIRES DU MUSÉUM NATIONAL D'HISTOIRE NATURELLE

SÉRIE A
ZOOLOGIE
TOME 154

Roland HOUART

Institut Royal des Sciences
Naturelles de Belgique
29, rue Vautier
1040 Bruxelles
Belgique

The genus Chicoreus and related genera (Gastropoda: Muricidae) in the Indo-West Pacific

ÉDITIONS
DU MUSÉUM
PARIS

1992

SOMMAIRE
CONTENTS

	Pages
ABSTRACT	9
RÉSUMÉ	11
INTRODUCTION	13
Material, methods and text conventions	13
Excluded taxa	13
Geographical distribution	14
Taxonomic characters	14
Protoconch	14
Operculum	25
Radula	25
Habitat	25
ABBREVIATIONS	32
SYSTEMATICS	34
Genus *Chicoreus*	34
Subgenus *Chicoreus*	36
Subgenus *Triplex*	46
Subgenus *Rhizophorimurex*	106
Subgenus *Siratus*	109
Subgenus *Chicopinnatus*	112
Genus *Chicomurex*	115
Genus *Naquetia*	125
Fossil species	134
ACKNOWLEDGEMENTS	142
REFERENCES	179
INDEX	185

ABSTRACT

HOUART, R., 1992. *THE GENUS* CHICOREUS *AND RELATED GENERA (GASTROPODA : MURICIDAE) IN THE INDO-WEST PACIFIC. Mém. Mus. natn. Hist. nat.,* (A), **154** : 1-188. Paris ISBN : 2-85653-194-6.
Published October 20th, 1992.

The genus *Chicoreus* Montfort, 1810 is divided into five subgenera : *Chicoreus (s.s.)* ; *Triplex* Perry, 1810, *Siratus* Jousseaume, 1880, *Rhizophorimurex* Oyama, 1950, and *Chicopinnatus* n. subgen. *Naquetia* Jousseaume, 1880 and *Chicomurex* Arakawa, 1964 are treated as separate genera, due to differences in radula morphology. A new subgenus of *Chicoreus*, *Chicopinnatus* is introduced for three species formerly classified in the genus *Pterynotus* Swainson, 1833.

All of the species are systematically revised and illustrated. The Recent species are redescribed and their distributions are mapped. Sixty-three Recent species and one subspecies are recognized : 7 are referable to *Chicoreus (s.s.)*, 39 to *Triplex*, 2 to *Siratus*, 1 to *Rhizophorimurex*, 3 to *Chicopinnatus*, 5 to *Naquetia*, and 7 to *Chicomurex*. One new species is named : *Chicomurex protoglobosus*. Of the 19 fossil species, 17 are assigned to *Triplex* and 2 to *Chicomurex*.

RÉSUMÉ

HOUART, R., 1992. *THE GENUS* CHICOREUS *AND RELATED GENERA (GASTROPODA : MURICIDAE) IN THE INDO-WEST PACIFIC*. Mém. Mus. natn. Hist. nat., (A), **154** : 1-188. Paris ISBN : 2-85653-194-6.
Publié le 20 octobre 1992.

Révision des espèces indo-pacifiques de *Chicoreus* et genres voisins (Gastropoda, Muricidae).

Le genre *Chicoreus* est divisé en 5 sous-genres : *Chicoreus* (*s.s.*) Montfort, 1810, *Triplex* Perry, 1810, *Siratus* Jousseaume, 1880, *Rhizophorimurex* Oyama, 1950 et *Chicopinnatus* n. subgen. Deux genres, habituellement reconnus comme sous-genres de *Chicoreus* : *Naquetia* Jousseaume, 1880 et *Chicomurex* Arakawa, 1964, sont traités en tant que genre sur base de différences radulaires.

63 espèces et une sous-espèce récentes sont reconnues : 52 appartiennent au genre *Chicoreus*, dont 7 à *Chicoreus* (*s.s.*), 39 à *Triplex*, 2 à *Siratus*, 1 à *Rhizophorimurex* et 3 à *Chicopinnatus*, 5 appartiennent au genre *Naquetia* et 7 à *Chicomurex*. Une nouvelle espèce de *Chicomurex* est décrite. Des 19 espèces fossiles, 17 appartiennent à *Triplex* et 2 à *Chicomurex*.

Le sous-genre *Chicoreus* (*s.s.*) est caractérisé par une coquille à ouverture large possédant une dent labiale plus ou moins développée, absente chez les autres sous-genres et chez les genres *Chicomurex* et *Naquetia*. Dans la plupart des cas la coquille est grande, atteignant une hauteur de plus de 300 mm chez *Chicoreus ramosus* (Linné, 1758).

Le sous-genre *Triplex* Perry, 1810 est subdivisé en 7 groupes, chaque groupe contenant des espèces proches au niveau de la structure de la coquille. La hauteur des coquilles ne dépassent pas 130 mm, la sculpture axiale est représentée par 3 varices arrondies, garnies d'épines généralement foliacées.

Le sous-genre *Siratus* Jousseaume, 1880, bien représenté dans l'Océan Atlantique Occidental, ne connaît que deux espèces dans l'Océan Pacifique. Celles-ci atteignent cependant une plus grande taille que les espèces atlantiques, sans toutefois dépasser la hauteur de 184 mm. Les varices sont généralement ornées d'une longue épine carinale et d'une expansion foliacée.

Le sous-genre *Rhizophorimurex* Oyama, 1950 est monotypique et se distingue par une espèce à coquille de taille moyenne, présentant 3 varices arrondies, non épineuses, sur le dernier tour, un bord columellaire adhérant à la coquille et un canal siphonal court, large et non épineux.

Chicopinnatus n. subgen. est créé pour 3 espèces classées auparavant chez le genre *Pterynotus* et dont les coquilles se distinguent par un canal siphonal long et mince, une ouverture striée intérieurement et non denticulée comme chez *Pterynotus*, et par des premiers tours de téléoconque dont la sculpture axiale diffère de celle de *Pterynotus*, mais se rapproche de *Chicoreus* (*s.l.*).

Le genre *Chicomurex* Arakawa, 1964 regroupe des espèces dont la coquille ne dépasse pas 85 mm de haut et présente 3 varices plus ou moins épineuses, légèrement foliacées adapicalement sur le dernier tour de téléoconque.

Le genre *Naquetia* Jousseaume, 1880 est représenté par des espèces dont la coquille se distingue par une spire haute, comportant 3 varices arrondies, parfois légèrement foliacées sur le dernier tour.

Certaines espèces appartenant très probablement au genre *Chicoreus* ou à un des autres genres étudiés, mais qui ne furent jamais, ou très mal illustrées et pour lesquelles aucun matériel type ne fut retrouvé, sont reprises comme *nomina dubia*. La liste de ces espèces est inclue dans l'introduction.

Toutes les espèces sont revues et illustrées systématiquement. Les espèces récentes sont redécrites.

La description des espèces est basée sur des spécimens adultes, et le matériel type est illustré pour la majorité des espèces.

La distribution géographique, indiquée à l'aide de cartes, est basée sur du matériel appartenant à différents musées, à des collections privées et à la collection de l'auteur.

La radula fut étudiée chez 17 espèces. Elle présente une dent rachidienne et une dent bilatérale unicuspide. La dent rachidienne supporte 5 denticules.

Deux types de radulas furent observés, l'un regroupant les espèces de *Chicoreus* (*s.l.*) (Figs 92-115, 129) et l'autre, les espèces des genres *Chicomurex* et *Naquetia* (Figs 116-128). Les espèces de *Chicomurex* et *Naquetia* possèdent une radula avec une dent rachidienne supportant un grand denticule triangulaire central et des rangées de denticules très rapprochées les unes des autres. Les espèces de *Chicoreus* (*s.l.*) au contraire, présentent une radula avec un denticule central plus étroit et allongé ainsi que des rangées de denticules plus espacées.

Deux autres observations importantes furent mises en évidence grâce à l'étude de la radula :
— Une séparation au point de vue spécifique peut être réalisée chez certaines espèces de *Triplex* classées ici dans un même groupe, alors qu'en général la morphologie de la radula est très souvent inutilisable pour une séparation spécifique fiable chez les Muricidae.
— L'existence de différences structurelles de la radula chez une même espèce, juvénile et adulte.

La protoconque est reconnue comme un élément essentiel pour l'identification des espèces. Deux formes de protoconques sont observées chez les Muricidae : celles appartenant aux espèces connaissant un développement larvaire planctotrophe sont multispirales et de forme conique, avec des tours légèrement convexes, tandis que celles appartenant aux espèces à développement non planctotrophe sont généralement paucispirales à tours arrondis. Les espèces à développement non planctotrophe peuvent à leur tour être subdivisées en deux groupes d'espèces : celles connaissant un développement lecitotrophe et celles à métamorphose intracapsulaire.

On observe une très large distribution géographique chez les espèces à développement larvaire planctotrophe, tandis que les espèces à développement non planctotrophe connaissent une distribution géographique soit restreinte s'il s'agit d'un développement intracapsulaire, soit élargie s'il s'agit d'une larve lecitotrophe. Ceci est dû au fait que les espèces à larve lécitotrophe présentent en général une protoconque assez petite et légère, leur permettant de vivre quelques temps comme des espèces planctotrophes, sans pour autant se nourrir du plancton, mais expliquant ainsi leur plus large distribution géographique.

L'opercule de toutes les espèces étudiées est généralement ovale ou arrondi, avec un nucleus apical ou subapical.

INTRODUCTION

Species of *Chicoreus* and related genera have always been the subject of taxonomic confusion.
Some species are very stable in sculpture, colour, spine ornamentation, with only slightly discernable differences between different populations (e.g. *C. trivialis, C. capucinus*). Others are extremely variable, especially *C. brunneus* and *C. microphyllus*, and to a lesser degree, such species as *C. torrefactus, C. banksii* and *C. axicornis*. This variability has resulted in the appearance of many different names to designate a same species.

This revision is an attempt to refine classification and facilitate identification of all Recent and fossil Indo-West Pacific species.

For revisions of the western Atlantic relatives, all of which belong in *Chicoreus* (s.l.), see CLENCH & PÉREZ FARFANTE (1945), VOKES (1965 and 1967), FAIR (1976) and RADWIN & D'ATTILIO (1976). The muricid fauna of the eastern Atlantic is currently being studied by the author, and the species tentatively included in the genus *Chicoreus* will be revised.

MATERIAL, METHODS AND TEXT CONVENTIONS

All Recent species are systematically redescribed. The description is based on adult shells, and dimensions are given for shells at maturity. Varicial spines are numbered consecutively abapically so that the adapical (i.e., shoulder) spine is considered as first spine and so on. Most of the available type material is illustrated, as well as other representative specimens, including inter- and intra-populational variants. Additional references are given from many publications.

Fossil species are not systematically redescribed but all were studied and carefully compared; all of them are illustrated and reviewed.

Types and other material were borrowed from many institutions (see abbreviations) but no extensive lists of material examined were compiled except for uncommon or rarer species.

EXCLUDED TAXA

Some species that clearly belong in *Chicoreus* were inadequately described and were either unillustrated or very poorly illustrated. Such species for which no type material exists are specifically unrecognizable and are thus *nomina dubia*.

Generic names in parenthese immediately following species names are as originally cited.

abortiva (Triplex) Perry, 1811 : pl. 6, fig. 5 (*nomen dubium*)
capucina (Purpura) Röding, 1798 : 143 (*nomen dubium*)
carneola (Purpura) Röding, 1798 : 142 (*nomen dubium*)
cornudama (Purpura) Röding, 1798 : 142 (*nomen nudum*) (questionably a species of *Chicoreus*)
elongata (Purpura) Link, 1807 : 121 (*nomen dubium*)
lactuca (Purpura) Röding, 1798 : 141 (*nomen dubium*) (composite species)
rosana (Purpura) Schumacher, 1817 : 212 (*nomen dubium*)

tirondus (Murex) de Gregorio, 1885 : 252 (*nomen dubium*)
tubulatus (Murex) Mörch, 1852 : 97 (*nomen dubium*)
variegata (Murex) Mörch, 1852 : 96 (*nomen dubium*)

GEOGRAPHICAL DISTRIBUTION

The distribution maps are based on material housed in museums, in private collections and in the personal collection of the author, for which the locality data is trustworthy.

Also included on distribution maps are records from literature associated with good illustrations or records based on trustworthy or verified personal communications. Published records unaccompanied by illustrations or accompanied by ambiguous illustrations, or that otherwise seem doubtful are excluded.

TAXONOMIC CHARACTERS

Through numerous observations, it has been possible to determinate the shell characters that are most useful for species discrimination. Most differences such as coloration, number of whorls, and size of shell spines are of little or no importance. On the other hand it seems that major differences in shell shape, and in the number and arrangement of spines could be of a greater importance. Major differences in protoconch, aperture and spiral sculpture are usually decisive.

PROTOCONCH
Figs 1-85

A knowledge of the protoconch is often essential for accurate identification of species. With an intact protoconch, the presence of a first teleoconch whorl, and with good locality data, it is usually possible to determine a species with considerable exactitude. On the other hand, it is sometimes very difficult or impossible to be absolutely sure of an identification without knowledge of the protoconch, even with an adult shell. The terminal varix of the protoconch is intraspecifically variable in species with non planktotrophic larval development and is thus useful for species discrimination in this case.

Two kinds of larval development are observed in the Muricidae : planktothrophic and non planktotrophic. Protoconchs of species presumed to have planktotrophic development are acutely conical and comprise 3 or more whorls, usually of the sinusigera type. In the non planktotrophic larval development distinction must be made between lecithotrophic development and intracapsular metamorphosis.

It might be expected that the species with planktotrophic larval development have a wide geographical distribution while those with non planktotrophic development (most of the species of *Chicoreus*) would have more limited distributions. While the first supposition is often correct, the second is not always true. Some species with non planktotrophic larval development have restricted geographical ranges (e.g. *C. trivialis, C. denudatus, C. longicornis*) while others (e.g. *C. palmarosae, C. microphyllus*) have much wider geographical distribution than might be expected. As suggested by PONDER & VOKES (1988), those species with small protoconchs probably have lecithotrophic larval development, spending a short time in the water column, which explain their wider geographical distribution, compared to those with intracapsular metamorphosis which are limited to a restricted geographical area. The size of the veliger at hatching is certainly another factor to take into account (BOUCHET, 1987).

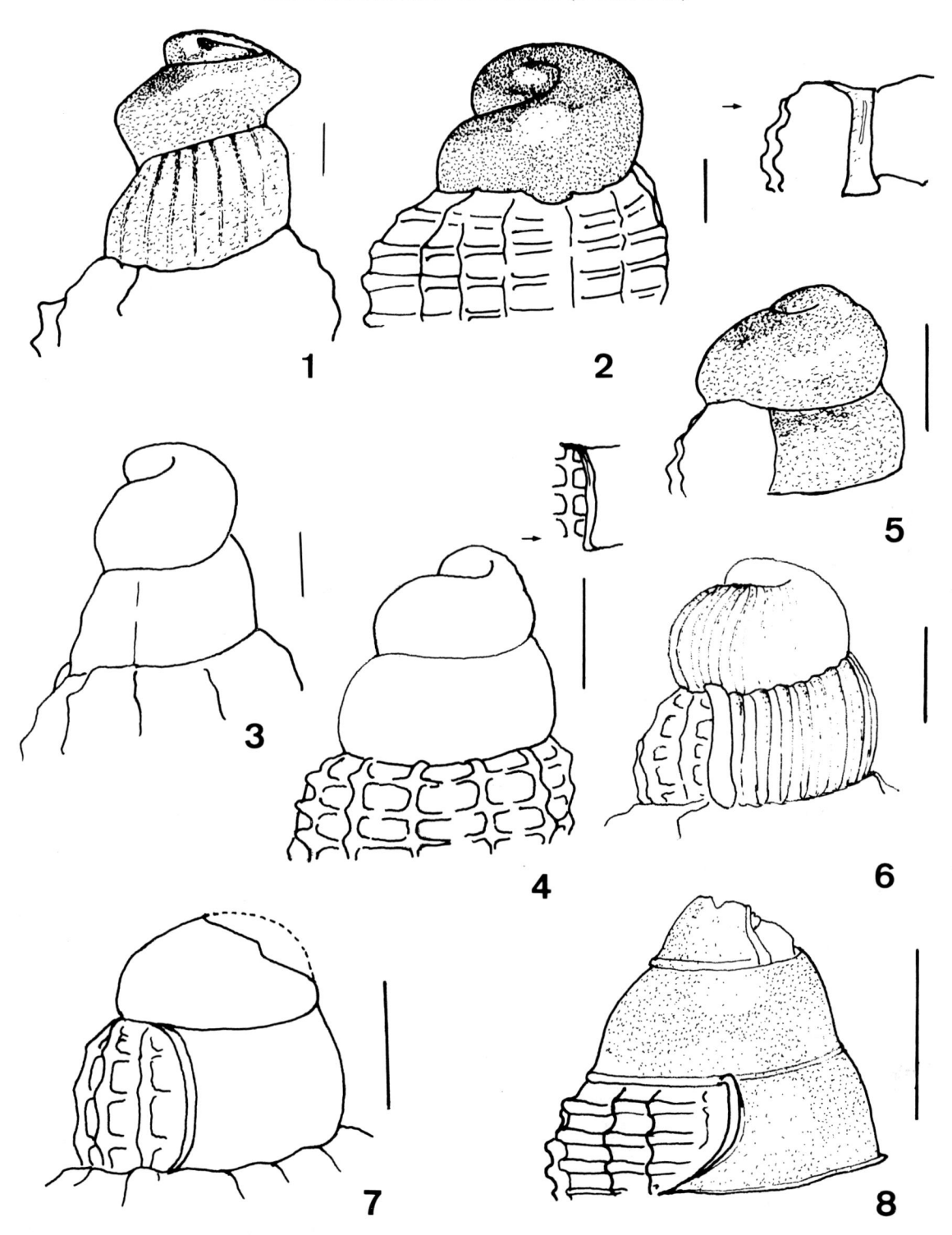

FIGS 1-8. — Protoconchs of *Chicoreus* sp. (scale bars : 0.5 mm)
1, *Chicoreus cornucervi* (Röding). Arafura Sea (RH).
2, *C. austramosus* Vokes. Port Alfred, South Africa (AMS).
3, *C. litos* Vokes. south east of Durban Bluff (paratype, RH).
4-5, *C. ramosus* (Linné). 4, Tulear, Madagascar (MNHN). 5, Seychelles (RH).
6, *C. bundharmai* Houart. Banjarmasin, South Borneo (paratype, Zoological Mus. Bogoriense, Bogor).
7, *C. virgineus* (Röding). Gulf of Aden (WPU).
8, *C. maurus* (Broderip). Tahiti (J. TRONDLE).

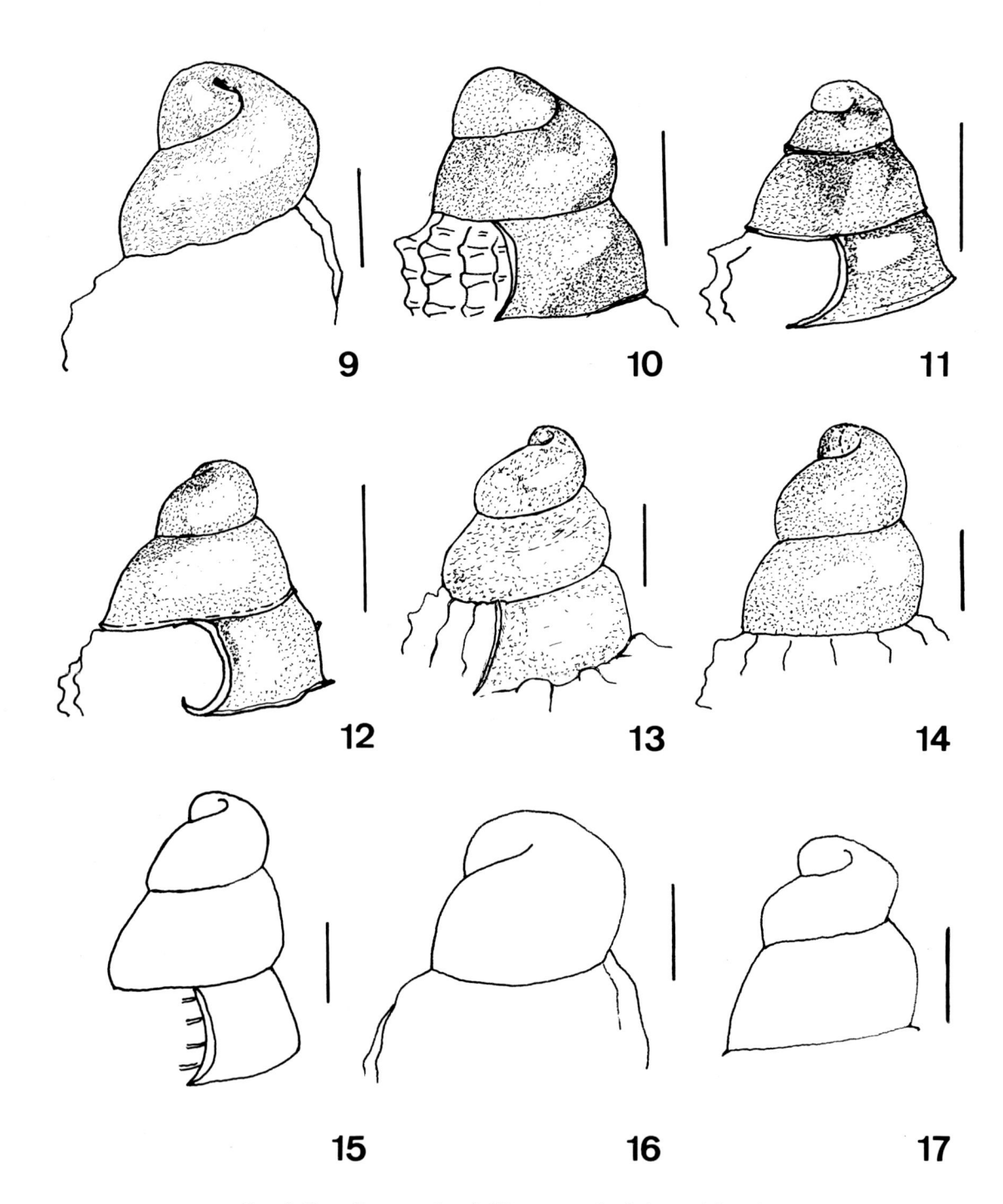

FIGS 9-17. — Protoconchs of *Chicoreus* sp. (scale bars : 0.5 mm)
9, *Chicoreus palmarosae* (Lamarck). Philippines (RH).
10, *C. dovi* Houart. Tanzania (RH).
11, *C. saulii* (Sowerby). Philippines (RH).
12, *C. torrefactus* (Sowerby). Seychelles (RH).
13-17, *C. microphyllus* (Lamarck). 13, New Caledonia (RH). 14-15, Queensland, Australia (AMS), illustrated Fig. 319. 16, Philippines (RH), illustrated Fig. 318. 17, Queensland, Australia (AMS), illustrated Fig. 301.

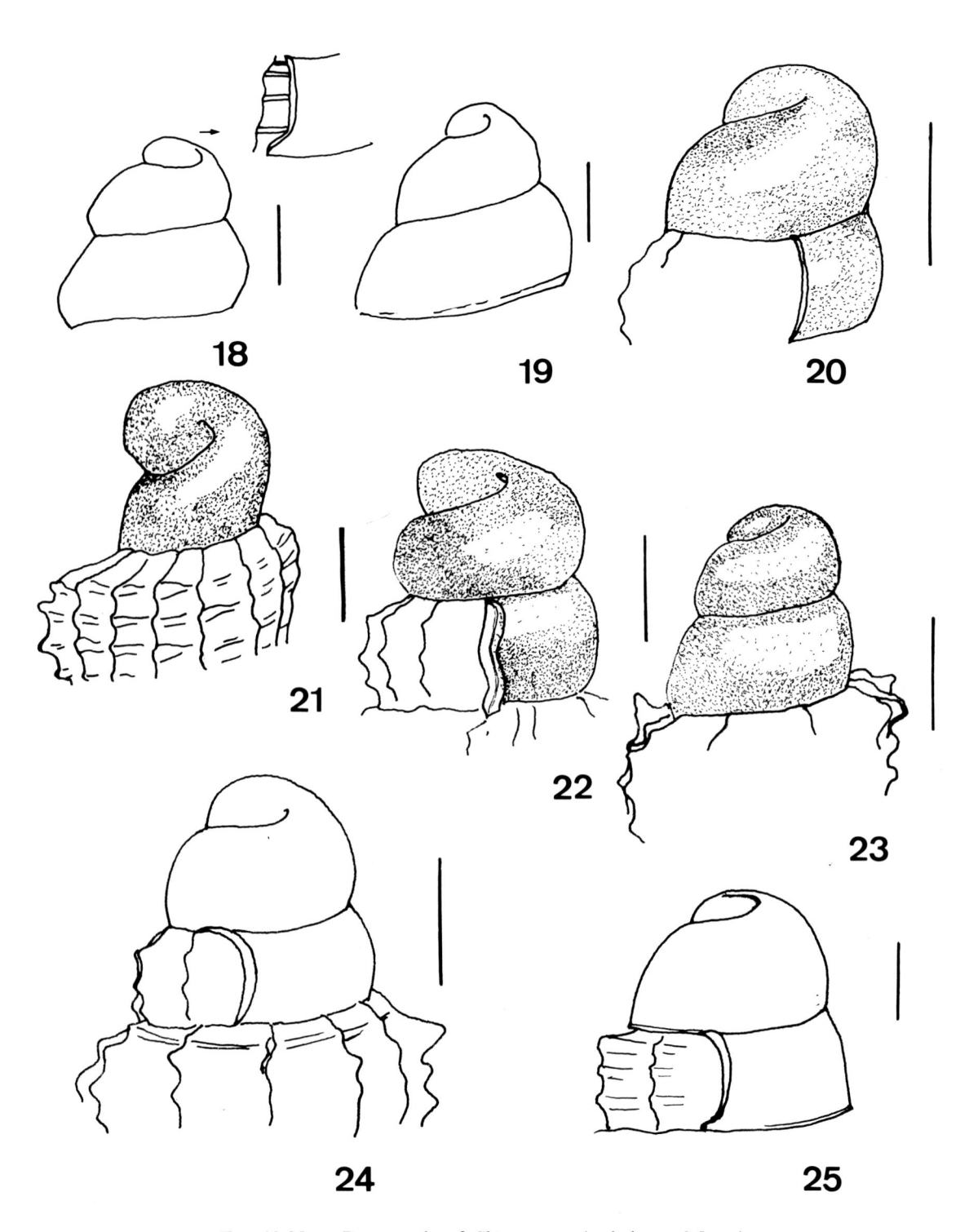

FIGS 18-25. — Protoconchs of *Chicoreus* sp. (scale bars : 0.5 mm)
18-19, *Chicoreus microphyllus* (Lamarck). 18, Fiji (AMS). 19, Fiji (AMS).
20, *C. strigatus* (Reeve). Moluccas (RH).
21, *C. paini* Houart. Papua New Guinea (AMS).
22, *C. trivialis* (A. Adams), W. Australia (coll. R. ISAACS).
23-25, *C. axicornis* (Lamarck). 23, Papua New Guinea (RH). 24, Papua New Guinea (RH). 25, Sri Lanka (RH).

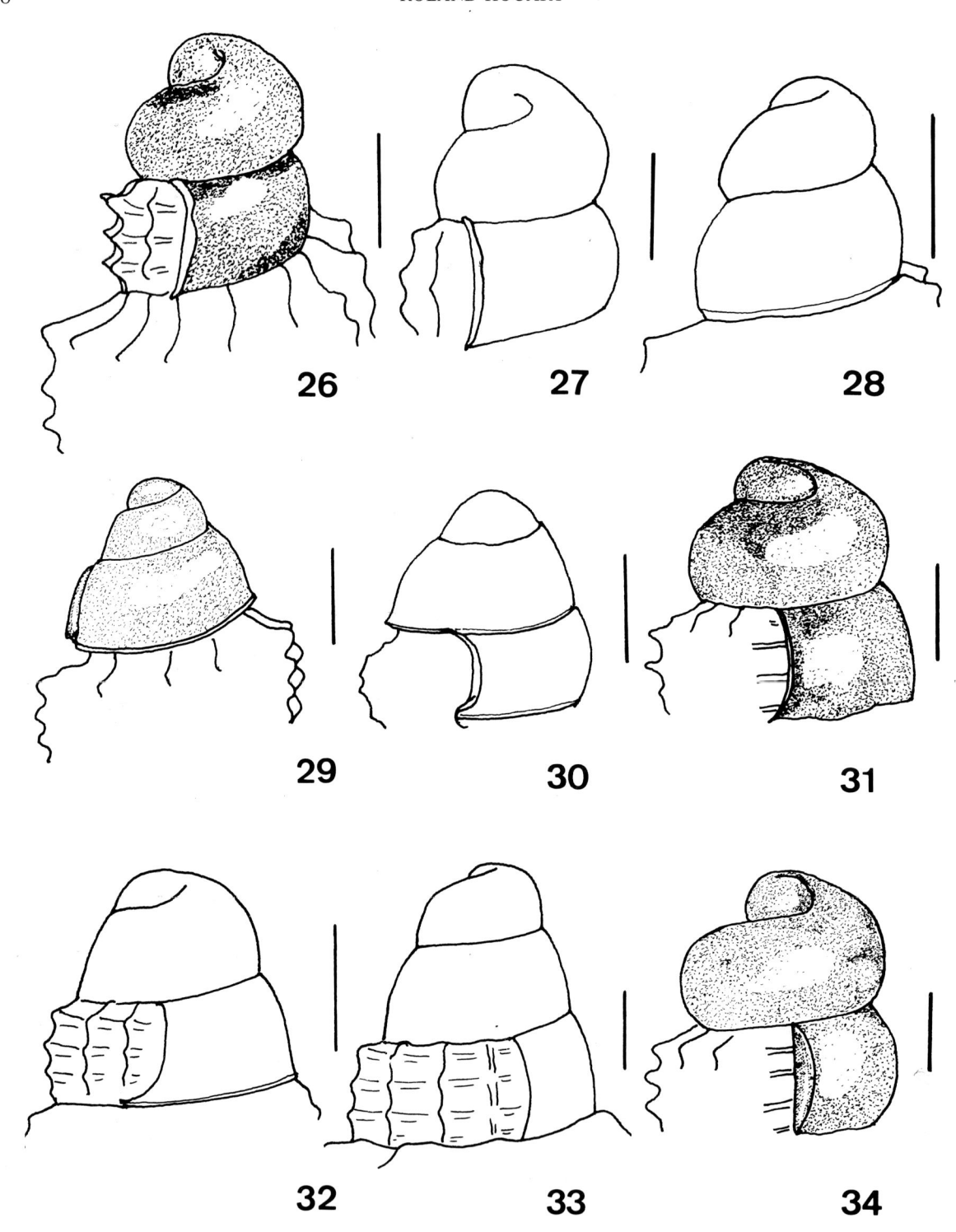

FIGS 26-34. — Protoconchs of *Chicoreus* sp. (scale bars : 0.5 mm)
26-28, *Chicoreus banksii* (Sowerby). 26, Queensland, Australia (AMS). 27, Queensland, Australia (RH). 28, New Caledonia (RH).
29-30, *C. bourguignati* (Poirier), Seychelles (RH).
31-33, *C. brunneus* (Link). 31, New caledonia (RH). 32, Chesterfields (MNHN), illustrated Fig. 351. 33, New Caledonia (RH).
34, *C. groschi* Vokes. Mozambique (RH).

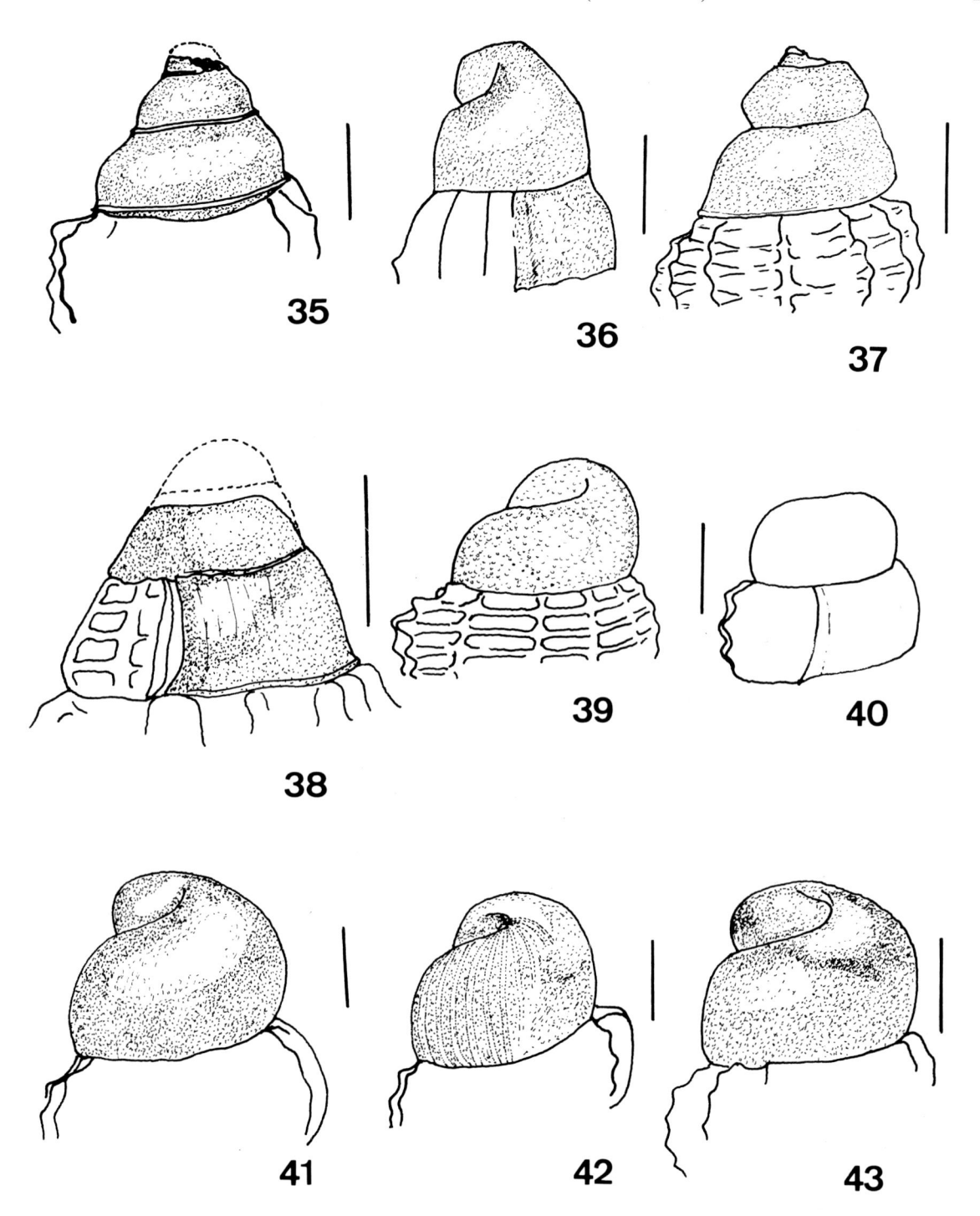

FIGS 35-43. — Protoconchs of *Chicoreus* sp. (scale bars : 0.5 mm)
35, *Chicoreus cnissodus* (Euthyme). Taiwan (RH).
36, *C. capucinus* (Lamarck). Queensland, Australia (AMS).
37, *C. thomasi* (Crosse). Marquesas (RH).
38, *C. rubescens* (Broderip). Wallis (coll. J. COLOMB).
39-40, *C. corrugatus* (Sowerby). Gulf of Suez (MNHN).
41, *C. damicornis* (Hedley). NSW, Australia (RH).
42, *C. denudatus* (Perry). NSW, Australia (RH).
43, *C. territus* (Reeve). Queensland, Australia (AMS).

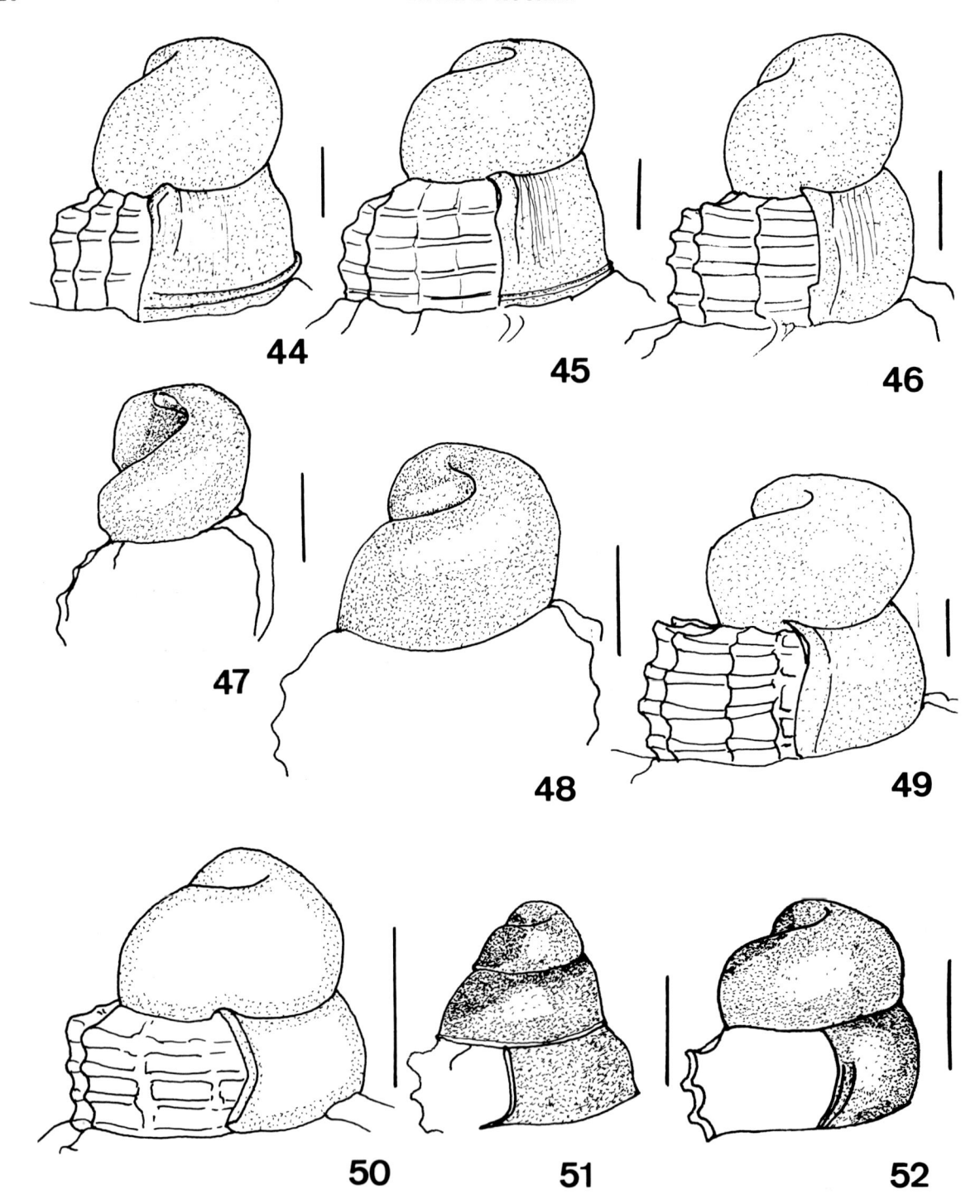

FIGS 44-52. — Protoconchs of *Chicoreus* sp. (scale bars : 0.5 mm)
44, *Chicoreus boucheti* Houart. New Caledonia (MNHN).
45, *C. subpalmatus* Houart. New Caledonia (MNHN).
46, *C. paucifrondosus* Houart. New Caledonia (MNHN).
47, *C. cervicornis* (Lamarck). Australia (RH).
48, *C. longicornis* (Dunker). Queensland, Australia (RH).
49, *C. fosterorum* Houart. S. Africa (paratype, coll. C. GLASS & R. FOSTER).
50, *C. zululandensis* Houart. S. Africa (holotype, NM).
51, *C. aculeatus* (Lamarck). Philippines (RH).
52, *C. cloveri* Houart. Mauritius (paratype, RH).

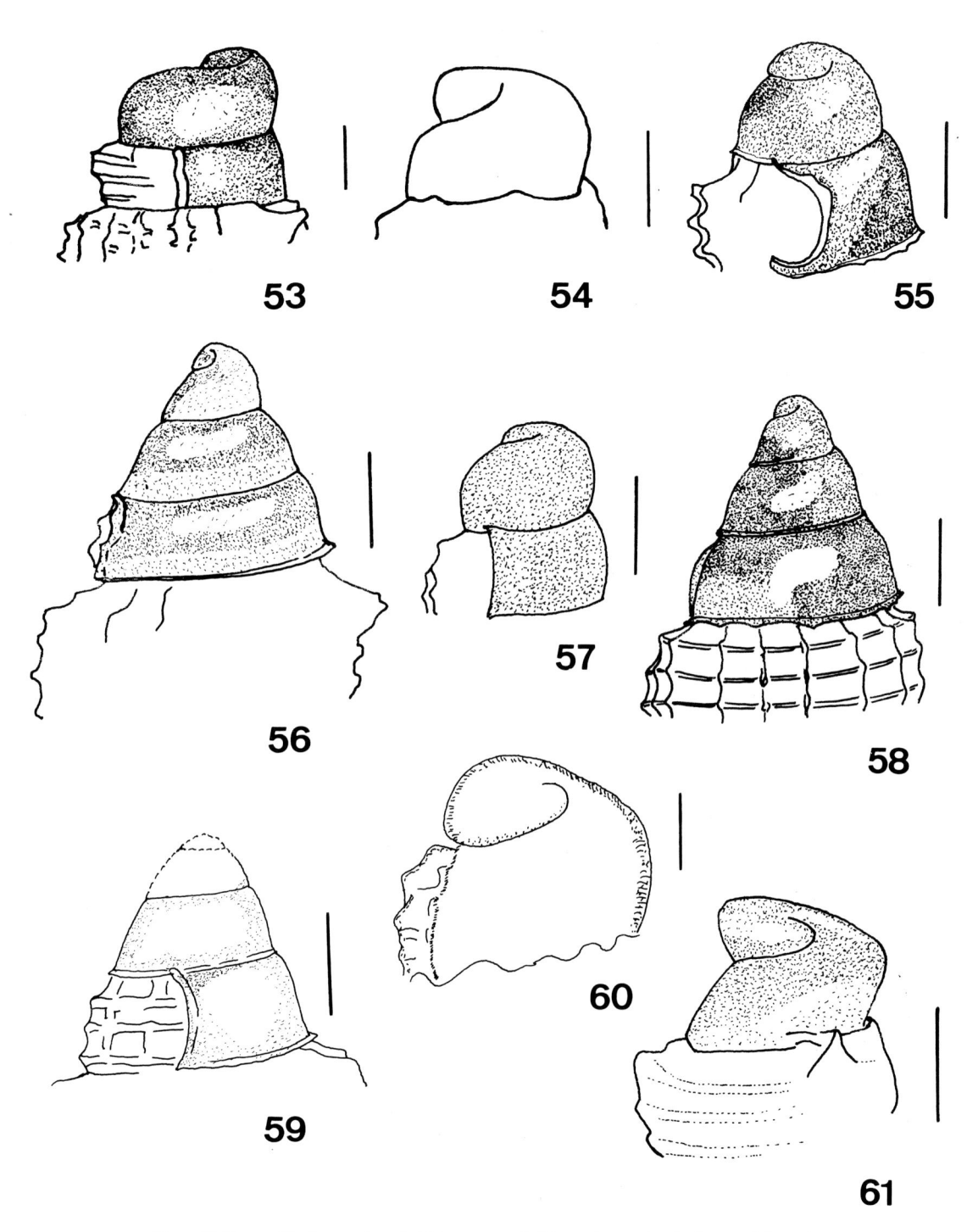

FIGS 53-61. — Protoconchs of *Chicoreus* and *Naquetia* sp. (scale bars : 0.5 mm)
53-54, *Chicoreus crosnieri* Houart. Madagascar (holotype, MNHN).
55, *C. nobilis* Shikama. Philippines (coll. A. LESAGE).
56, *C. rossiteri* (Crosse). Philippines (RH).
57, *C. ryukyuensis* Shikama. Okinawa (KPM), illustrated Fig. 253.
58, *Naquetia barclayi* (Reeve). Philippines (RH).
59, *N. triqueter* (Born). Papua New Guinea (RH).
60, *N. fosteri* D'Attilio & Hertz. Eilat, Red Sea (paratype, coll. C. GLASS & R. FOSTER, reproduced from D'ATTILIO & HERTZ, 1989, with author's permission).
61, *N. cumingii* (A. Adams). Philippines (RH).

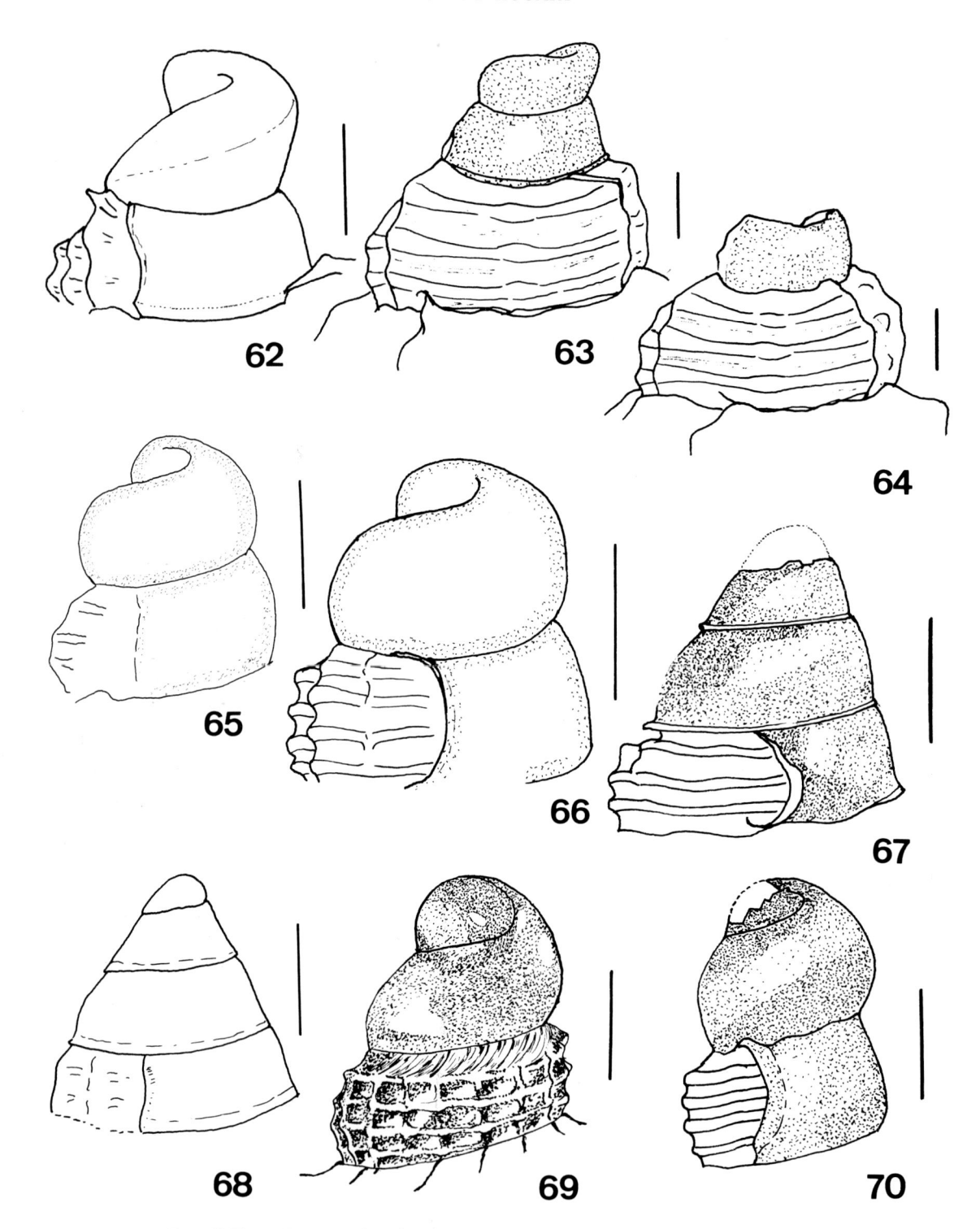

FIGS 62-70. — Protoconchs of *Naquetia* and *Chicomurex* sp. (scale bars : 0.5 mm)
62-64, *Naquetia cumingii* (A. Adams). 62, Philippines (coll. C. GLASS & R. FOSTER). 63, Moluccas (RH). 64, Dahlak, Red Sea (RH).
65-66, *N. vokesae* (Houart). 65, Tanzania (paratype, IRSNB). 66, Zululand, S. Africa (NM).
67, *Chicomurex superbus* (Sowerby). Taiwan (RH).
68, *C. venustulus* (Rehder & Wilson). Marquesas (paratype, USNM).
69-70, *C. turschi* Houart. 69, Papua New guinea (paratype, RH). 70, Philippines (coll. J. COLOMB), illustrated Fig. 432.

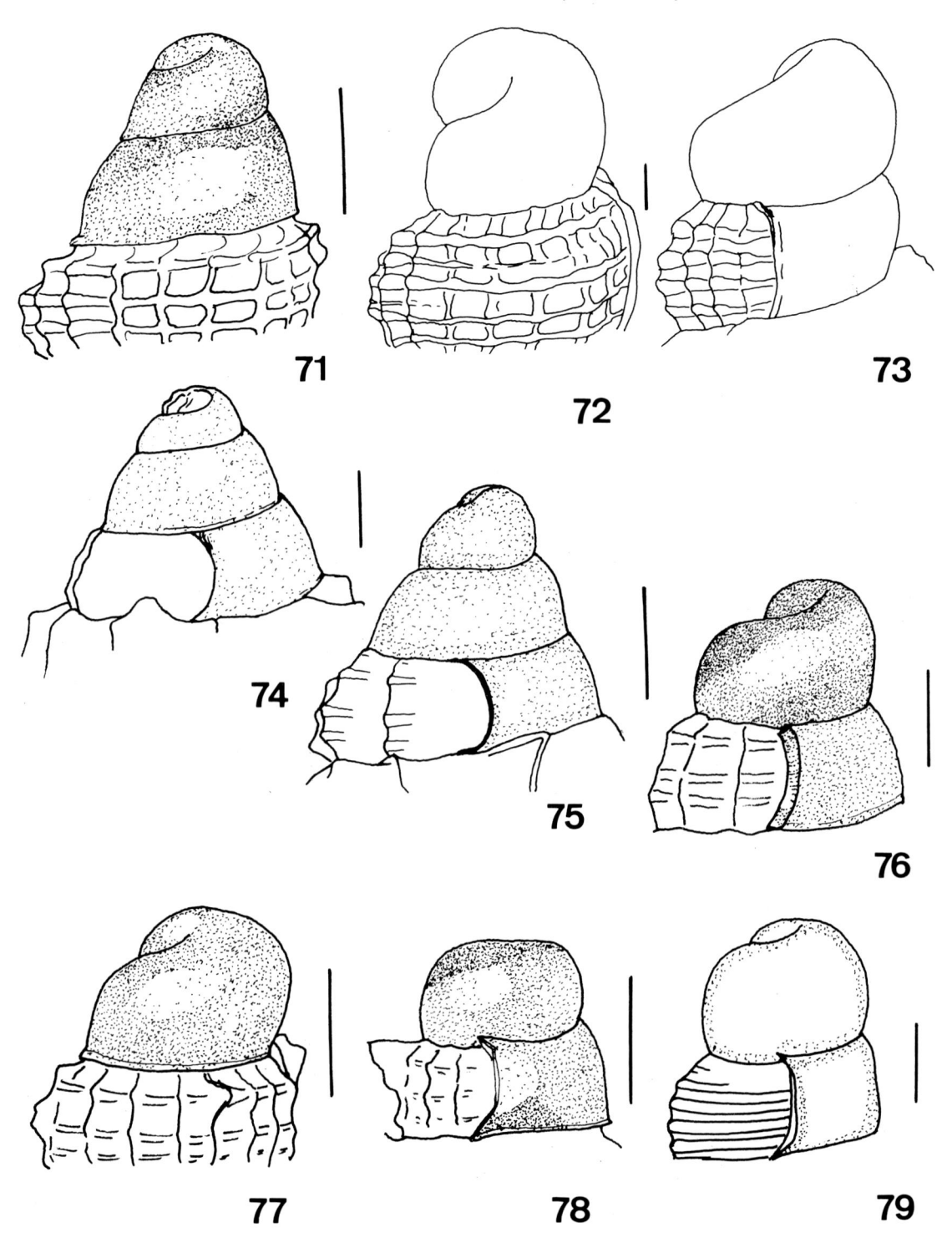

FIGS 71-79. — Protoconchs of *Chicomurex* and *Chicoreus* sp. (scale bars : 0.5 mm)
71, *Chicomurex laciniatus* (Sowerby), Philippines (RH).
72-73, *C. protoglobosus* n.sp., New Caledonia (holotype, MNHN).
74, *Chicoreus alabaster* (Reeve), Philippines (coll. G. POPPE).
75, *C. pliciferoides* (Kuroda), Philippines (RH).
76, *C. guillei* (Houart), Reunion (holotype, MNHN).
77-78, *C. orchidiflorus* (Shikama), Philippines (RH).
79, *C. dennanti* (Tate), Victoria, Australia (paralectotype, SAM).

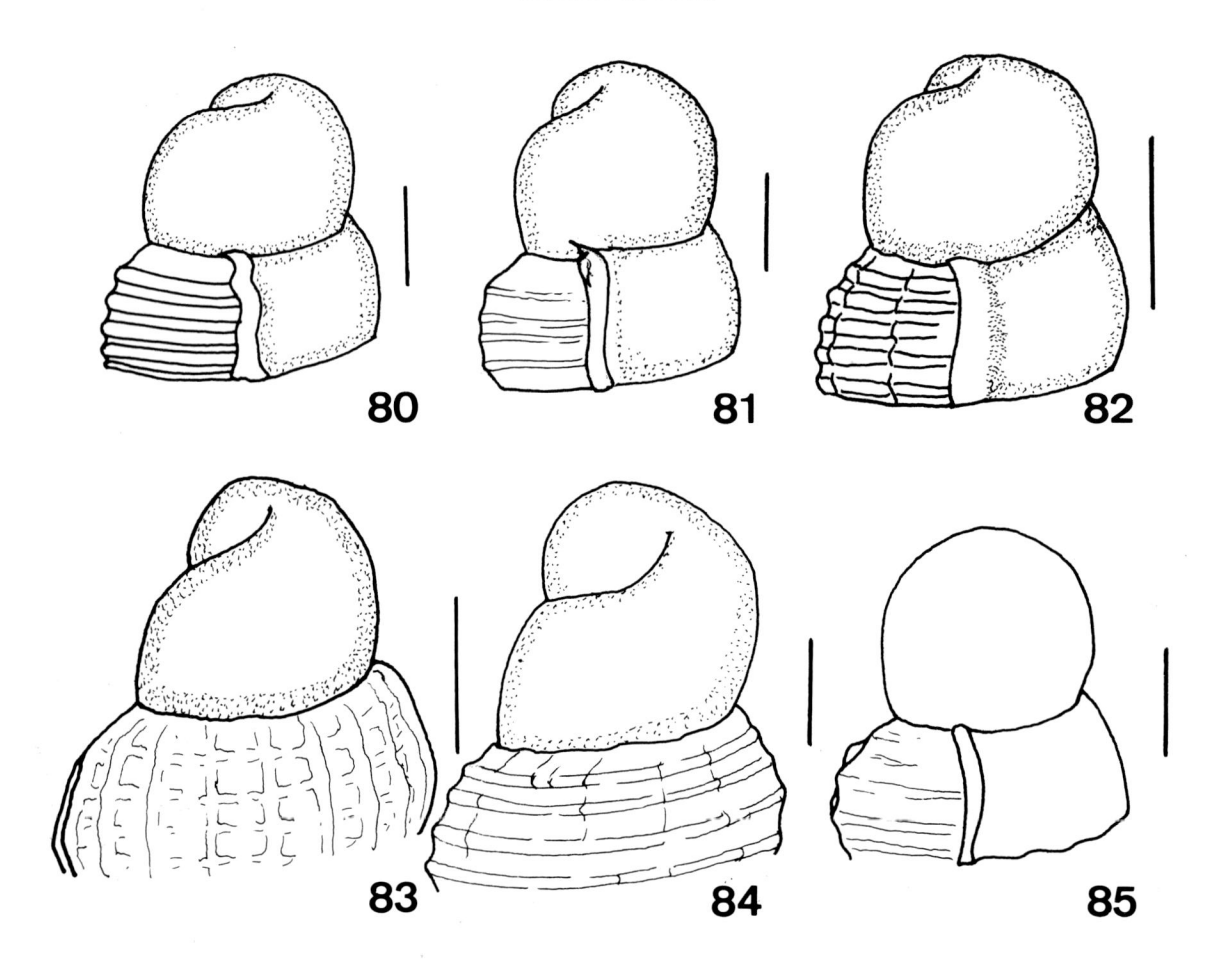

FIGS 80-85. — Protoconchs of *Chicoreus* and *Chicomurex* sp. (scale bars : 0.5 mm)
80, *Chicoreus basicinctus* (Tate), South Australia (paralectotype, SAM).
81, *C.* cf. *amblyceras*, Mornington, Victoria, Australia (MV).
82, *C. amblyceras* (Tate), Victoria, Australia (paralectotype G, SAM).
83, *C. juttingae* (Beets), Borneo (paratype, RML).
84-85, *Chicomurex lophoessus* (Tate), Mornington, Victoria, Australia (MV).

SHUTO (1983) studied the larval development and geographical distribution of some of the Indo-West Pacific *Murex* species. The most reliable explanation retained by SHUTO is an evolutionary change of the style of larval development, from planktonic to non-planktonic in Neogene time. Unfortunately, SHUTO erroneously supposed that all species of *Murex* (*s.s.*) have non planktotrophic development for in fact many of the species cited by him have planktotrophic larvae.

RADWIN & D'ATTILIO (1976) indicated that species with wide distribution patterns " **must have been arrived at via low, stepwise migration over a great many generations, each generation producting a small extension and consolidation of the species range** ", an interpretation somewhat accordant with that of Ponder & Vokes (see above).

In conclusion, although it is nearly impossible to tell lecithotrophic from intracapsular development from protoconch morphology only, it is almost certain that a species with a large, heavy, and irregular shaped protoconch with 1 1/4 to 1 3/4 whorls have intracapsular metamorphosis. On the other hand, a species with small paucispiral protoconch of 1 1/4 to 2 1/2 whorls may have lecithotrophic or intracapsular development, but the non planktotrophic species with small protoconch and extensive geographical range have certainly lecithotrophic larval development, lecitotrophes living a more or less short time as planktonic species (but without feeding in the plankton), which explains their wide geographical range.

OPERCULUM
Figs 86-91

The operculum of all the *Chicoreus* (*s.l.*) species (i.e. all species here studied) is usually ovate or roundly ovate with an apical or subapical nucleus surrounded by numerous incremental concentric ridges. There is little intraspecific variation in shape. It seems to have only very limited value for species discrimination, although it is possible that the sculpture of the inner side may prove to be significant.

RADULA
Figs 92-129

Radular characters were studied for 17 species. The radula of all species described herein consists of a broad, serrated rachidian tooth with a single unicuspid lateral tooth on each side. The rachidian bears five cusps of which the central cusp is generally the longest. Intermediate denticles were not observed in any radula studied.

Two basic types of radulae were observed, one grouping the species of *Chicomurex* and *Naquetia* (Figs 116-128), the other the species of *Chicoreus* and its subgenera (Figs 92-115, 129). *Chicomurex* and *Naquetia* have a radular ribbon with front to back crowded rows of teeth with a large, triangular central cusp, while *Chicoreus* (*s.s.*) and subgenera have more widely separate rows of teeth with long and more slender central cusp. *Chicomurex* and *Naquetia* are related to *Phyllonotus* Swainson, 1833, judging from the similarity of their radula to that of *Phyllonotus pomum* (Gmelin, 1791) (BANDEL, 1984 : pl. 11, figs. 5 & 7). For this reason, I prefer to consider *Chicomurex* and *Naquetia* as separate from *Chicoreus* (*s.l.*), although their shells are closely similar.

The radulae of species of the subgenus *Triplex* may be separated into two loosely defined groups : one with more or less long and slender central and lateral cusps (e.g. *C. rossiteri*, *C. boucheti*) (Figs 106-107), the other with shorter, more broadly triangular cusps (e.g. *C. nobilis*, *C. banksii*, figs 100-101). This is a useful character for some specific discriminations. It may be used for example, to distinguish *C. nobilis* from *C. rossiteri* (Figs 110-113), which are here considered to belong to the same group on the basis of shell morphology alone.

Another significant observation in *Triplex* is the apparent change in radula morphology between juvenile and adult specimens. The radula of a juvenile specimen of *C. torrefactus* (Figs 94-95) shows a short and prominent central cusp, similar to that in an adult specimens of *C. brunneus* (Figs 102-103), while less well developed, while the radula of an adult *C. torrefactus* (Figs. 96-97) shows a longer and less prominent central cusp. The radula of *C. capucinus* (Lamarck) also has a more or less prominent central cusp (Fig. 105), although not apparent in Ponder & Vokes's illustration of the radula (1988 : 137, fig. E).

HABITAT

Like other muricids, most *Chicoreus* (*s.l.*) species live among rocks and coral reefs, but some live on sandy or muddy substrata. They occur generally from the intertidal zone to about 70 meters depth, though a few species range deeper than 200 meters.

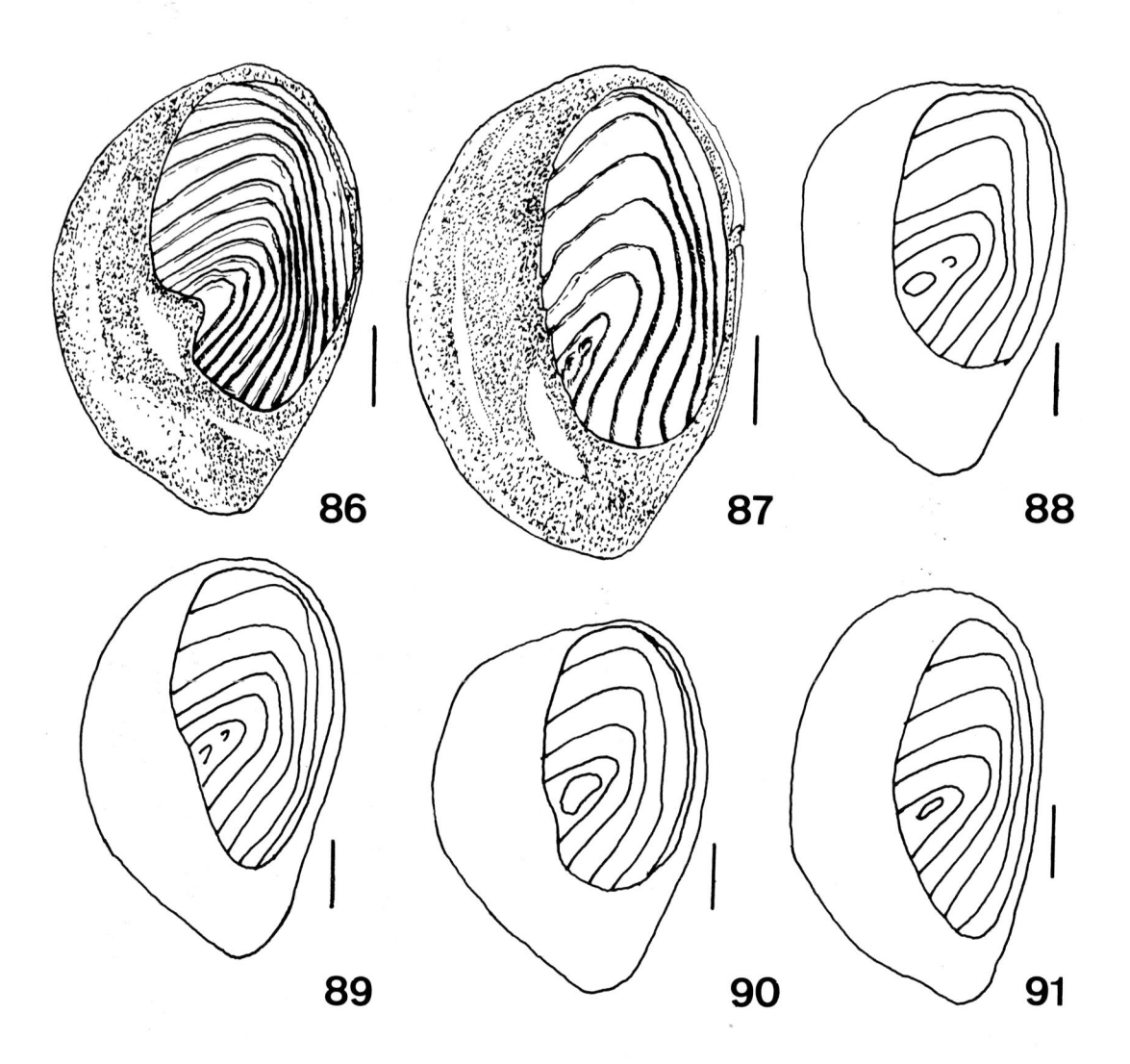

FIGS 86-91. — Opercula (scale bars : natural size)
86, Interior sculpture of operculum of *Chicoreus strigatus* (Reeve).
87-91, Interior sculpture of operculum of *Chicoreus microphyllus* (Lamarck).

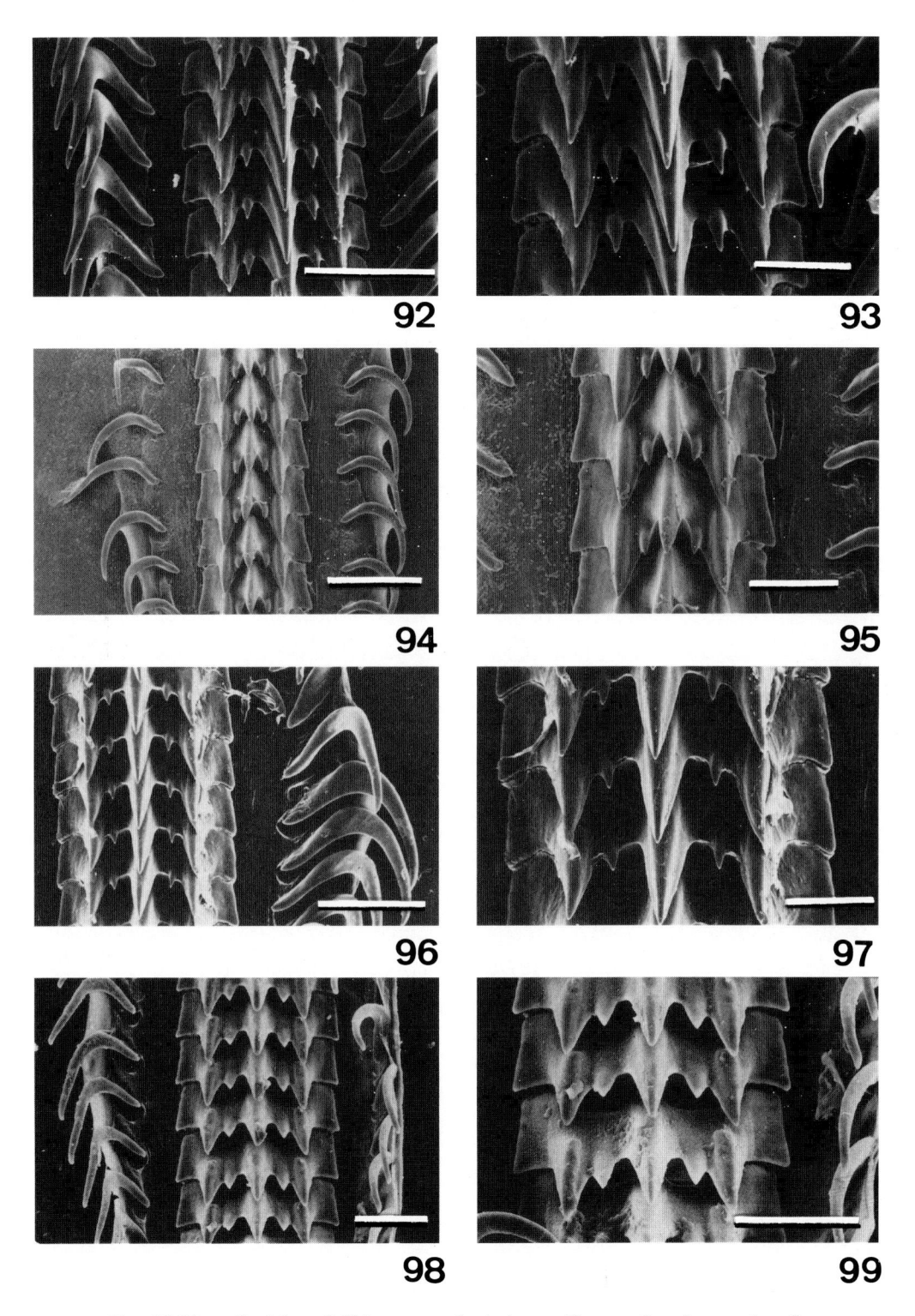

FIGS 92-99. — Radulae of *Chicoreus* sp. (scale bars : 50 μm ; otherwise mentioned)
92-93, *Chicoreus virgineus* (Röding). Gulf of Aden (WPU).
94-97, *C. torrefactus* (Sowerby). 94-95, Papua New Guinea (juvenile, 33 mm) (IRSNB). 96-97, Papua New Guinea (adult, 73 mm) (RH) (94 and 96 : scale bar 100 μm).
98-99, *C. paini* Houart. Solomon Is (paratype, RH).

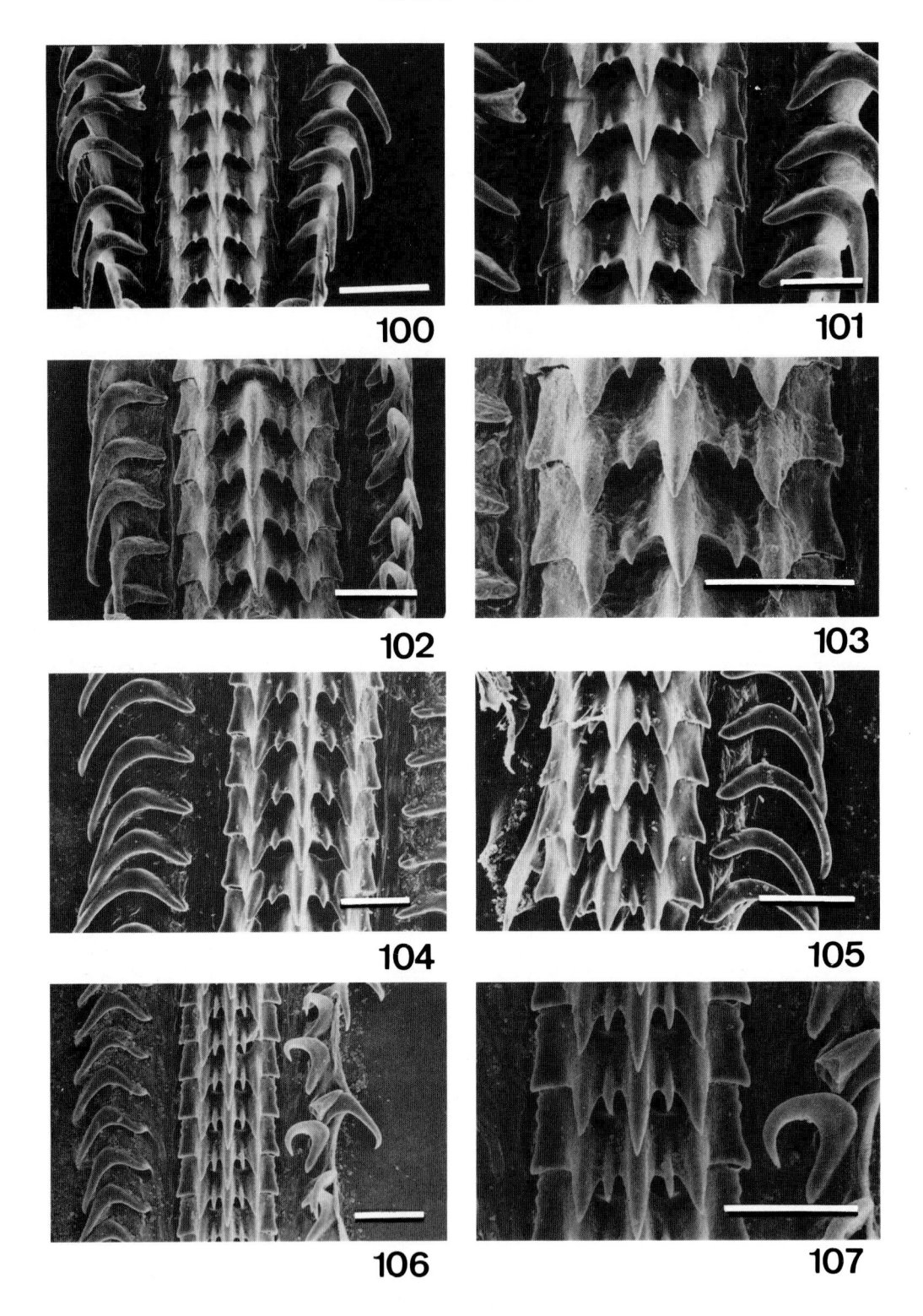

FIGS 100-107. — Radulae of *Chicoreus* sp. (scale bars : 50 μm ; otherwise mentioned)
100-101, *Chicoreus banksii* (Sowerby). New Caledonia (RH) (100 : scale bar 100 μm).
102-104, *C. brunneus* (Link). 102-103, New Caledonia (small form), illustrated Fig. 354 (RH). 104, New Caledonia (typical form) (RH).
105, *C. capucinus* (Lamarck). N.E. Queensland, Australia (QM).
106-107, *C. boucheti* Houart. New Caledonia, illustrated Fig. 399 (MNHN).

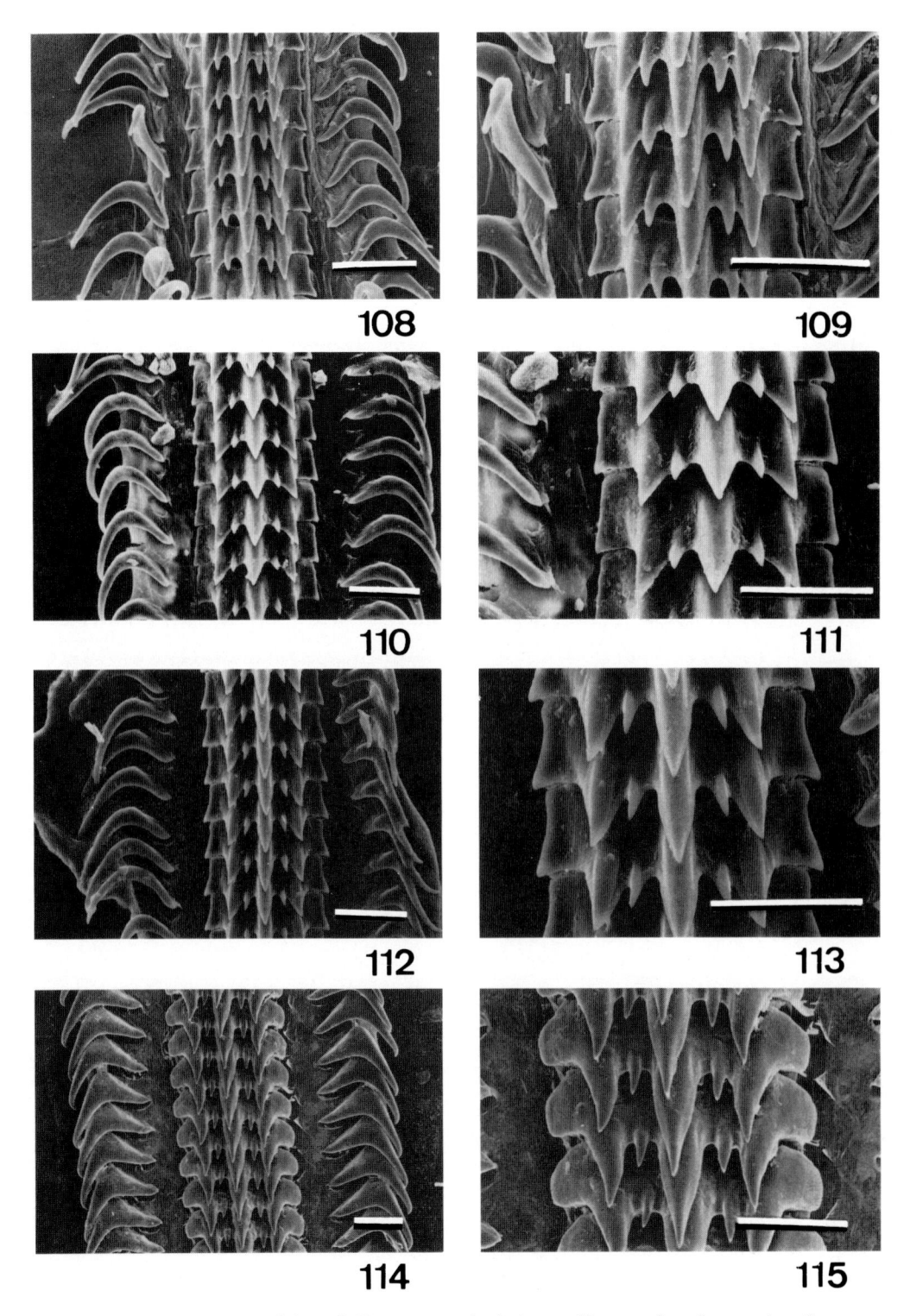

FIGS 108-115. — Radulae of *Chicoreus* sp. (scale bars : 50 µm ; otherwise mentioned)
108-109, *Chicoreus subpalmatus* Houart. New Caledonia (MNHN).
110-111, *C. nobilis* Shikama. Coral Sea (MNHN).
112-113, *C. rossiteri* (Crosse). New Caledonia (RH).
114-115, *C. pliciferoides* Kuroda. New Caledonia (MNHN) (114 : scale bar 100 µm).

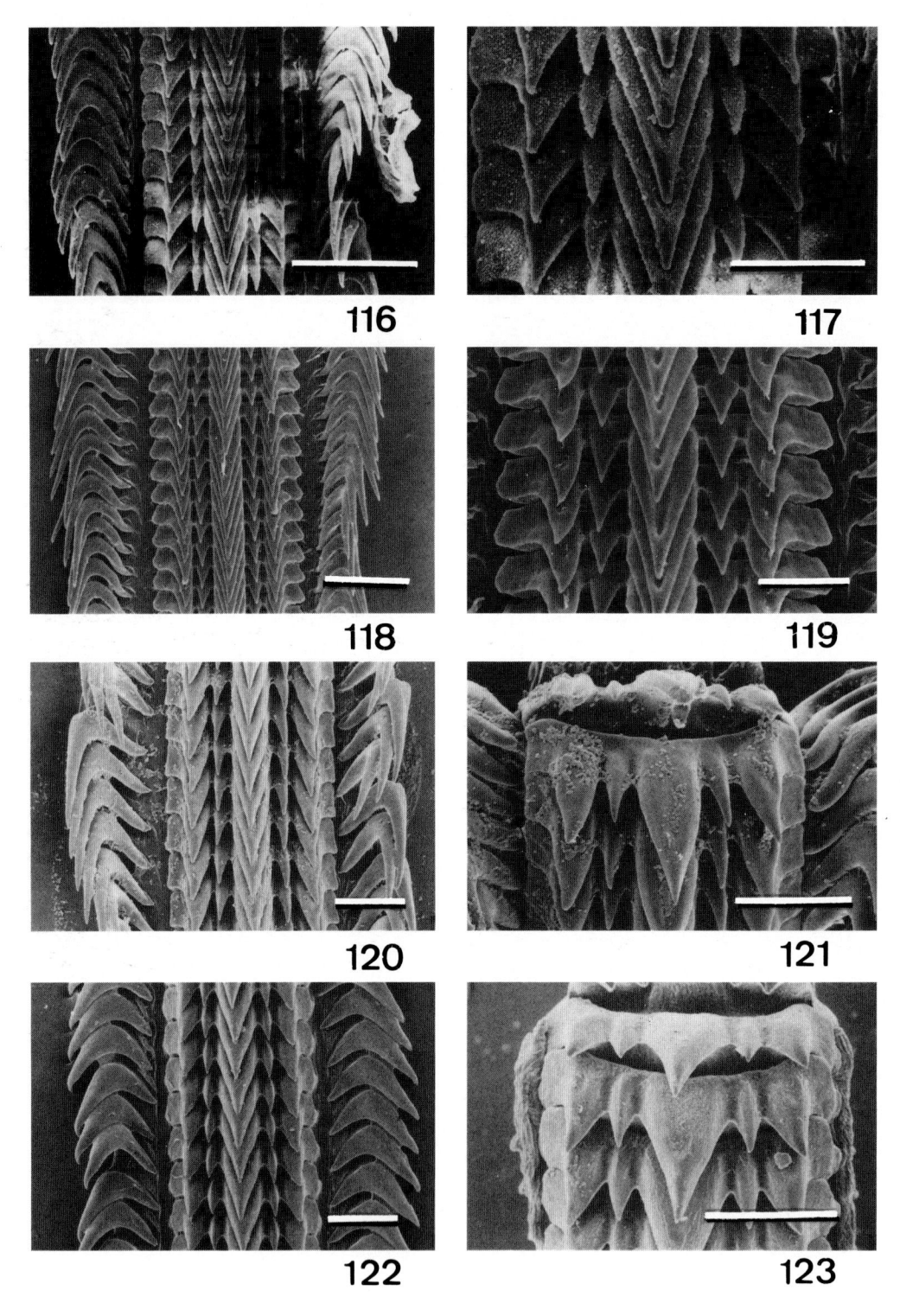

FIGS 116-123. — Radulae of *Chicomurex* sp. (scale bars : 50 µm ; otherwise mentioned)
116-119, *Chicomurex venustulus* (Rehder & Wilson). 116-117, Taiwan (RH). 118-119, New Caledonia (MNHN) (116 and 118 : scale bar 100 µm).
120-121, *C. laciniatus* (Sowerby). Papua New Guinea (IRSNB).
122-123, *C. turschi* (Houart). Papua New Guinea (IRSNB).

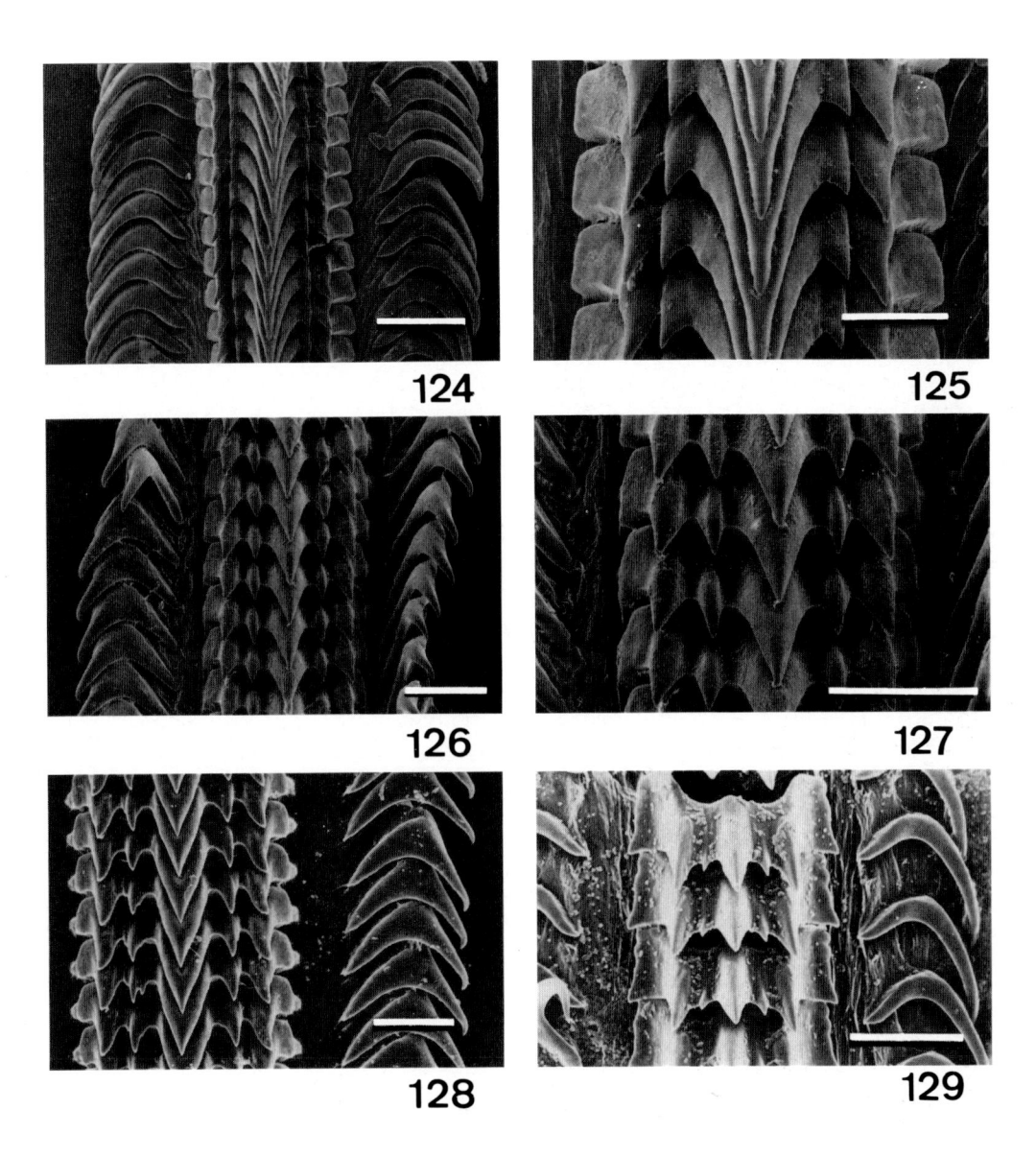

FIGS 124-129. — Radulae of *Chicomurex*, *Naquetia* and *Chicoreus* sp.
(scale bars : 50 µm ; otherwise mentioned)
124-125, *Chicomurex superbus* (Sowerby). Coral Sea, illustrated Fig. 424 (MNHN) (124 : scale bar 100 µm).
126-127, *C. protoglobosus* n.sp. New Caledonia, illustrated Fig. 427 (holotype, MNHN).
128, *Naquetia triqueter* (Born). Papua New Guinea (IRSNB).
129, *Chicoreus orchidiflorus* (Shikama). Philippines (RH).

ABBREVIATIONS

AIM : Auckland Institute and Museum, Auckland, New Zealand
AMS : Australian Museum, Sydney, Australia
ANSP : Academy of Natural Sciences of Philadelphia, USA
BMNH : The Natural History Museum, London, UK
SMF : Forschungsinstitut Senckenberg, Frankfurt, Germany
GSI : Geological Survey of India, Calcutta, India
HLD : Hessisches Landesmuseum, Darmstadt, Germany
HUJ : Hebrew University, Jerusalem, Israel
IGPS : Institute of Geology and Paleontology, Tohoku University, Sendai, Japan
IMG : Istituto e Museo di Geologia, Palermo, Italy
IMT : Institute of Malacology, Tokyo, Japan
IRSNB : Institut Royal des Sciences Naturelles de Belgique, Bruxelles, Belgium
KPM : Kanagawa Prefectural Museum, Yokohama, Japan
LM : Löbbecke Museum und Aquarium, Düsseldorf, Germany
MCSN : Museo Civico di Storia Naturale, Genova, Italy
MGM : Mineralogisch-Geologisch Museum, Delft, The Netherlands
MHNG : Muséum d'Histoire Naturelle, Genève, Switzerland
MHNM : Muséum d'Histoire Naturelle, Marseille, France
MNHN : Muséum National d'Histoire Naturelle, Paris, France
MV : Museum of Victoria, Melbourne, Australia
NHMB : Naturhistorisches Museum, Basel, Switzerland
NHMW : Naturhistorisches Museum, Wien, Austria
NM : Natal Museum, Pietermaritzburg, South Africa
NMW : National Museum of Wales, Cardiff, UK
NSMT : National Science Museum, Tokyo, Japan
NZGS : New Zealand Geological Survey, Lower Hutt, New Zealand
OM : Otago Museum, Dunedin, New Zealand
QM : Queensland Museum, South Brisbane, Australia
RH : R. Houart collection
RML : Rijksmuseum Geologie en Paleontologie, Leiden, The Netherlands
RMNH : Rijksmuseum voor Natuurlijke Historie, Leiden, The Netherlands
SAM : South Australian Museum, Adelaide, Australia
SDNHM and SDSNH : San Diego Natural History Museum, San Diego, USA
TU : Tulane University, New Orleans, USA
TUF : Tokyo University of Fisheries, Tokyo, Japan
UMZ : University Museum of Zoology, Cambridge, UK
UO : University of Otago, Dunedin, New Zealand
USNM : National Museum of Natural History, Washington D.C., USA
WAM : Western Australian Museum, Perth, Australia
WPU : Wilhelm-Pieck Universität, Rostock, Germany
YCM : Yokosuka City Museum, Yokosuka, Kanagawa, Japan
ZMA : Instituut voor Taxonomische Zoölogie, Zoölogisch Museum, Amsterdam, The Netherlands

ZMB : Museum für Naturkunde der Humboldt Universität zu Berlin, Zoologisches Museum, Germany
ZSI : Zoological Survey of India, Calcutta, India
ZSM : Zoologischen Staatssammlung, München, Germany
lv : live-taken specimen(s)
dd : empty shell(s)

SYSTEMATICS

Genus *CHICOREUS* Montfort, 1810

Subgenus *CHICOREUS* Montfort, 1810

Chicoreus Montfort, 1810 : 610. Type-species (by original designation) : *Murex ramosus* Linné, 1758.

Frondosaria Schlüter, 1838 ; 20. Type-species (by subsequent designation, VOKES, 1967 : 7) : *Frondosaria inflata* (Lamarck, 1822) = *Murex ramosus* Linné, 1758.
Euphyllon Jousseaume, 1880 : 335. Type-species (by original designation) : *Murex monodon* Sowerby, 1825 [= *Purpura cornucervi* Röding, 1798].

DESCRIPTION. Shell up to 327 mm in length ; contour roughly trigonal, with three frondose varices. Other axial sculpture consisting generally of one or two intervaricial axial ridges or nodes ; suture well impressed ; aperture rounded, bearing a small to large labral tooth. Inner side of outer lip lirate for a short distance within ; anal notch of shallow to moderate depth ; siphonal canal generally medium-sized to long. Generally whitish.

GEOGRAPHICAL DISTRIBUTION. Indo-West Pacific including the Red Sea.

REMARKS. FRANÇOIS (1891) reported that the labral tooth of *C. ramosus* is used as a tool to open and to maintain opened the valves of a bivalve (*Arca* sp.) during feeding. I know of no additional reports pertaining to the use of the labral tooth as a tool in this or other species of *Chicoreus* but PONDER & VOKES (1988 : 7) provided references to the function of the labral tooth in the genus *Acanthina*.

Subgenus *TRIPLEX* Perry, 1810

Triplex Perry, 1810, pl. 23. Type-species (by monotypy) : *Triplex foliatus* Perry, 1810 (= *Murex palmarosae* Lamarck, 1822).

Pirtus de Gregorio, 1884 : 257. Type-species (by monotypy) : *Murex (Pirtus) fiatus* de Gregorio, 1884 (= *Murex dujardini* Tournouer, 1875).
Torvamurex Iredale, 1936 : 323. Type-species (by original designation) : *Triplex denudata* Perry, 1811.

DESCRIPTION. Shell medium-sized, up to 130 mm in length ; fusiform, bearing 3 rounded and mostly frondose varices. Other axial sculpture consisting of 1 to 3 ridges ; whorls suture well impressed ; aperture roundly-ovate to ovate ; outer lip crenulate or weakly undulate, no labral tooth, striated for a short distance within ; anal notch shallow to deep. Siphonal canal medium-sized. Colour generally brownish.

GEOGRAPHICAL DISTRIBUTION. Indo-West Pacific uncluding the Red Sea ; western and eastern Atlantic.

REMARKS. Both names *Chicoreus* and *Triplex* were proposed in 1810. The exact date of Montfort work is not known however IREDALE (1915 : 457) has shown that it was reviewed in the Göttinger Anzeiger, issued May 28, 1810 and thus precedes PERRY'S work which did not appear until June (VOKES, 1964 : 8).

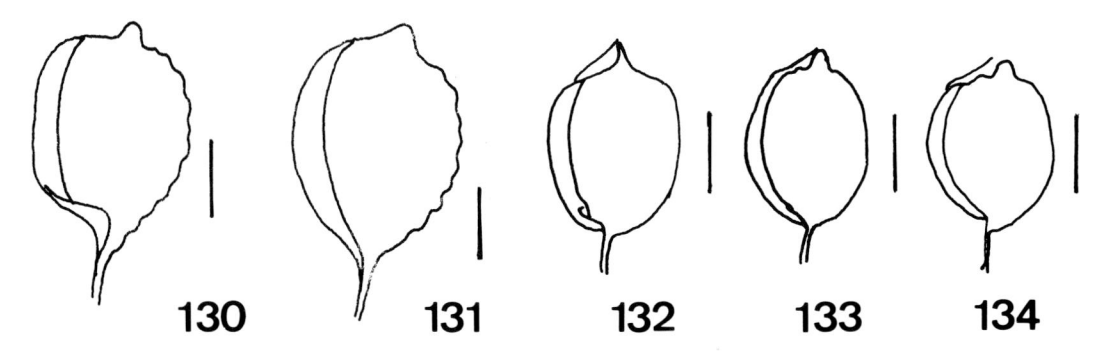

FIGS 130-134. — Apertures (scale bars of 91 and 92 : natural size ; other figures : 0.5 mm)
130, *Chicoreus ryukyuensis* Shikama.
131, *C. cloveri* Houart.
132, *C. torrefactus* (Sowerby).
133, *C. microphyllus* (Lamarck).
134, *C. dovi* Houart.

Subgenus *SIRATUS* Jousseaume, 1880

Siratus Jousseaume, 1880 : 335. Type-species (by original designation) : *Purpura sirat* "Adanson" Jousseaume, 1880 (= *Murex senegalensis* Gmelin, 1791).

DESCRIPTION. Shell large, up to 184 mm in length, with three winglike varices with one short to long shoulder spine. Spire high. Aperture rounded ; anal sulcus shallow and broad, outer lip striate for short distance within. Siphonal canal long or of moderate length.

GEOGRAPHICAL DISTRIBUTION. West Pacific Ocean and West Atlantic Ocean. More common in latter.

Subgenus *RHIZOPHORIMUREX* Oyama, 1950

Rhizophorimurex Oyama, 1950 : 10. Type-species (by original designation) : *Murex capuchinus* (sic) Lamarck, 1822.

DESCRIPTION. Shell medium-sized, up to 124 mm in length, fusiform, with 3 rounded, almost spineless varices. Aperture ovate. Columellar lip fully adherent, anal notch shallow. Siphonal canal short and broad, spineless, occasionally with short webbed expansion abapically. Colour brownish.

GEOGRAPHICAL DISTRIBUTION. North-East Indian Ocean and West Pacific Ocean.

Subgenus *CHICOPINNATUS* n. subgen.

Type-species : *Pterynotus orchidiflorus* Shikama, 1972.

ETYMOLOGY. *Chico* : from *Chicoreus* ; *pinnatus* : winged.

DESCRIPTION. Three wing-like varices and 1 to 3 axial ridges on last whorl. Aperture roundly-ovate to ovate with crenulated outer lip. Inner side of outer lip lirate. Anal notch moderately deep. Siphonal canal long, spiny, narrowly open.

GEOGRAPHICAL DISTRIBUTION. Indo West-Pacific.

REMARKS. The three species included in this new subgenus have been variously classified in *Pterynotus* (*s.s.*) and other genera, but early teleoconch whorls are entirely different from *Pterynotus*, and essentially similar to those in *Chicoreus* (s.l.). In *Pterynotus* the first teleoconch whorl is ornamented with 6 small varices that progressively disappear on the second and third whorl, leaving 3 varices with 1 intervaricial node per whorl. In *Chicopinnatus* there are 8-12 small axial ridges on the first teleoconch whorl, some of which become webbed varices on subsequent whorls, the others remaining as intervaricial ridges. The long, slender siphonal canal and the internally lirate rather than dentate outer lip are additional differences between *Chicopinnatus* and *Pterynotus*.

Chicopinnatus differs from *Chicoreus* (s.l.) by its spineless, webbed varices.

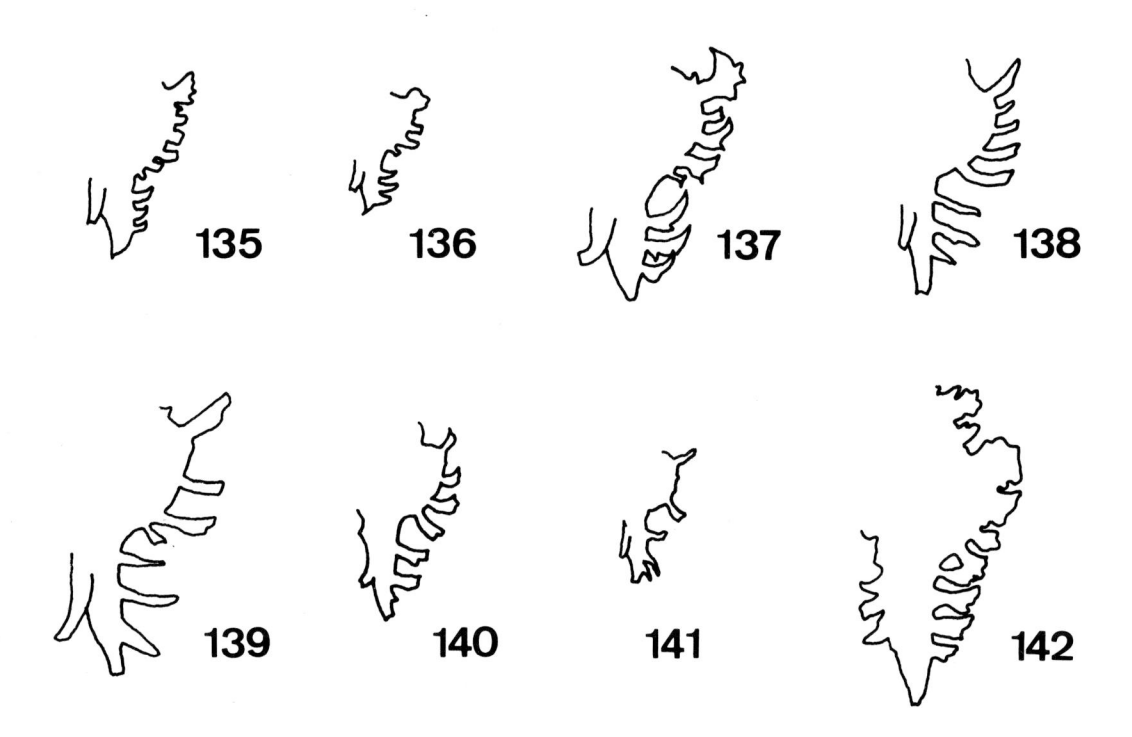

FIGS 135-142. — Spines ornamentation (approximately natural size)
135, *Chicoreus ryukyuensis* Shikama.
136, *C. cloveri* Houart.
137, *C. nobilis* Shikama.
138, *C. rossiteri* (Crosse).
139, *C. aculeatus* (Lamarck).
140, *C. zululandensis* Houart.
141, *C. crosnieri* Houart.
142, *C. fosterorum* Houart.

Genus **CHICOREUS** Montfort, 1810

Subgenus **CHICOREUS** (*s.s.*)

Chicoreus (Chicoreus) asianus Kuroda, 1942
Figs 143, 144, 145

Murex sinensis Reeve, 1845 : pl. 6, fig. 24.
Chicoreus asianus Kuroda, 1942 : 80 (new name for *Murex sinensis* Reeve, 1845, non Gmelin, 1791).
Chicoreus orientalis Zhang, 1965 : 18, pl. 2, fig. 2.

ADDITIONAL REFERENCES

Chicoreus asianus. — KIRA, 1965 ; 61, pl. 23, fig. 15 ; ZHANG, 1965 : 18, pl. 1, fig. 5 ; KURODA, HABE & OYAMA, 1971 : 139, pl. 40, fig. 1 ; KAICHER, 1973 : card 133 ; RADWIN & D'ATTILIO, 1976 : 32, pl. 6, fig. 8 ; EISENBERG, 1981 : 87, pl. 69, fig. 11 ; ABBOTT & DANCE, 1982 : 136, text fig. ; LAI, 1987 : 63, pl. 30, fig. 4.
Murex (Chicoreus) asianus. — SMITH, 1953 : 7, pl. 7, fig. 2 [not pl. 3, fig. 2 = *Chicoreus (Triplex) brevifrons* (Lamarck, 1822)].
Chicoreus (Chicoreus) asianus — FAIR, 1976 : 22, pl. 6, fig. 82.
NOT *Murex asianus.* — SMITH, 1953 : pl. 3, fig. 2 [= *Chicoreus (Triplex) brevifrons* (Lamarck, 1822)].

TYPE LOCALITIES. *M. sinensis* : China ; *C. orientalis* : Chiu-Po, Canton.

TYPE MATERIAL. *M. sinensis* : none ; *C. orientalis* : unknown.

MATERIAL EXAMINED. c. 50 specimens from throughout the distributional range.

DISTRIBUTION. (Fig. 143) China : Che-Kiang, Xiamen, Canton ; Hong Kong ; Taiwan ; South Korea ; Japan, Boso Peninsula as north limit. Depth range : 5-20 m, on rocks or muddy rocks.

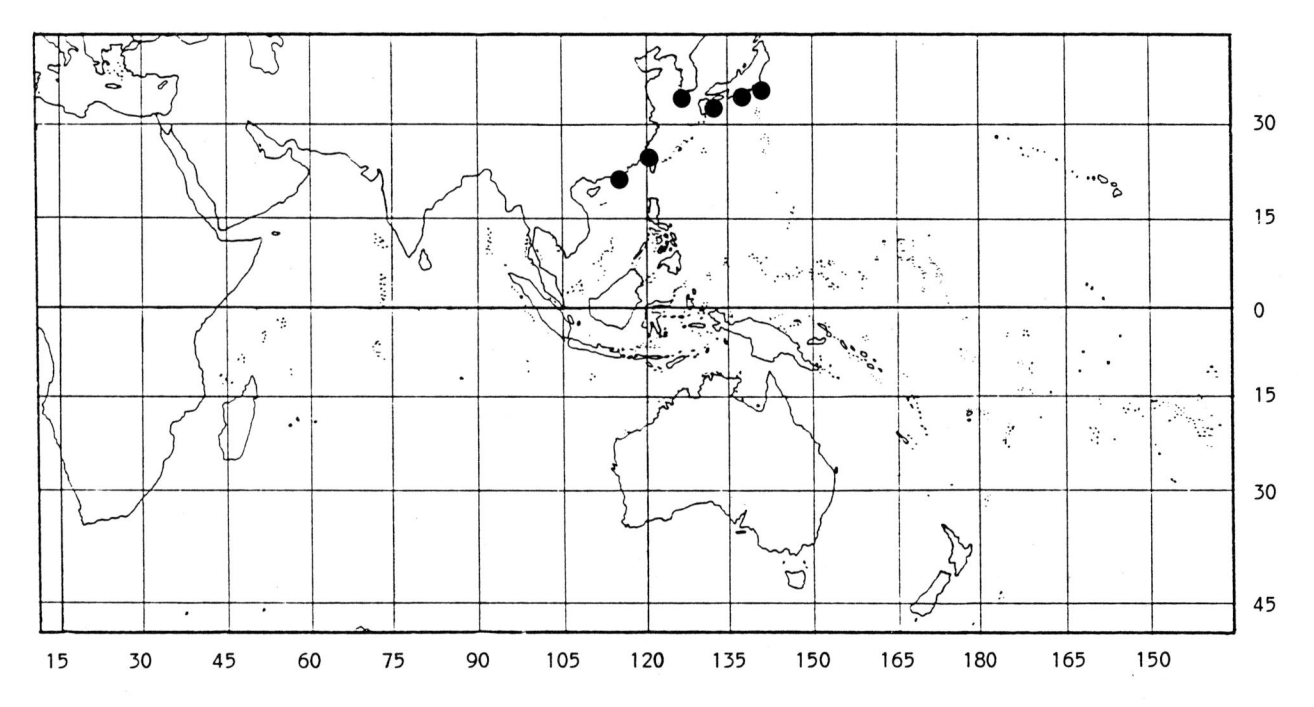

FIG. 143. — Distribution of *Chicoreus asianus* Kuroda.

DESCRIPTION. Shell up to 125 mm in length. Spire high, protoconch unknown (eroded or broken in all the specimens studied), but apparently paucispiral with rounded whorls ; up to 8 rounded teleoconch whorls, suture deeply impressed.
Last teleoconch whorl with 3 frondose varices, each with 5 major spines and intermediate spinelets. Shoulder spine longest. Other axial sculpture consisting of 2 or 3 low axial ridges (usually 2). Spiral sculpture consisting of small cords connecting the varicial spines, and numerous finer threads between each cord.

Aperture broad and rounded. Columellar lip smooth, rim adherent or slightly erect. Anal notch deep, relatively narrow. Outer lip denticulate, bearing a relatively strong labral tooth adapically ; interior of outer lip strongly lirate for short distance within. Siphonal canal broad, moderately long, narrowly open, bent abaperturally, with 3 straight open spines.
Light to dark brown with darker traces on the spiral sculpture. Aperture white, edge of the columellar lip usually light pink.

REMARKS. When REEVE (1845, sp. 24) named *Murex sinensis* he overlooked an earlier use of this binomen by GMELIN (1791), KURODA (1942) synonymised *Murex elongatus* Lamarck, 1822 (not Lightfoot, 1786) and *Murex sinensis* Reeve, 1845 and proposed *Chicoreus asianus* as a replacement

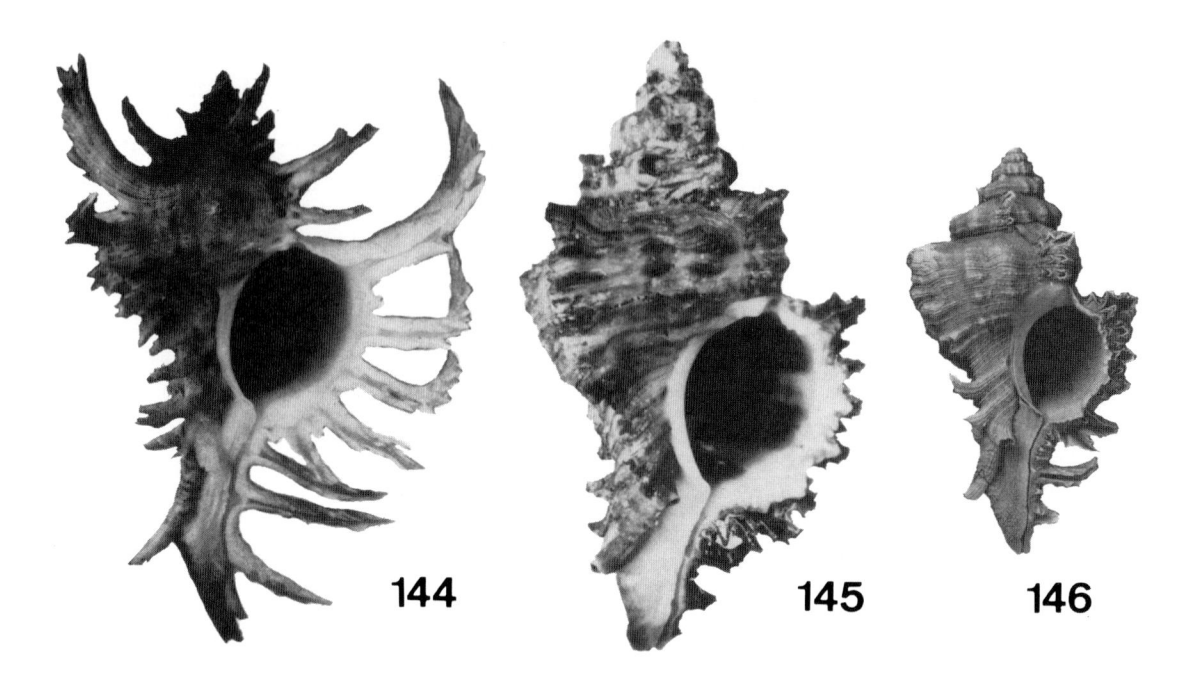

FIGS 144-146. — Subgenus *C. (Chicoreus)*.
144-145, *C. asianus* Kuroda. 144, Wakayama, Japan, 106 mm (RH). 145, Hong Kong, 72 mm (MNHN).
146, *C. austramosus* Vokes. Off Natal, South Africa, 53.5 mm (holotype, NM 5348/T2141. Courtesy of E.H. VOKES).

name. The new name is applicable only to *Murex sinensis*, however, because *Murex elongatus* Lamarck is undoubtedly *Murex brevifrons* Lamarck, 1822, a West-Indian species of *Chicoreus (Triplex)*. Some forms of *C. brevifrons* closely resemble *C. asianus* but the shell is easily separable by the presence of a labral tooth in *C. asianus* (none in *C. brevifrons*). Judging from the original illustration (type material not seen), and the description, kindly translated for me by T.C. Lan, *C. orientalis* is indistinguishable from *Murex sinensis* and is thus subjective synonym of *C. asianus*.

Chicoreus (Chicoreus) austramosus Vokes, 1978
Figs 2, 146, 147, 271

Chicoreus (C.) austramosus Vokes, 1978 : 388, pl. 4, figs 1-2.

ADDITIONAL REFERENCES

Chicoreus austramosus. — KAICHER, 1979 : card 1995 (holotype); HOUART, 1981c : 16 (text fig.); KILBURN & RIPPEY, 1982 : 81, pl. 18, fig. 2; RIPPINGALE, 1987 : 3, fig. 3.

TYPE LOCALITY. Off Natal (*ex pisce*).

TYPE MATERIAL. Holotype NM 5348/T2141.

OTHER MATERIAL EXAMINED. Transkei coast, S. Africa, RH (3 dd); Pondoland, South Africa, RH (1 dd); Transkei, on reef, 35 m, South Africa, RH (1 lv); off Transkei coast, 30-35 m, South Africa, coll. Glass & Foster (2 lv); Port Alfred, South Africa, AMS C147585 (3 lv).

DISTRIBUTION. (Fig. 147). Off Durban to Western Transkei, southern Africa. Depth range : 30-35 m.

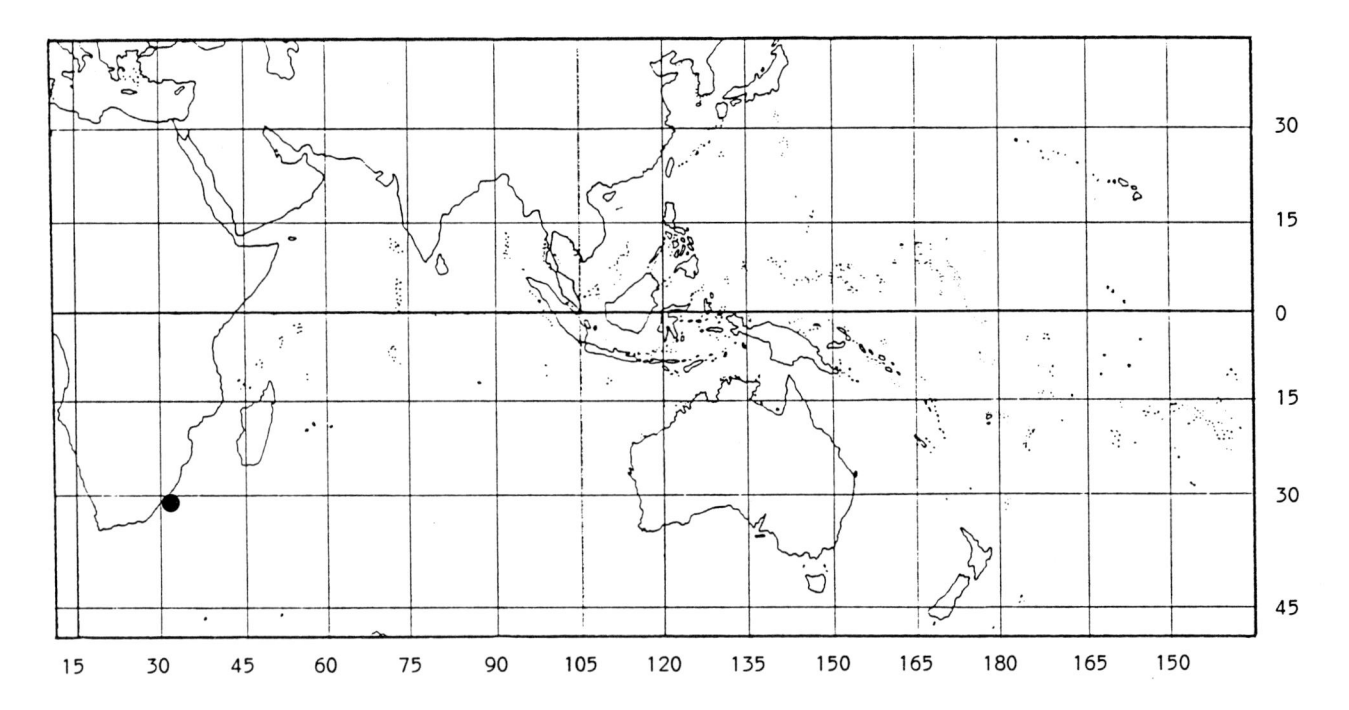

FIG. 147. — Distribution of *Chicoreus austramosus* Vokes.

DESCRIPTION. Shell up to 65 mm in length, heavy. Spire high, with 2 protoconch whorls and up to 7 teleoconch whorls. Suture slightly appressed. Protoconch large and bulbous.

Last teleoconch whorl with 3 rounded varices, each with 5 short foliated spines, spines being slightly longer at shoulder and on most adapical part of each varice. Other axial sculpture consisting of 2 intervaricial nodes (rarely one). Spiral sculpture consisting of cords interconnecting varical fronds, and numerous spiral threads in each interspace.

Aperture rounded. Columellar lip smooth, adherent, rim slightly erect abapically. Anal notch deep, broad, delineated by small callus. Outer lip denticulate, bearing a labral tooth adapically. Siphonal canal moderately long, narrowly open, abaperturally bent at tip, ornamented with 2 upcurved spines.

Light brown suffused with some darker blotches to light or bright orange. Aperture white, sometimes suffused with pink.

REMARKS. The species was originally compared with *C. ramosus* from which it differs in being much smaller, more narrowly fusiform and in having a different protoconch, twice the diameter of that of *C. ramosus*. It is nearer to *C. asianus* but differs from that species in being smaller and in having fewer spines on the siphonal canal. *C. asianus* has a comparatively higher spire and the varicial spines are usually larger. *C. asianus* is apparently restricted to southeast Asia.

Chicoreus (Chicoreus) bundharmai Houart, 1992
Figs 6, 148, 280-281

Chicoreus (Chicoreus) bundharmai Houart, 1992 : 27, figs 1, 5-8.

TYPE LOCALITY. Banjarmasin, South Borneo, 03°22' S, 114°33' E, 20 m.

TYPE MATERIAL. Holotype MNHN.

MATERIAL EXAMINED. Type material, and 7 paratypes.

DISTRIBUTION. (Fig. 148). South Borneo, Java Sea, 20-40 m.

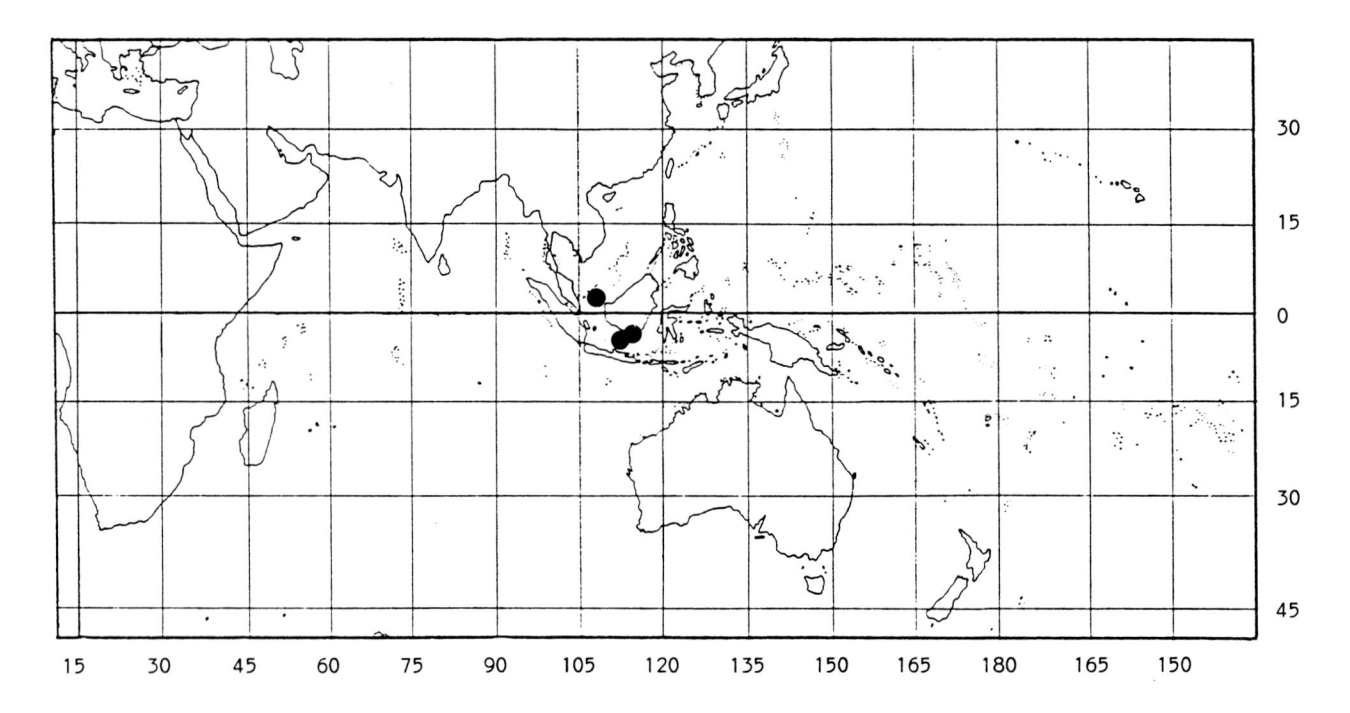

FIG. 148. — Distribution of *Chicoreus bundharmai* Houart.

DESCRIPTION. Shell medium sized for the genus, 51-68 mm in length at maturity, frondose. Spire moderately high with 2.25 protoconch whorls and up to 6 weakly shouldered, broad teleoconch whorls. Protoconch whorls rounded, sculptured with strong axial ribs, more apparent on last whorl; terminal varix strong, thick, almost straight. Suture impressed.

First and second teleoconch whorls with 10 axial ribs, third whorl with axial ribs and forming varices, fourth, fifth and sixth teleoconch whorls with 4 varices. Each varix of last whorl with 5 medium sized spines, adapical (shoulder) spine longest, other spines decreasing in length abapically. Short intermediate spine between shoulder spine and second abapical spine. Spiral sculpture throughout consisting of numerous cords of varying strength.

Aperture rounded. Columellar lip smooth, weakly erect abapically, adherent adapically. Anal notch shallow, broad. Outer lip erect, crenulate; strong, large, narrow labral tooth abapically. Siphonal canal of moderate length, narrow, straight, abaperturally bent at tip, narrowly open, with 3 short open spines.

Light brown with darker spiral cords, and brown peripheral band at adapical part of whorls. Aperture whitish with pink rim.

REMARKS. *Chicoreus bundharmai* is close to *C. ramosus* (Linné, 1758) from which it differs primarly in its longer, more prominent labral tooth, in having 4 varices instead of 3 on the last teleoconch whorl, and in having a strongly axially sculptured, instead of smooth, protoconch (Fig. 6). *Chicoreus bundharmai* is also similar to *C. cornucervi* (Röding, 1798) but differs in having 4 varices on the last teleoconch whorl, in attaining smaller size with the same number of teleoconch whorls (average size of *C. cornucervi* is of 90 mm in length), while the varicial spines are generally shorter. The protoconch in *C. cornucervi* is twice as large and strongly keeled.

Chicoreus (Chicoreus) cornucervi (Röding, 1798)

Figs 1, 149, 277, 278

Purpura cornucervi Röding, 1798 : 142.
Murex monodon Sowerby, 1825 : 19.
Murex aranea de Blainville *in* Kiener, 1842 : 34, pl. 36, fig. 1.

ADDITIONAL REFERENCES

Murex (Euphyllon) cornucervi. — SHIKAMA, 1963 : 70, pl. 53, fig. 3.
Euphyllon cornucervi. — CERNOHORSKY, 1967b : 124, pl. 26, fig. 155 ; HINTON, 1972 : 36, pl. 18, fig. 5-7.
Chicoreus cornucervi. — D'ATTILIO, 1967 : 6, fig. 5 ; WILSON & GILLETT, 1971 : 86, pl. 58, fig. 5 ; KAICHER, 1973 : card 155 ; RADWIN & D'ATTILIO, 1976 : 36, pl. 4, fig. 11 ; HINTON, 1979 : pl. 26, fig. 4 ; ABBOTT & DANCE, 1982 : 138, text fig. ; WELLS & BRYCE, 1985 : 88, pl. 25, fig. 284 ; SHORT & POTTER, 1987 : 56, pl. 27, fig. 15.
Chicoreus (Chicoreus) cornucervi. — FAIR, 1976 : 33, pl. 8, fig. 101.
Murex (Chicoreus) cornucervi. — EISENBERG, 1981 : 89, pl. 71, fig. 3.
Murex cornucervi. — LEEHMAN, 1981 : 9, text fig.

TYPE LOCALITIES. *P. cornucervi* and *M. monodon* : East Indish Sea (MARTINI, 1777) ; *M. aranea* : Indian Ocean.

TYPE MATERIAL. None.

OTHER MATERIAL EXAMINED. c. 40 specimens from throughout the distributional range.

DISTRIBUTION. (Fig. 149). North-West Cape, West Australia ; Dampier to Darwin ; Torres Strait ; northern Australia ; Arafura Sea ; south-western New Guinea.

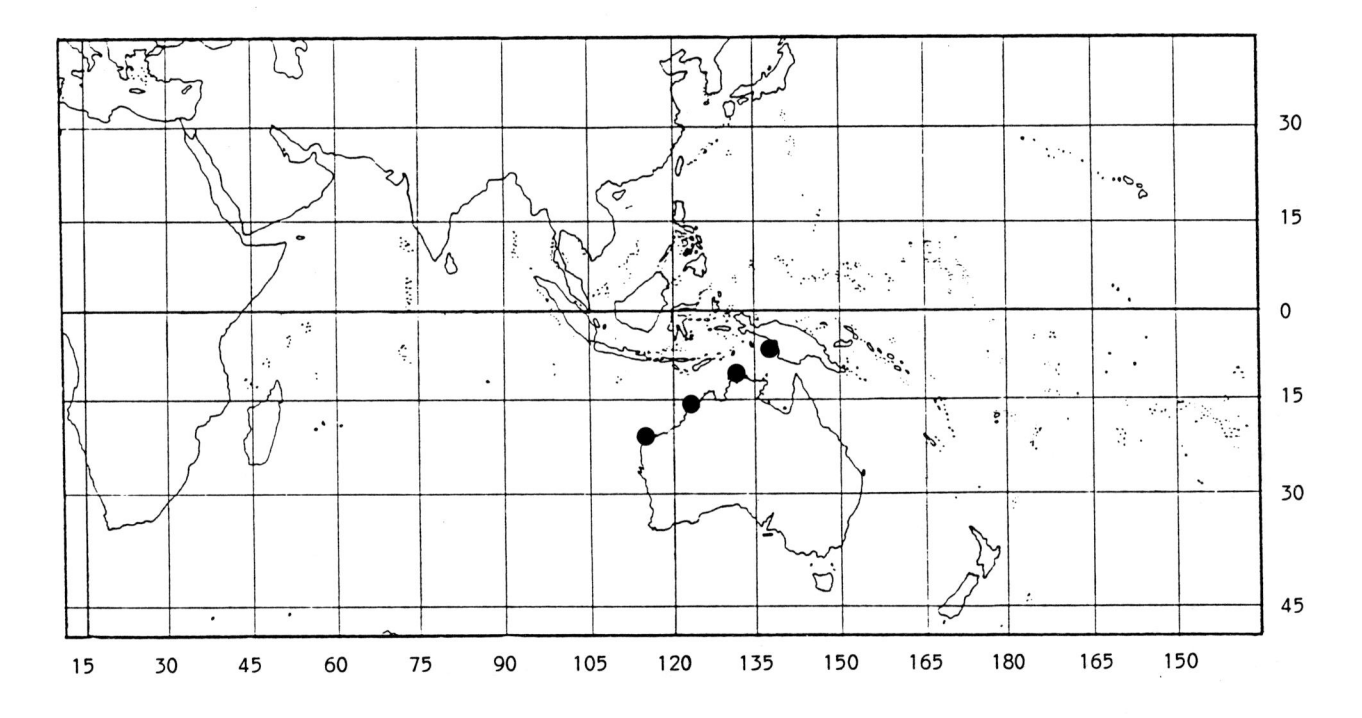

FIG. 149. — Distribution of *Chicoreus cornucervi* (Röding).

DESCRIPTION. Shell up to 140 mm in length, whorls rounded. Spire high, with 2 1/2 protoconch whorls and up to 6 rounded teleoconch whorls. Suture deeply impressed. Protoconch large, first whorl flattened, last whorl sculptured with low axial ridges.

Last teleoconch whorl with 3 low rounded varices, each ornamented with 4 or 5 frondose spines and intermediate spinelets, 3 adapical spines very long and often strongly bent abaperturally ; 1 or 2 smaller abapical spines. No other axial sculpture on last whorl. Spire whorls with one low intervaricial axial node. Spiral sculpture consisting of spiral cords connecting varicial spines, and numerous threads in each interspace.

Aperture rounded. Columellar lip smooth, rim slightly detached. Anal notch broad, shallow. Outer lip denticulate, with a strong labral tooth abapically. Siphonal canal broad, relatively long, narrowly open, bent abaperturally, with 2 frondose spines, shoulder spine very long, recurved, second spine short.

Whitish or light brown to nearly black. Aperture bluish-white, rim light pink.

REMARKS. One of the most easily recognizable species because of its strongly curved fronds and large protoconch, the latter suggesting intracapsular metamorphosis, which may explain its restricted geographical range.

Chicoreus (Chicoreus) litos Vokes, 1978
Figs 3, 150, 273, 279

Chicoreus (Chicoreus) litos Vokes, 1978 ; 390, pl. 5, fig. 1.

ADDITIONAL REFERENCES

Murex axicornis. — BARNARD, 1959 : 196 ; KENSLEY, 1973 : 140, fig. 472 (not *Murex axicornis* Lamarck, 1822).
Chicoreus litos. — KAICHER, 1979 : card 1973 (holotype).

TYPE LOCALITY. North-east of Beira, Mozambique, 75 meters.

TYPE MATERIAL. Holotype NM G8656/T2130.

OTHER MATERIAL EXAMINED. Aliwal Shoal, South Coast of Natal, 30°15′ S, 30°49′ E, 45 m., coll. GLASS & FOSTER (1 lv) ; trawled South east of Durban Bluff, 220 m, ex paratype NM T2146, RH (1 dd) ; trawled off southern Mozambique, RH (1 dd).

DISTRIBUTION. (Fig. 150). Natal, southern Africa ; Mozambique.

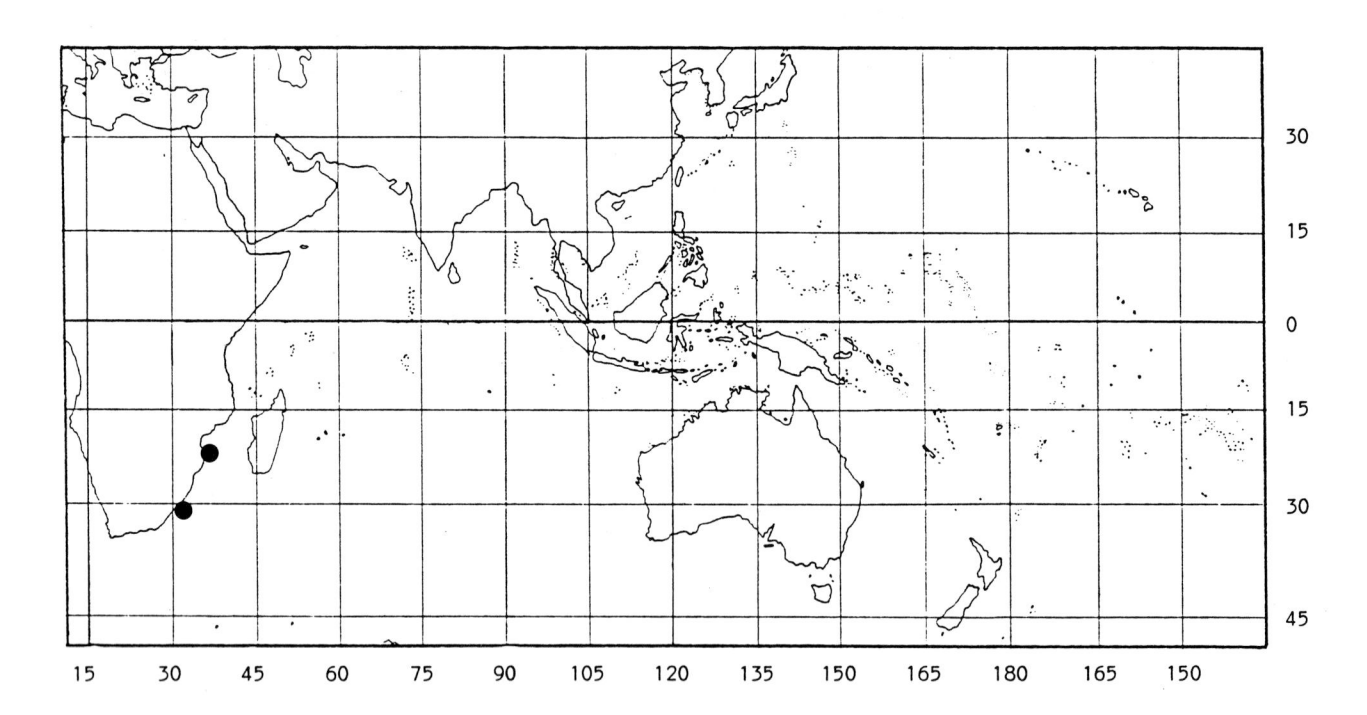

FIG. 150. — Distribution of *Chicoreus litos* Vokes.

DESCRIPTION. [In part after VOKES (1978) with some changes].

Shell up to 78 mm in length, stout. Spire high, with 2 1/2 protoconch whorls and up to 6 teleoconch whorls. Suture appressed. Protoconch bulbous and extended.

Last whorl bearing 3 spinose varices. Shoulder spine longer than 3 abapical spines. Abapical spines sometimes connected by a web, and with small intermediate spinelets. Other axial sculpture consisting of a single intervaricial node. Spiral sculpture consisting initially of 4 cords, gradually blending into intercalated threadlets to give an overall pattern of numerous equal sized threads that cover entire surface.

Aperture rounded. Columellar lip smooth, erect at abapical extremity. Outer lip crenulate, with a small labral tooth abapically. Siphonal canal wide, elongate, abaperturally recurved at distal end, ornamented with 2 short, straight spines, the most adapical spine longer.

White to tan, darker on summits of intervaricial nodes and spiral threads. Aperture white.

REMARKS. *C. litos* was formerly confused with *C. axicornis* both by BARNARD (1959 : 196, fig. 41 C) and KENSLEY (1973 : fig. 472), but *C. axicornis* differs in having a more angulate shell, a more bulbous protoconch, and in lacking a labral tooth. No closely related species are known.

Chicoreus (Chicoreus) ramosus (Linné, 1758)
Figs 4-5, 151, 270, 272, 276

Murex ramosus Linné, 1758 : 747.

Purpura incarnata Röding, 1798 : 142.
Purpura fusiformis Röding, 1798 : 144.
Murex inflatus Lamarck, 1822 : 160.
Murex frondosus Mörch, 1852 : 97.
Murex fortispinna François, 1891 : 240, fig. 241.

ADDITIONAL REFERENCES

Murex (Chicoreus) ramosus. — SMITH, 1953 ; 7, pl.22, fig. 5 ; EISENBERG, 1981 : 93, pl. 75, fig. 1.
Chicoreus ramosus. — SPRY, 1961 : 19, pl. 2, fig. 134 ; KIRA, 1965 : 61, pl. 23, fig. 17 ; ZHANG, 1965 : 17, pl. 2, fig. 4 ; CERNOHORSKY, 1967a : 120, text fig. 5, pl. 14, fig. 11 ; CERNOHORSKY, 1967b : 122, pl. 25, fig. 152 ; D'ATTILIO, 1967 : 6, fig. 6 ; WILSON & GILLETT, 1971 : 86, pl. 58, fig. 1 ; HINTON, 1972 : 34, pl. 17, fig. 12 ; ASTARY, 1973 : 7 ; KAICHER, 1973 : card 132 ; SALVAT & RIVES, 1975 : 311, fig. 191 ; RADWIN & D'ATTILIO, 1976 : 40, fig. 21, pl. 4, fig. 8 ; HINTON, 1979, pl. 26, fig. 1 ; ABBOTT & DANCE, 1982 : 138, text fig. ; BOSCH & BOSCH, 1982 : 89, text fig. ; KILBURN & RIPPEY, 1982 : 81, pl. 18, fig. 1 ; SMYTHE, 1982 : 59 ; SHARABATI, 1984, pl. 18, fig. 9 ; LAI, 1987 : 61, text figs A & B ; SHORT & POTTER, 1987 : 56, pl. 27, fig. 11 ; DRIVAS & JAY, 1988 : 68, text fig.
Murex ramosus. — KENNELLY, 1964 : 69, pl. 17, fig. 85 ; BROST & COALE, 1971 : 69, pl. 15, figs 1, 2 ; KENSLEY, 1973 : 140, fig. 476.
Chicoreus (Chicoreus) ramosus. — FAIR, 1976 : 71, pl. 9, fig. 121 ; VOKES, 1978 : 387, pl. 4, fig. 3 ; SPRINGSTEEN & LEOBRERA, 1986 : 134, pl. 36, fig. 11.

TYPE LOCALITIES. *M. ramosus* : " Jamaica " ; *P. incarnata* and *M. inflatus* : Red Sea (MARTINI, 1777) ; *P. fusiformis* : none ; *M. frondosus* : West India ; *M. fortispinna* : New Caledonia.

TYPE MATERIAL. None.

OTHER MATERIAL EXAMINED. More than 200 specimens from throughout the distributional range.

DISTRIBUTION. (Fig. 151). Throughout the Indo-West Pacific : The Red Sea to Durban Bay, southern Africa ; the Gulf of Oman ; Sri Lanka ; South China Sea to south of Japan ; north-western Australia ; eastern Queensland to the Marquesas (East limit).

DESCRIPTION. Shell up to 327 mm in length, stout. Spire low to moderately high, 2 protoconch whorls and up to 9 rounded teleoconch whorls. Suture slightly appressed. Protoconch whorls rounded.

Last whorl bearing 3 frondose varices, each ornamented with 5 major frondose spines and smaller intermediate spinelets ; shoulder spine strongest and often longest ; length of other spines variable. Other axial sculpture consisting of one narrow major intervaricial node with usually a second smaller one. Spiral sculpture consisting of cords connecting the varicial spines and minor cords that connect intermediate spinelets ; numerous additional spiral threads.

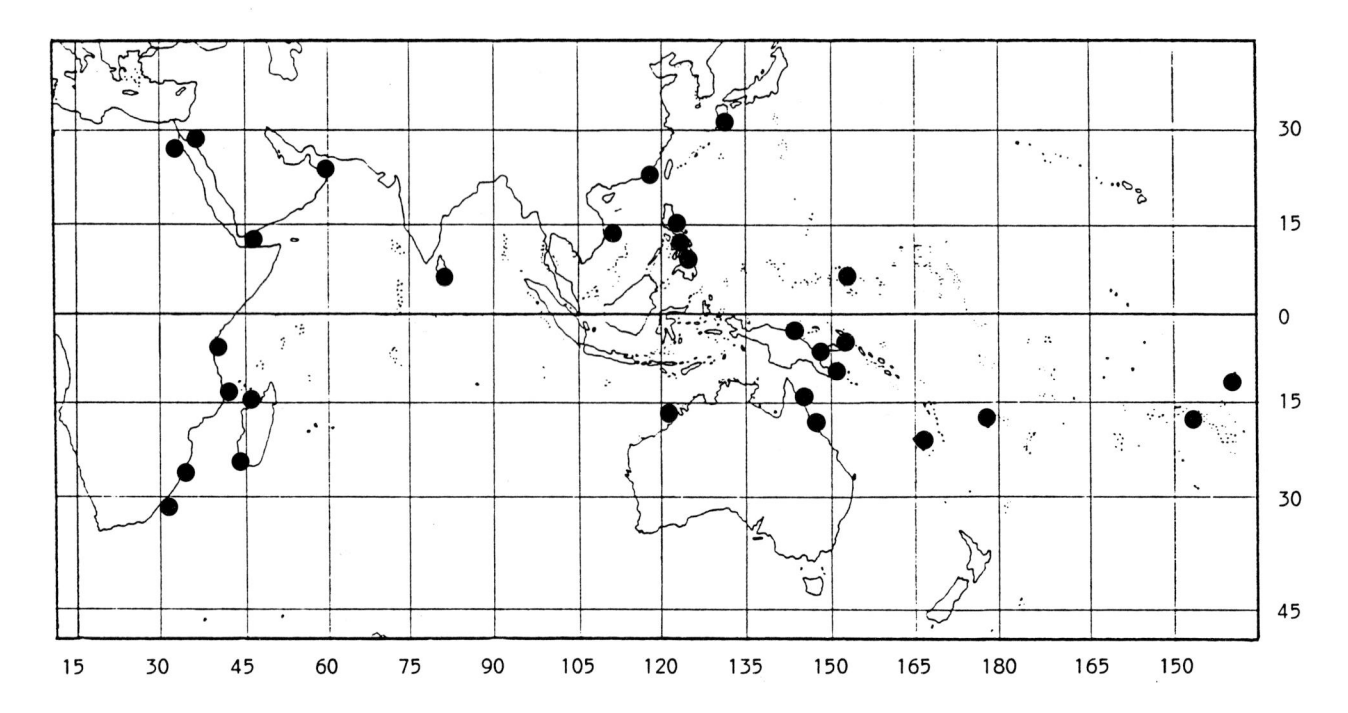

FIG. 151. — Distribution of *Chicoreus ramosus* (Linné).

Aperture large, rounded. Columellar lip smooth, rim adherent or slightly erect. Anal notch large, moderately deep, delineated by a callus. Outer lip crenulate, a relatively strong labral tooth abapically, interior smooth or lirate for short distance within. Siphonal canal broad, relatively long, narrowly open, bent abaperturally, ornamented with 2 or 3 frondose spines.

Withish with pinkish-red outer lip and columellar lips.

REMARKS. As pointed out by DODGE (1957 : 88), *C. ramosus* of LINNÉ is a composite species since it originally included the species *C. palmarosae, C. brevifrons, C. axicornis, C. brunneus* and probably others. The type-locality " Jamaica " probably refers to a specimen of *C. brevifrons*. LAMARCK (1822 : 160) proposed *Murex inflatus*, refering to MARTINI (1777 : figs 980 and 981), but he included also *M. ramosus* in synonymy of his species, and thus considered they were the same.

Three specimens labelled *Murex inflatus* are in the LAMARCK coll. (MHNG 1099/10-12) (FINET, *in litt*). All are Indian Ocean specimens of *C. ramosus*.

Since LAMARCK's time the name *C. ramosus* has been widely used for the large, common Indo-Pacific species. All other species included by Linné under that name have been named by subsequent authors.

South African shells (Durban Bay) are stunted and smaller, reaching a maximum length of 110 mm (KILBURN & RIPPEY, 1982 : 81). Despite this, specimens with intact protoconchs (Durban, NM) are indistinguishable from *C. ramosus*.

So-called dwarf forms of *C. ramosus* in collections are juveniles.

Chicoreus (Chicoreus) virgineus (Röding, 1798)
Figs 7, 92-93, 152-153, 154, 274-275

Purpura virgineus Röding, 1798 : 141.

Purpura rudis Link, 1807 : 121.
Murex anguliferus Lamarck, 1822 : 171.
Murex ferrugo Wood, 1828 : 15, pl. 5, fig. 16.

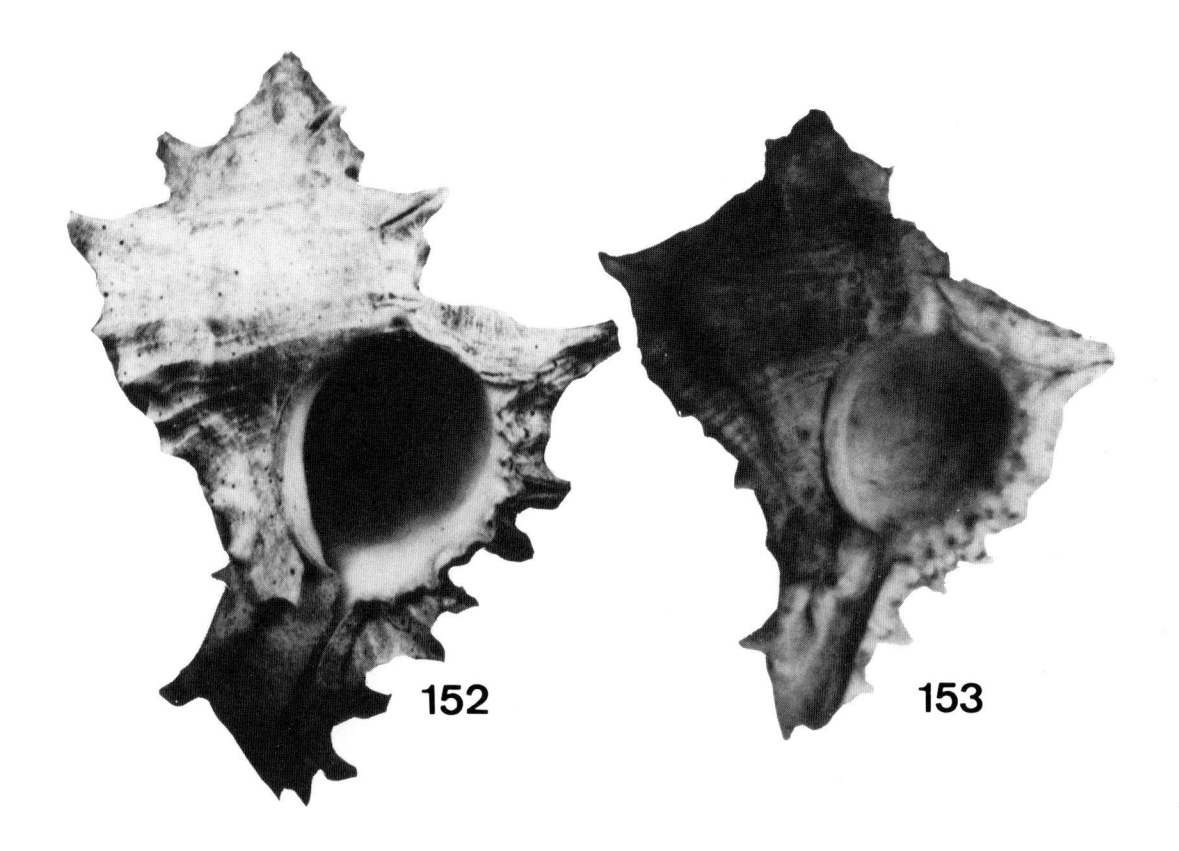

FIGS 152-153. — *Chicoreus (Chicoreus) virgineus* (Röding).
152, locality unknown, 110 mm (lectotype of *Murex erythraeus* Fischer, MNHN).
153, Red Sea, 80 mm (lectotype of *Murex anguliferus* Lamarck, MNHN).

Murex erythraeus Fischer, 1870 : 176.
Murex cyacantha Sowerby, 1879 : 11, fig. 160.
Murex ponderosus Sowerby, 1879 : 12, fig. 67.

ADDITIONAL REFERENCES

Murex (Chicoreus) anguliferus ponderosus. — SMITH, 1953 : 7, pl. 23, fig. 3.
Chicoreus virgineus. — KAICHER, 1973 : card 136 ; ABBOTT & DANCE, 1982 : 136, text fig. ; SHARABATI, 1984, pl. 18, fig. 6.
Chicoreus (Chicoreus) virgineus. — FAIR, 1976 : 85, pl. 6, fig. 81.
?*Siratus virgineus*. — RADWIN & D'ATTILIO, 1976 : 108, pl. 17, fig. 15.
Murex (?Siratus) virgineus. — EISENBERG, 1981 : 95, pl. 77, fig. 1.

TYPE LOCALITIES. *M. virgineus, M. rudis, M. ferrugo* : none ; *M. anguliferus* : Atlantic Ocean and on African coasts ; *M. erythraeus* : Suez ; *M. cyacantha* : ?Red Sea ; *M. ponderosus* : Ceylon.

TYPE MATERIAL. *M. anguliferus* : lectotype MNHN, here selected from 2 syntypes ; *M. erythraeus* : lectotype MNHN, here selected from 3 syntypes. No material located for the other names.

OTHER MATERIAL EXAMINED. c. 50 specimens from throughout the distributional range.
DISTRIBUTION (Fig. 154). Suez and off Dahlak, Red Sea ; Gulf of Aden ; Sri Lanka and the Bay of Bengal.

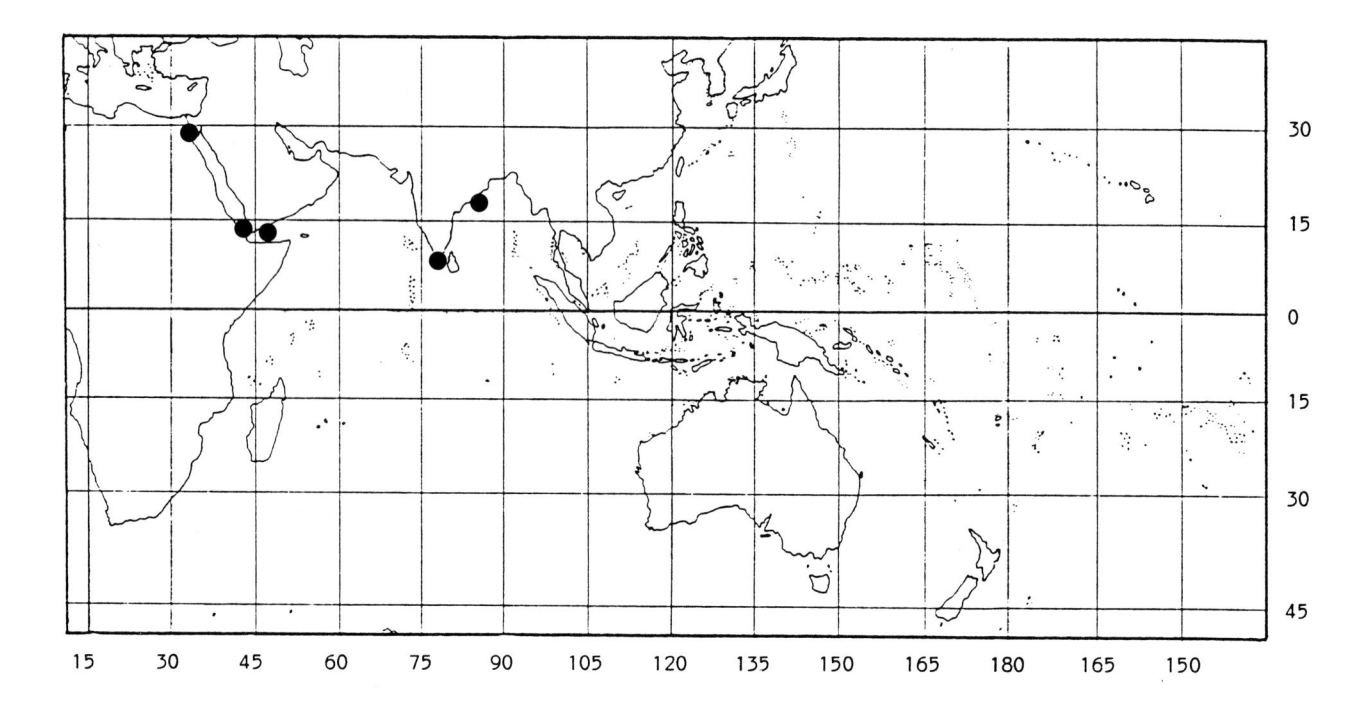

FIG. 154. — Distribution of *Chicoreus virgineus* (Röding).

DESCRIPTION. Shell up to 150 mm in length, heavy and stout. Spire moderately high, up to 8 teleoconch whorls and a protoconch of probably 2 whorls (partially broken). Suture appressed. Last preserved whorl of protoconch suggesting paucispiral nucleus.

Last whorl with three rounded and heavy varices (occasionally four). Intervaricial axial sculpture consisting of one single heavy node, mostly near the succeeding varix or sometimes fused with it. Spiral sculpture of 5 or 6 obsolete major cords, and numerous minor cords and threads in each interspace. Major cords giving rise to short, broadly open spines on the varices : abapical spines sometimes connected by webbing.

Aperture large, rounded. Columellar lip completely adherent, smooth. Anal notch large and shallow. Outer lip denticulate, very weakly lirate for short distance within, with a weak labral tooth abapically. Siphonal canal moderately long, broad, bent abaperturally, ornamented with 2 or 3 (rarely 4) open spines.

White to dark brown ; edges of columellar and outer lips generally pink.

Radula (Figs 92-93).

REMARKS. Somewhat resembling *C. ramosus*, in shell morphology, *C. virgineus* is heavier and less spinose with more rounded varices and a more strongly triangular shape. Although type material of *M. cyacantha*, *M. ponderosus* and *M. ferrugo* could not be located, the original illustrations leave no doubt about their synonymy with *C. virgineus*. *C. virgineus* was doubtfully included in *Siratus* by RADWIN & D'ATTILIO (1976) but the presence of a labral tooth (unknown in *Siratus* species) and the gross shell morphology suggest a better placement in *Chicoreus* (s.s.).

Genus **CHICOREUS** Montfort, 1810

Subgenus **TRIPLEX** Perry, 1810

SYSTEMATIC TREATMENT. Rather than arrange all of the species alphabetically, they are treated in groups of related species. A comparison table is added at the end of some of these groups where appropriate. The species are grouped as follows :

GROUP 1

Shell generally attaining medium to large size (length 60-133 mm), contour fusiform; mostly brownish with white to yellowish aperture; varicial spines frondose, short; usually 4 or 5 spines on varices of last whorl. Four species with planktotrophic larval development, one non planktotrophic and one unknown (broken or eroded in all specimens examined).

Chicoreus (T.) microphyllus is included in the comparison table at the end of this group to separate it from *Chicoreus (T.) torrefactus* with which it is sometimes confused or erroneously synonymised.

Included taxa : *Chicoreus (Triplex) dovi* Houart, 1984; *C. (T.) insularum* (Pilsbry, 1921); *C. (T.) maurus* (Broderip, 1832); *C. (T.) palmarosae* (Lamarck, 1822); *C. (T.) saulii* (Sowerby, 1834); *C. (T.) torrefactus* (Sowerby, 1841).

GROUP 2

Shell medium-sized, 23-90 mm in length. Brown with darker spiral threads. 5-7 frondose varicial spines on last whorl. Non planktotrophic larval development, except in *Chicoreus (T.) rubescens* (Broderip, 1833).

Included taxa : *Chicoreus (Triplex) microphyllus* (Lamarck, 1816); *C. (T.) paini* Houart, 1983; *C. (T.) strigatus* (Reeve, 1849); *C. (T.) rubescens* (Broderip, 1833); *C. (T.) trivialis* (A. Adams, 1854).

GROUP 3

Withish to dark brown or nearly black shells, though some other colour forms. From 30-90 mm in length. 2-7 frondose spines on varices of last whorl. Non planktotrophic larval development, except *Chicoreus (Triplex) bourguignati*.

Included taxa : *Chicoreus (Triplex) axicornis* (Lamarck, 1822); *C. (T.) banksii* (Sowerby, 1841); *C. (T.) bourguignati* (Poirier, 1883); *C. (T.) brunneus* (Link, 1807); *C. (T.) elisae* Bozzetti, 1991; *C. (T.) groschi* Vokes, 1978; *C. (T.) ryosukei* Shikama, 1978.

GROUP 4

Shell milky white to light brown, spiral sculpture usually darker. Outline rounded, 4-6 short to medium-sized frondose spines on varices of last whorl, aperture rounded, 36-88 mm in length. Planktotrophic larval development in *C. cnissodus*, unknown in *C. peledi*.

Included taxa : *Chicoreus (Triplex) cnissodus* (Euthyme, 1889); *C. (T.) peledi* Vokes, 1978.

GROUP 5

Shell relatively small, 30-70 mm in length, spines small and slightly frondose, adapically spines usually connected on each other by low varicial flange, whitish to brownish. Non planktotrophic larval development. Geographical range restricted.

Included taxa : *Chicoreus (Triplex) corrugatus corrugatus* (Sowerby, 1841); *C. (T.) corrugatus ethiopius* Vokes, 1978; *C. (T.) damicornis* (Hedley, 1903); *C. (T.) denudatus* (Perry, 1811); *C. (T.) territus* (Reeve, 1845); *C. (T.) thomasi* (Crosse, 1872).

GROUP 6

Shell relatively small, 25 to 75 mm in length. Protoconch paucispiral. Varicial fronds usually long, sometimes reduced to acute spines or sometimes interconnected. Whitish to yellowish or yellowish brown. Non planktotrophic larval development. Restricted geographical range.

Included taxa : *Chicoreus (Triplex) boucheti* Houart, 1983 ; *C. (T.) cervicornis* (Lamarck, 1822) ; *C. (T.) longicornis* (Dunker, 1864) ; *C. (T.) paucifrondosus* Houart, 1988 ; *C. (T.) subpalmatus* Houart, 1988.

GROUP 7

Shell relatively small, 20-53 mm in length, white, yellowish or pinkish ; short varicial frondose spines. Planktotrophic or non planktotrophic larval development.

Included taxa : *Chicoreus (Triplex) aculeatus* (Lamarck, 1822) ; *C. (T.) cloveri* Houart, 1985 ; *C. (T.) crosnieri* Houart, 1985 ; *C. (T.) fosterorum* Houart, 1989 ; *C. (T.) nobilis* Shikama, 1977 ; *C. (T.) rossiteri* (Crosse, 1872) ; *C. (T.) ryukyuensis* Shikama, 1978 ; *C. (T.) zululandensis* Houart, 1989.

GROUP 1

Chicoreus (Triplex) dovi Houart, 1984
Figs 10, 134, 155, 260, 292

Chicoreus dovi Houart, 1984 : 55, figs 1-4, pl. 1.

ADDITIONAL REFERENCES

Chicoreus dovi. — HOUART, 1986a : 9, text fig.

TYPE LOCALITY. Malindi, Kenya.

TYPE MATERIAL. Holotype IRSNB I. G. 26656/402.

OTHER MATERIAL EXAMINED. c. 20 specimens from throughout the distributional range.

DISTRIBUTION. (Fig. 155). South Somalia to Dar es Salaam (Tanzania). Depth range : 2-6 m, on coral reefs.

DESCRIPTION. Shell up to 109 mm in length, stout. Spire high, with 2 1/2 teleoconch whorls, and up to 8 teleoconch whorls. Suture slightly appressed. Protoconch whorls smooth, rounded.

Last whorl bearing 3 foliose varices, each with 5 major short, straight spines, an intermediate recurved spinelet between each major spine. Other intervaricial sculpture consisting of a moderately high nodulose rib, extending from suture to siphonal canal, mostly erect on carinal edge of body whorl. Spiral sculpture of numerous cords and intermediate threads in each interspace.

Aperture ovale. Columellar lip totally adherent, smooth with sometimes 1 or 2 folds adapically, near callus. Anal notch deep, large, " U " shaped. Outer lip erect, denticulate, lirate for short distance within. Siphonal canal moderately long for genus, open, slightly bent abaperturally at distal end, bearing 2 or 3 straight frondose spines.

Light brown with darker coloured spiral cords and threads. Aperture glossy white.

REMARKS. *C. dovi* differs from the superficially similar *C. torrefactus* in having a more globose last whorl, and a smaller siphonal canal. As in *C. torrefactus* it generally has foliaceous spines but they are narrower at their tips. The last whorl in adult specimens always exhibits a single intervaricial node, while *C. torrefactus* bears 1-3 nodes. *C. torrefactus* always has a yellow to cream-peach columellar lip while that of *C. dovi* is consistantly white. The shell of *C. torrefactus* always has a fold on the abapical part of the columella, while *C. dovi* has a straight columellar lip. In *C. torrefactus* the protoconch is conical, and of the sinugera type, consisting of 3 whorls with a small first whorl, denoting planktotrophic larval development. The protoconch in *C. dovi* is more globose, with fewer whorls suggesting non planktotrophic development. Moreover the terminal varix of the protoconch is differently shaped (Fig. 10).

C. dovi differs from *C. microphyllus* in shell contour and in the form of aperture and protoconch (Table 1).

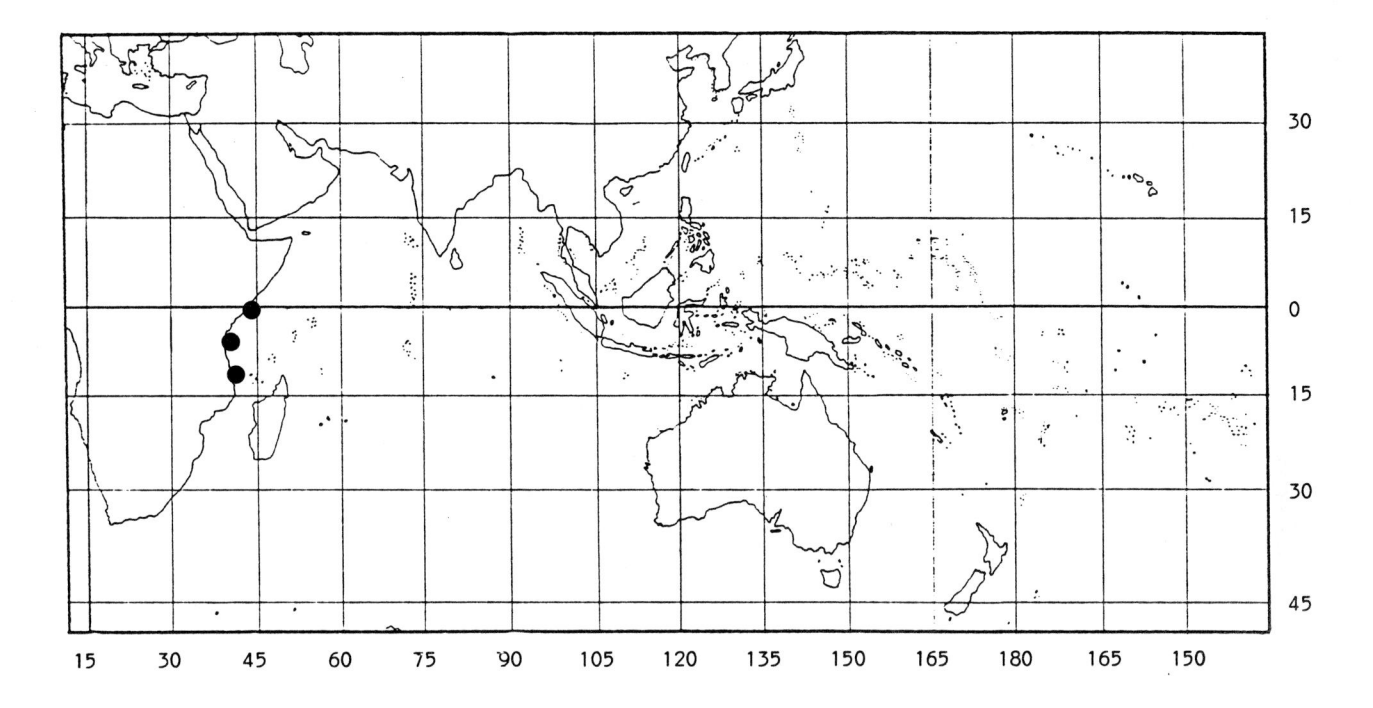

FIG. 155. — Distribution of *Chicoreus dovi* Houart.

TABLE 1. — Comparisons of *Chicoreus* species of group 1.

Characters	*C. torrefactus*	*C. microphyllus*	*C. dovi*	*C. palmarosae*	*C. saulii*	*C. maurus*	*C. insularum*
Protoconch	Conical, 3-3.5 whorls. Rounded, erect terminal varix	1.75-3 rounded whorls. Weakly curved, terminal varix	2.5 rounded whorls. Straight, weakly angulate terminal varix	1.45 rounded whorls. Terminal varix unknown	Conical, 3.25 whorls. Erect, weakly curved terminal varix	Conical, probably 3.25-3.5 whorls. Erect weakly curved terminal varix	unknown
Number of teleoconch whorls	7-11	7-9	7-8	9	9	9	7
Spines on last whorl	Usually 5 with intermediate spinelets	Usually 5 with intermediate spinelets	5 with intermediate spinelets	4, no intermediate spinelets	4 with small intermediate spinelets	4 with intermediate spinelets	5 with intermediate spinelets
Intervaricial axial sculpture	1 large and 1 shallow, or 1 strong, single node, rarely 3	2-4, generally 3	1 ridge	Generally 3 low axial ridges	1 heavy single node, or 2 weak nodes	1 prominent node	1 prominent node, occasionally with additional axial ridge
Aperture and columellar lips	Aperture ovate or subcircular. Outer lip curved. Columellar lip smooth, rarely bordered by few folds. Heavy fold on abapical end	Aperture ovate. Outer lip curved. Columellar lip mostly bordered by fine folds. Very shallow or no fold on abapical end	Aperture ovate. Outer lip curved. Columellar lip smooth. No fold on abapical end	Aperture ovate, weakly angulate. Columellar lip bordered by folds. Small fold abapically	Aperture ovate. Columellar lip smooth. No fold abapically	Aperture ovate or subcircular. Columellar lip weakly folded adapically. Small fold abapically	Broadly ovate. Columellar lip with weak node abapically and folds adapically. Small fold abapically
Adult shell length	60-123.8 mm	40-90 mm	70-109 mm	85-133 mm	85-124 mm	70-91 mm	71-101.5 mm

Chicoreus (Triplex) insularum (Pilsbry, 1921)
Figs 156, 286-287

Murex torrefactus insularum Pilsbry, 1921 : 319.

ADDITIONAL REFERENCES

Chicoreus insularum. — KAICHER, 1974 : card 536 ; RADWIN & D'ATTILIO, 1976, p. 38, pl. 5, fig. 4 ; KAY, 1979 : 236, fig. 83(j) ; EARLE, 1980 : 10, text fig. ; OKUTANI, 1983 : 8, pl. 24, fig. 11.
Chicoreus (Chicoreus) insularum. — FAIR, 1976 : 51, pl. 6, fig. 80 ; HOUART & PAIN, 1983a : 19, text figs : 17.
NOT *Chicoreus (Chicoreus) insularum*. — HOUART & PAIN, 1983a : 19, text fig. (syntype of *Murex mexicanus* Stearns) [= *Chicoreus (Triplex) maurus* (Broderip, 1833)].

TYPE LOCALITY. Off Waikiki, Oahu, 64-91 m.

TYPE MATERIAL. Holotype and 2 paratypes ANSP 47192.

OTHER MATERIAL EXAMINED. Ca. 10 specimens from throughout the distributional range.

DISTRIBUTION. (Fig. 156). Endemic to the Hawaiian Island Chain and the Midway Is, on coral reefs.

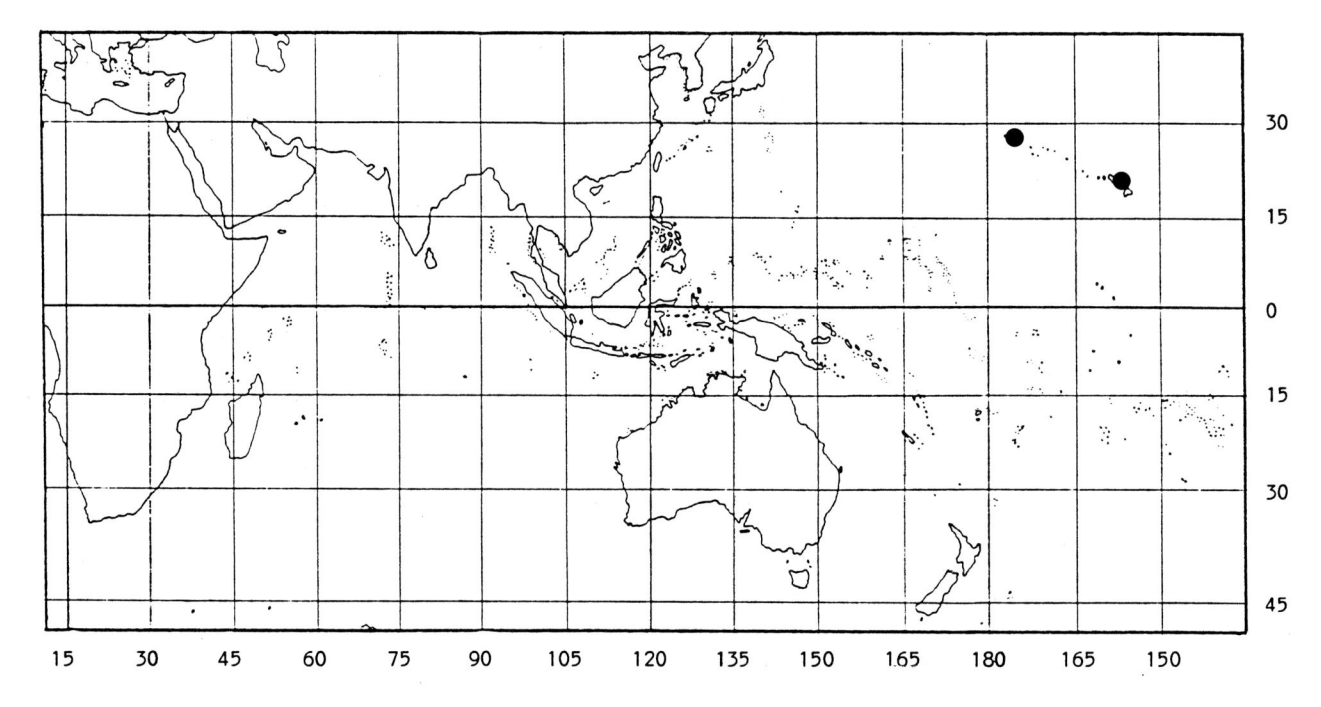

FIG. 156. — Distribution of *Chicoreus insularum* (Pilsbry).

DESCRIPTION. Shell up to 101.5 mm in length, stout. Spire moderately high, up to 7 convex, angulate teleoconch whorls. Suture appressed. Protoconch unknown.

Last whorl with 3 rounded varices, each ornamented with 5 frondose, short major spines and some intermediate spinelets. Shoulder spine longest, straight. Second and third abapically spines generally recurved abaperturally. Other axial sculpture consisting of one heavy intervaricial node and small associated axial ridge. Spiral sculpture consisting of 5 major cords and fine scabrous intermediate threads in each interspace.

Aperture rounded. Columellar lip rim partially erect and adherent adapically, with a shallow denticle on its adapical part ; anal notch well delineated by small callus. Outer lip crenulate, strongly lirate for short distance within. Siphonal canal short, tapering abapically, bent distally, with 2 or 3 straight spines.

Light brown with somewhat darker spiral sculpture. Aperture glossy white, sometimes fine yellow or pink line around adapical part.

REMARKS. *Murex colpos* Dall *in* Burch, 1955 was synonymized with *C. insularum* by HOUART & PAIN (1983a : 19) but after a more detailed study it is here synonymised with *C. maurus* (see below).

Chicoreus (Triplex) maurus (Broderip, 1833)
Figs 8, 157, 288-291, 298

Murex maurus Broderip, 1833 : 174.

Murex steeriae Reeve, 1845, pl. 8, fig. 28.
Murex mexicanus Stearns, 1893 : 345 (non *M. mexicanus* Petit de la Saussaye, 1852).
Murex colpos Dall *in* Burch, 1955 : 12 (new name for *Murex mexicanus* Stearns, 1894, non Petit de la Saussaye, 1852).

ADDITIONAL REFERENCES

Murex (Chicoreus) maurus. — SMITH, 1953 : 5 (in part), pl. 8, fig. 7 ; EISENBERG, 1981 : 91, pl. 73, fig. 5.
Chicoreus maurus. — RADWIN & D'ATTILIO, 1976 : 39, pl. 5, fig. 5 ; CERNOHORSKY, 1978a : 72, figs 15-17 (syntypes), 18-19 ; CERNOHORSKY, 1978b : 65, pl. 18, fig. 2 ; ABBOTT & DANCE, 1982 : 136, text fig. ; RIPPINGALE, 1987 : 7, fig. 16.
Chicoreus (Chicoreus) maurus. — HOUART & PAIN, 1983b : 3, text figs p. 4 (syntype of *Murex maurus* and holotype of *Murex steeriae* (Reeve).
Murex (Chicoreus) palma-rosae mexicanus. — SMITH, 1953 : 13, pl. 11, fig. 7 (holotype).
Murex steeriae. — ASTARY, 1973 : 7, text fig.
Chicoreus steeriae. — KAICHER, 1973 : card 148 ; SALVAT & RIVES : 311, fig. 192.
Chicoreus (Chicoreus) steeriae. — FAIR, 1976 : 78, pl. 8, fig. 106.
Chicoreus torrefactus. — SALVAT & RIVES, 1975 : 311, fig. 193 (not *Murex torrefactus* Sowerby, 1841).
Chicoreus (Chicoreus) insularum. — HOUART & PAIN, 1983a : 19 (holotype of *Murex mexicanus* Stearns, 1893) (not *Murex insularum* Pilbry, 1921).
NOT *Murex (Chicoreus) maurus.* — SMITH, 1963, pl. 3, fig. 7.
NOT *Chicoreus maurus.* — KAICHER, 1973, card 148 [= *Chicoreus (Triplex) torrefactus* (Sowerby, 1841)].
NOT *Chicoreus (Chicoreus) maurus.* — FAIR, 1976 :57, pl. 8, fig. 103 ; VOKES, 1978 : 384, pl. 3, fig. 1 [= *Chicoreus (Triplex) torrefactus* (Sowerby, 1841)].

TYPE LOCALITIES. *Murex maurus* : Anaa I., Tuamotu Archipelago, Pacific Ocean ; *Murex steeriae* : unknown ; *Murex mexicanus* : Gulf of California.

TYPE MATERIAL. *M. maurus* : lectotype BMNH 197473, here selected from 3 syntypes ; *M. steeriae* : holotype UMZ ; *Murex mexicanus* (= *M. colpos*) : holotype USNM 46803.

OTHER MATERIAL EXAMINED : ca. 20 specimens from throughout the distributional range.

DISTRIBUTION. (Fig. 157). North-east of New Caledonia ; Tahiti, Tuamotu Archipelago, and Marquesas Is. Depth range : 1-15 m.

DESCRIPTION. Shell up to 91 mm in length, stout. Spire moderately high, with a conical protoconch of probably 3 1/4-3 1/2 whorls (damaged) and up to 9 rounded, somewhat angulate, teleoconch whorls. Suture slightly appressed.

Last whorl with 3 strong, rounded varices, each with 4 short to medium-sized, moderately frondose spines. Shoulder spines usually broad, long, abapical 3 slightly frondose ; abapical spine shortest ; one intermediate spinelet between each pair of major spines. Other axial sculpture consisting of a single strong intervaricial node, very rarely with an associated shallow axial ridge. Spiral sculpture of numerous scabrous cords and threads.

Aperture broadly ovate. Columellar lip adherent, with weak node abapically, a strong callus, followed by 2 or 3 small ridges abapically. Anal notch deep, fairly broad, " U " shaped. Outer lip crenulate, partially strongly ridged within. Siphonal canal short, tapering abapically, distally recurved, with 2 or 3 straight foliated spines.

Colour of alternating bands of dark violet and white or very light brown. Fronds and apertural margin stained with pink or pale violet. Aperture glossy white.

REMARKS. Judging from a juvenile specimen from Tahiti (J. TRÖNDLÉ coll.) with an almost perfect protoconch, this species is evidently related to *C. saulii* and indeed if protoconchs prove to be absolutely identical (Figs 8 and 11), *C. maurus* may prove to be a local population of *C. saulii*, although by priority, the name *C. maurus* will have to be used over *saulii*.

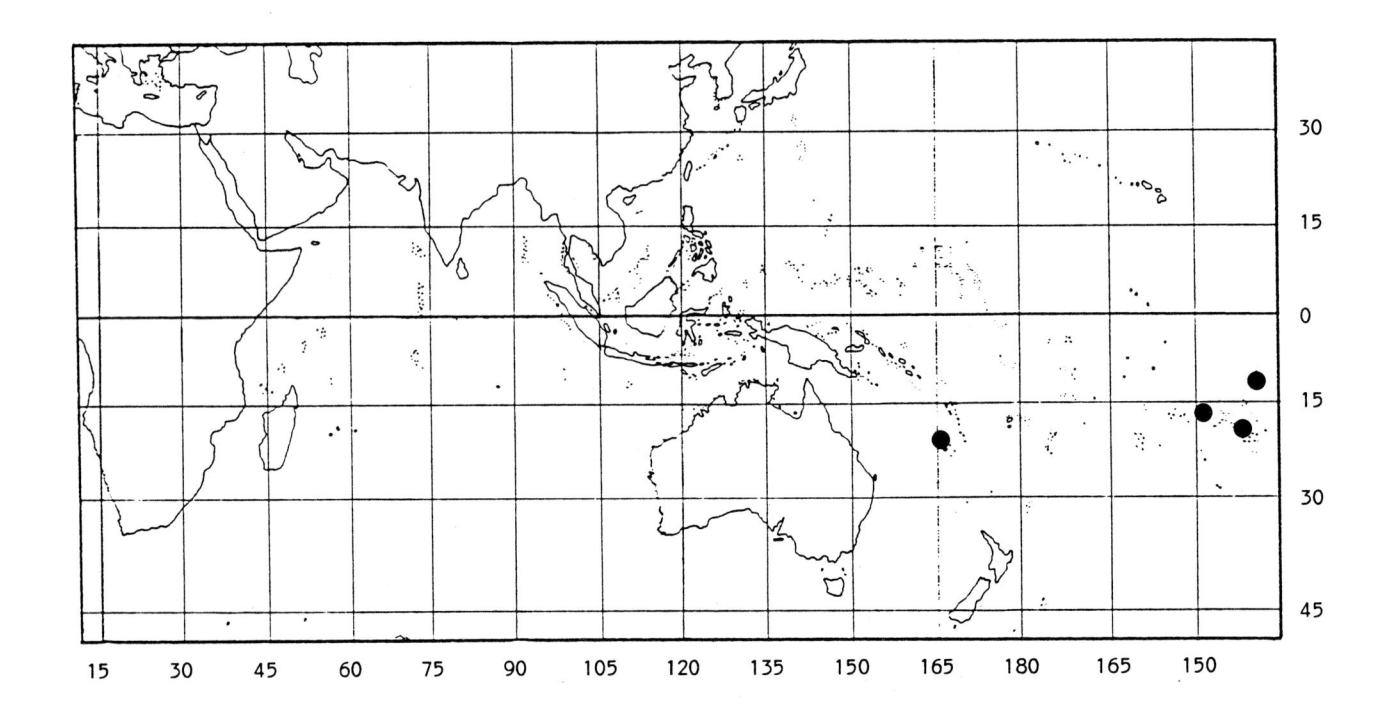

FIG. 157. — Distribution of *Chicoreus maurus* (Broderip).

MIENIS (1984 : 14) stated the type-locality, given by HOUART & PAIN (1983 : 4), to be erroneous. These authors were fully aware that Anaa Island is situated in the Tuamotu Archipelago, however, and were merely mentioning the locality given by Broderip on the original label (i.e. " Annaa Is. Philippines "). The type-locality of *C. maurus* is most probably Anaa Island, Tuamotus, as stated by CERNOHORSKY (1978 : 73).

The lectotype here selected has previously been illustrated by CERNOHORSKY (1978 : 73, fig. 15) and by HOUART & PAIN (1983b : 3, text fig. p. 4), and is probably the specimen illustrated by SOWERBY (1834). The largest paralectotype measures 70.1 × 39 mm. Several authors have erroneously illustrated the South African taxon known as *Chicoreus kilburni* Houart & Pain, 1982 (a form of *C. torrefactus*) as *Chicoreus maurus*.

The name *Murex steeriae* is based on a long-spined adult specimen of *C. maurus*.

Although HOUART & PAIN (1983 : 19) considered that *Murex colpos* Dall in Burch, 1955 was a synonym of *C. insularum*, the worn holotype exhibits the outer lip lirations sculpture and coloration that are typical of *C. maurus*. The type-locality of *M. mexicanus* is certainly erroneous.

Chicoreus (Triplex) palmarosae (Lamarck, 1822)
Figs 9, 158, 293-294

Murex palmarosae Lamarck, 1822 : 161.

Triplex rosaria Perry, 1811 : pl. 6, fig. 3 [secondary homonym of *Purpura rosarium* Röding, 1798 rejected before 1961 (ICZN, art. 59 b i)].
Triplex foliatus Perry, 1810, pl. 23 [suppressed in favor of *Murex palmarosae* Lamarck, 1822 (ICZN opinion 911)].
Murex argyna Mörch, 1852 : 97.

ADDITIONAL REFERENCES

Murex (Chicoreus) palmarosae. — KIENER, 1842, pl. 17, fig. 1 ; pl. 18, fig. 1. SMITH, 1953 : 5, pl. 3, fig. 1 ; SHIKAMA, 1963 : 70, pl. 53, fig. 5 ; EISENBERG, 1981 : 91, pl. 73, fig. 20.
Chicoreus palmarosae. — HINTON, 1972 : 34, pl. 17, fig. 13 ; KAICHER, 1973 : card 142 ; RADWIN & D'ATTILIO, 1976 : 40, pl. 5, fig. 2 ; HINTON, 1979 : 26, fig. 2 ; ABBOTT & DANCE, 1982 : 1236, text fig. ; OKUTANI, 1983 : 8, pl. 22, fig. 4 ; LAI, 1987 : 61,pl. 29, fig. 2 ; DRIVAS & JAY, 1988 : 68, pl. 19, fig. 4.
Chicoreus (Chicoreus) palmarosae. — FAIR, 1976 : 65, pl. 8, fig. 104 ; VOKES, 1978 : 386.
Murex (Chicoreus) maurus sauliae. — SMITH, 1953 : 5, pl. 3, fig. 9 (not *Murex sauliae* Sowerby, 1834).
Chicoreus rosarius. — ARAKAWA, 1964 : 361, pl. 21, figs 11-12 (radula) ; CERNOHORSKY, 1978b : 65, pl. 18, fig. 1.
Chicoreus (Triplex) rosarius. — KIRA, 1965 : 60, pl. 23, fig. 13.
Chicoreus (Chicoreus) rosarius. — SPRINGSTEEN & LEOBRERA, 1986 : 134, pl. 36, fig. 14.

TYPE LOCALITIES. *T. foliatus* : unknown ; *T. rosaria* : Island of Ceylon (Sri Lanka) ; *M. palmarosae* : ?Indian Ocean ; *M. argyna* : unknown.

TYPE MATERIAL. No type material known.

OTHER MATERIAL EXAMINED : c. 60 specimens from throughout the geographical range.

DISTRIBUTION. (Fig. 158). Natal, southeastern Africa (NM D9872) ; south-western Indian Ocean (Reunion and Mauritius Is ; the Seychelles) ; Sri Lanka ; the Philippine Is ; south of Japan, and the Solomons Is. Depth Range : Intertidal to 90 m.

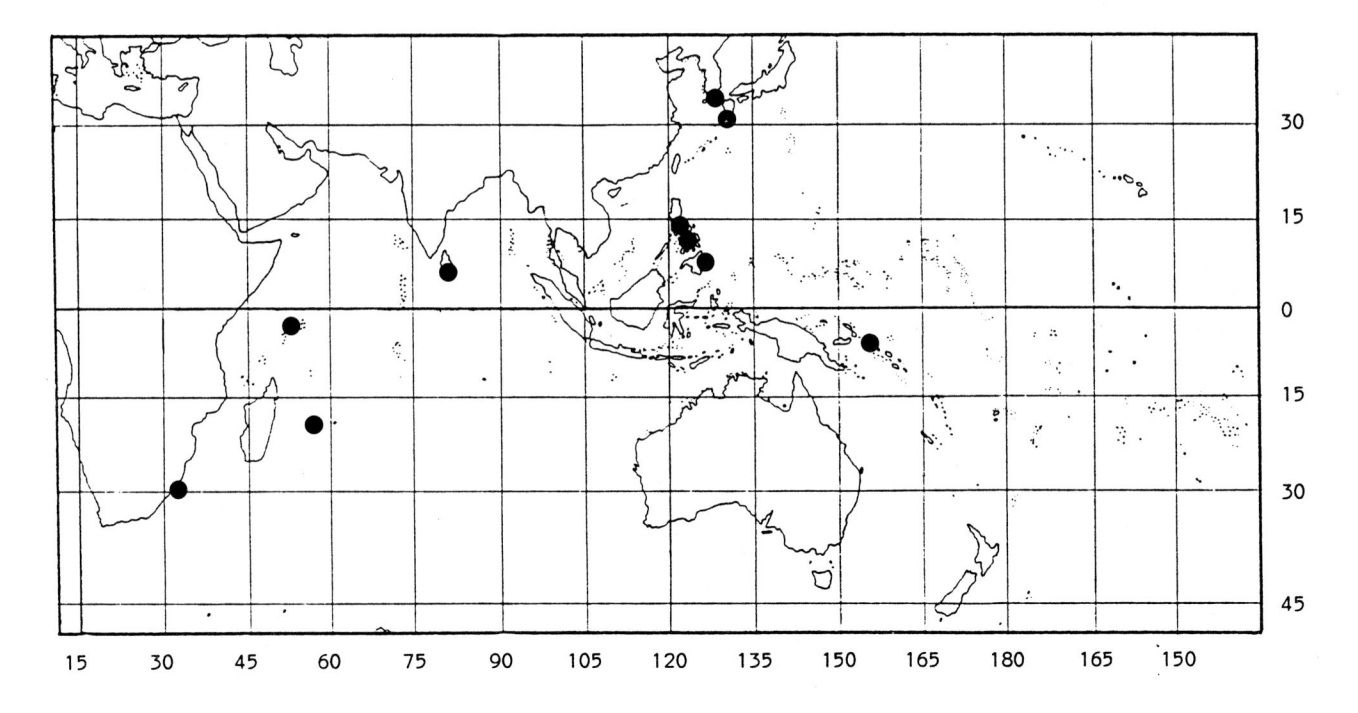

FIG. 158. — Distribution of *Chicoreus palmarosae* (Lamarck).

DESCRIPTION. Shell up to 133 mm in length, elongate and stout. Spire high, with 1 3/4 protoconch whorls and up to 9 elongate teleoconch whorls. Suture slightly appressed. Protoconch whorls glossy, smooth.

Last teleoconch whorl bearing 3 frondose varices, each with 4 short to relatively long, frondose spines. Shoulder spines longest, an appreciable gap separating it from 3 shorter, straight frondose spines ; no intermediate spinelets. Other axial sculpture consisting of 2 or 3 nodose axial ridges. Spiral sculpture of 9 or 10 scabrous spiral cords and numerous intermediate scabrous threads in each interspace.

Aperture broadly ovate. Columellar lip adherent, smooth inside but bearing numerous, somewhat elongate nodes on its edge. Strong callus delineating deep, large anal notch. Outer lip crenulate, strongly lirate for short distance within. Siphonal canal relatively long, straight, narrowly open, abaperturally bent at tip, bearing 2 or 3 frondose spines.

Cream or ochre to pale or dark brown. Spiral cords and threads darker. Interior of spines pinkish or violet in Indian Ocean and Japanese specimens, otherwise brownish. Aperture white bordered with dark brown; columellar denticles white, brown between.

REMARKS. No type material has been located. There is a specimen designated as a possible type of *M. palmarosae* in MHNG. This specimen, however, actually represents *C. torrefactus* and does not agree with Lamarck's description, which clearly refers to the species long known as *C. palmarosae*.

Widely distributed, it is rather variable in shell morphology and colour (Figs 293-294). To my knowledge, specimens with pink or violet stained frond occur only in the Indian Ocean and off Japan. Specimens from the Philippines are uniformly brown and have shorter spines. The most characteristic features for this species are the sculptured columellar lip edge, and the absence of intermediate spinelets between the varicial fronds.

Chicoreus (Triplex) saulii (Sowerby, 1834)
Figs 11, 159, 295, 303

Murex saulii Sowerby, 1841 : pl. 190, fig. 77; 1841b : 141.

ADDITIONAL REFERENCES

Murex (Chicoreus) saulii. — SHIKAMA, 1963 : 70, pl. 53, fig. 4; EISENBERG, 1981 : 93, pl. 75, fig. 12.
Chicoreus (Triplex) saulii. — KIRA, 1965 : 204, pl. 70, fig. 1.
Chicoreus saulii. — CERNOHORSKY, 1967a : 122, pl. 15, fig. 12; CERNOHORSKY, 1967b : 122, pl. 24, fig. 143; KAICHER, 1974 : card 501; RADWIN & D'ATTILIO, 1976 : 42, pl. 5, fig. 8; MC DONALD, 1979 : 7, fig. 8; ABBOTT & DANCE, 1982 : 136, text fig.; LAI, 1987 : 61, pl. 29, fig. 4; RIPPINGALE, 1987 : 11, fig. 25; DRIVAS & JAY, 1988 : 68, pl. 19, fig. 5.
Chicoreus (Chicoreus) saulii. — FAIR, 1976 : 75, pl. 8, fig. 99; SPRINGSTEEN & LEOBRERA, 1986 : 134, pl. 36, fig. 13.
NOT *Murex (Chicoreus) maurus sauliae*. — SMITH, 1953 : 5, pl. 3, fig. 9 [= *Chicoreus (Triplex) palmarosae* (Lamarck, 1822)].
NOT *Chicoreus saulii*. — HINTON, 1979 : 26, fig. 6 [= *Chicoreus (Triplex) torrefactus* (Sowerby, 1841)].

TYPE LOCALITY. Capul I., Philippine Islands.

TYPE MATERIAL. Holotype UMZ.

OTHER MATERIAL EXAMINED : c. 50 specimens from throughout the geographical range.

DISTRIBUTION. (Fig. 159). Central western Indian Ocean and north-western Pacific Ocean. Reunion and Mauritius Is, Comoros Archipelago, the Seychelles Is, Indonesia, Cocos I., New Britain, the Philippine Is, southern Japan, and the Marshall Is (Kwajalein), on coral reefs. Depth Range : 15-140 m.

DESCRIPTION. Shell up to 124 mm in length, stout. Spire high, acute, with 3 to 3 1/4 protoconch whorls and up to 10 elongate teleoconch whorls. Suture appressed. Protoconch glossy, conical.

Last whorl bearing 3 rounded, frondose varices, each ornamented with 4 moderately short frondose open spines. Shoulder spine generally longest, abapical second spine shortest; intermediate spinelets short, frondose. Other axial sculpture consisting of 1 or 2 low axial ridges or nodes. Spiral sculpture consisting of numerous crenulate cords and threads over entire surface.

Aperture ovate. Columellar lip smooth, adherent. Anal notch deep, narrow, delineated by strong callus. Outer lip crenulate, lirate for short distance within. Siphonal canal moderately long, narrowly open, tip abaperturally bent; bearing 3, rarely 4, frondose, open spines.

Light brown to pale orange-brown, spiral cords darker, interior of spines mostly pink. Aperture glossy white with deep pink rim.

REMARKS. This easily recognizable species has a much wider distribution than indicated by RADWIN & D'ATTILIO (1976 : 42) and exhibits very little geographic variation. It is sometimes confused by collectors with the purple form of *C. torrefactus*, but is easily separable from that species by its more elongate shell, paler colour and heavy pink line around the aperture.

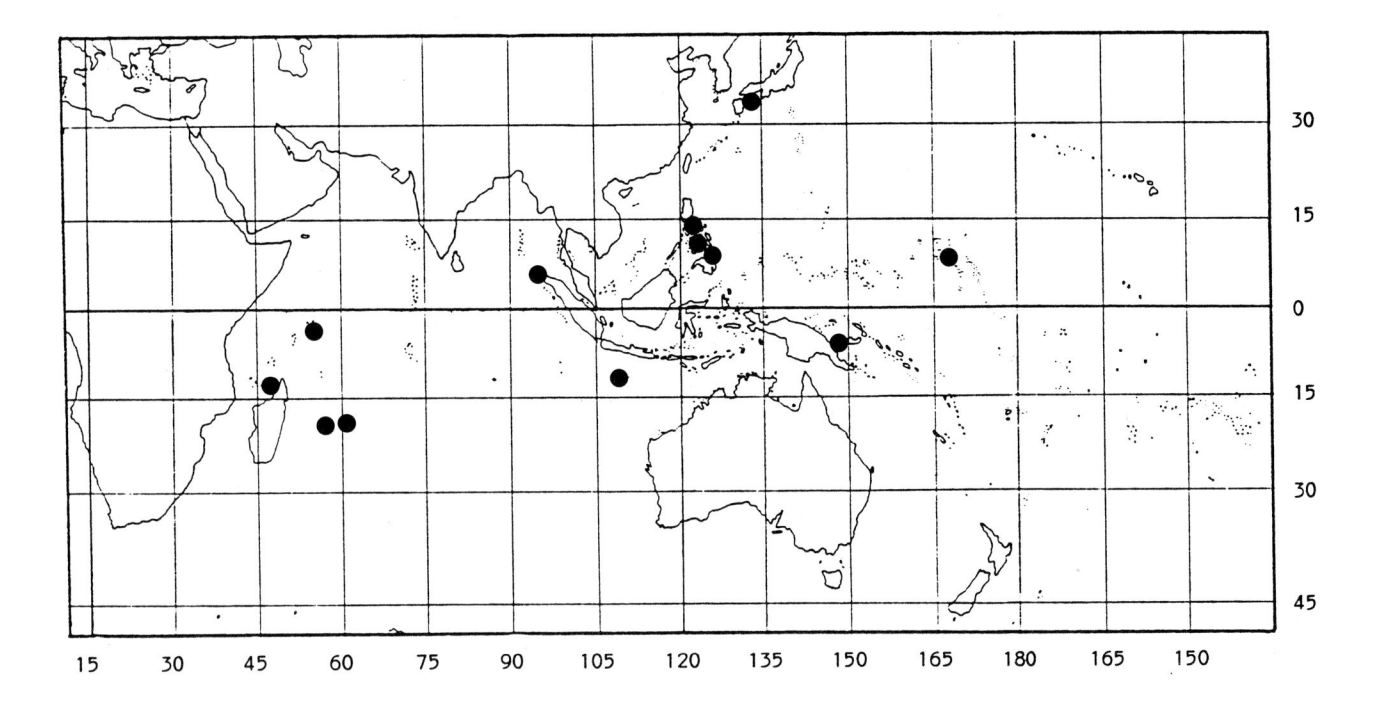

Fig. 159. — Distribution of *Chicoreus saulii* (Sowerby).

The shell differs from that of *C. palmarosae* in having intermediate spinelets, a smooth columellar lip, and heavy pink line around the aperture.

Chicoreus (Triplex) torrefactus (Sowerby, 1841)
Figs 12, 94-97, 132, 160-162, 163, 282-285, 296-297, 306-308

Murex torrefactus Sowerby, 1841 : pl. 199, fig. 120 ; 1841b : 141.

Murex rubiginosus Reeve, 1845 : pl. 8, fig. 32.
Murex affinis Reeve, 1846 : pl. 35, fig. 182 (not *M. affinis* Gmelin, 1791).
Murex benedictinus Löbbecke, 1879 : 79.
Murex rochebruni Poirier, 1883 : 57, pl. 5, fig. 1.
Chicoreus (Chicoreus) kilburni Houart & Pain, 1982 : 51, pl. 3, figs 1-4.

ADDITIONAL REFERENCES

Murex (Chicoreus) maurus. — SMITH, 1953 : pl. 3, fig. 7 (only) (not *Murex maurus* Broderip, 1833).
Murex maurus. — KENNELLY, 1964 : 69, pl. 17, fig. 84 ; KENSLEY, 1973 : 140, fig. 475 (not *Murex maurus* Broderip, 1833).
Chicoreus maurus. — KAICHER, 1973 : card 148 ; RICHARDS, 1981 : 55, pl. 28, fig. 217 ; KILBURN & RIPPEY, 1982 : 81, pl. 18, fig. 3 ; FREEMAN, 1986 : 9 (text fig. only) (not *Murex maurus* Broderip, 1833).
Chicoreus (Chicoreus) maurus. — FAIR, 1976 : 57, pl. 8, fig. 107 ; VOKES, 1978 : 384, pl. 3, fig. 1 (not *Murex maurus* Broderip, 1833).
Murex (Chicoreus) torrefactus. — SMITH, 1953 : 6, pl. 6, fig. 17.
Chicoreus torrefactus. — SPRY, 1961 : 19, pl. 4, fig. 135 ; ZHANG, 1965 : 20, pl. 1, figs 2, 3 ; CERNOHORSKY, 1967b : 122, pl. 25, fig. 153 ; WILSON & GILLETT, 1971 : 86, pl. 58, fig. 3 ; HINTON, 1972 : 36, pl. 18, fig. 4 ; ASTARY, 1973 : 7 ; KAICHER, 1974 : card 498 ; HINTON, 1979 : pl. 26, fig. 3 ; ABBOTT & DANCE, 1982 : 136, text fig. ; MUHLHAUSSER & ALF, 1983 : fig. 2 (in part) ; FREEMAN, 1986 : 9, text fig. ; LAI, 1987 : 61, pl. 29, fig. 1 ; DRIVAS & JAY, 1988 : 68, pl. 19, fig. 3.
Chicoreus (Triplex) torrefactus. — KIRA, 1965 : 61, pl. 23, fig. 14.
Murex torrefactus. — BROST & COALE, 1971 : 69, pl. 15, fig. 3.

FIGS 160-162. — *Chicoreus (Triplex) torrefactus* (Sowerby).
160, Ticao, Philippines, 90.5 mm (neotype, MNHN).
161, Natal, South Africa, 81.5 mm (holotype of *Chicoreus kilburni* Houart & Pain, IRSNB 26429/385).
162, Indian Ocean, 88.7 mm (holotype of *Murex benedictinus* Löbbecke, LM).

Chicoreus (Chicoreus) torrefactus. — FAIR, 1976 : 81, pl. 8, fig. 103 ; VOKES, 1978 : 384, pl. 3, fig. 4 ; HOUART & PAIN, 1982 : 51, pl. 3, fig. 6 (neotype) ; HOUART & PAIN, 1983a : 17, figs 1, 2 (syntypes of *Murex rubiginosus* Reeve), figs 3, 4 (syntypes of *Murex affinis* Reeve), text fig. p. 16 (protoconch), text fig. p. 18 (syntype of *Murex rochebruni* Poirier) ; HOUART, 1984b : 57, fig. 2, 4, pl. 1, fig. 4.
Chicoreus carneolus. — CERNOHORSKY, 1967a : 119, pl. 14, fig. 8 (not *Purpura carneola* Röding, 1798).
Chicoreus rubiginosus. — WILSON & GILLETT, 1971 : 86, pl. 58, fig. 4 ; KAICHER, 1973 : card 150 ; RADWIN & D'ATTILIO, 1976 : 42, pl. 6, fig. 10 ; ABBOTT & DANCE, 1982 : 137, text fig. ; RIPPINGALE, 1987 : 11, fig. 26.
Chicoreus (Chicoreus) rubiginosus. — FAIR, 1976 : 73, pl. 8, fig. 111 ; SPRINGSTEEN & LEOBRERA, 1986 : 134, pl. 36, fig. 12.
Chicoreus palmiferus. — HINTON, 1972 : 36, pl. 18, figs 2, 3 (not *Murex palmiferus* Sowerby, 1841).
Chicoreus microphyllus. — RADWIN & D'ATTILIO, 1976 : 39 (in part), pl. 5, fig. 7 ; MC DONALD, 1979 : 7, fig. 9 ; EISENBERG, 1981 : 91, pl. 73, fig. 9 ; WELLS & BRYCE, 1985 : 86, pl. 25, fig. 282 (not *Murex microphyllus* Lamarck, 1816).
Chicoreus (Chicoreus) microphyllus. — SPRINGSTEEN & LEOBRERA, 1986 : 135 (in part), pl. 36, fig. 16b (only) (not *Murex microphyllus* Lamarck, 1816).
Chicoreus saulii. — HINTON, 1979 : pl. 26, fig. 6 (not *Murex saulii* Sowerby, 1834).
Chicoreus (Chicoreus) kilburni. — HOUART & PAIN, 1983b : 5, text fig. (holotype and paratypes).
Chicoreus kilburni. — HOUART, 1984a : 12, text fig. (paratype) ; FREEMAN, 1986 : 9, text fig. ; RIPPINGALE, 1987 : 7, fig. 15.
NOT *Chicoreus torrefactus*. — SALVAT & RIVES, 1975 : 311, fig. 193 [= *Chicoreus (Triplex) maurus* (Broderip, 1833)].
NOT *Chicoreus (Chicoreus) benedictinus*. — FAIR, 1976 : 25, pl. 9, fig. 116 [= *Chicoreus (Triplex) corrugatus ethiopius* Vokes, 1978].
NOT *Chicoreus benedictinus*. — RADWIN & D'ATTILIO, 1976 : 34, pl. 23, fig. 10 [= *Chicoreus (Triplex) peledi* Vokes, 1978].

TYPE LOCALITIES. *M. torrefactus* : Ticao, Philippines ; *M. rubiginosus* : Philippine Islands ; *M. affinis* : unknown ; *M. benedictinus* : ?Indian Ocean ; *M. rochebruni* : Diego-Suarez ; *C. kilburni* : Durban Bay, Natal, South Africa.

TYPE MATERIAL. *M. torrefactus* : Neotype MNHN (designated by HOUART & PAIN, 1982) ; *M. rubiginosus* : lectotype BMNH 197475, here selected from 7 syntypes (see discussion) ; *M. affinis* : lectotype BMNH 197499, here selected from 2 syntypes ; *M. benedictinus* : holotype LM ; *M. rochebruni* : holotype MNHN ; *C. kilburni* : holotype IRSNB 26429/385.

OTHER MATERIAL EXAMINED : more than 300 specimens from throughout the geographical range.

DISTRIBUTION. (Fig. 163). Throughout the Indo-West Pacific, from South-Western Indian Ocean, Durban Bay, South Africa to Mozambique ; Madagascar, Réunion, Seychelles, India, North-Western Australia, Philippine Is to south of Japan (as North limit), Kwajalein Atoll, New Guinea, New Caledonia, and Fiji.

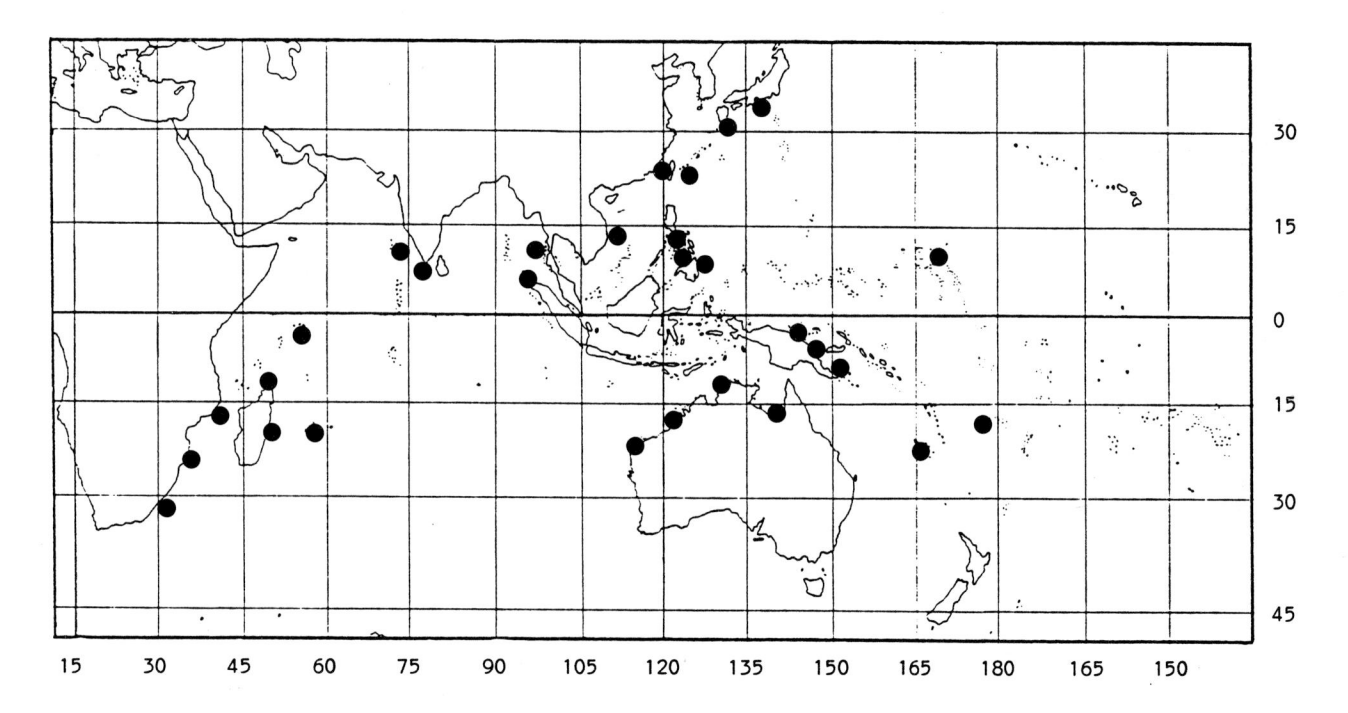

FIG. 163. Distribution of *Chicoreus torrefactus* (Sowerby).

DESCRIPTION. Shell up to 123.8 mm in length, stout. Spire high, acute, with 3 protoconch whorls and up to 10 convex teleoconch whorls. Suture ranging from slightly impressed to well appressed. Protoconch glossy, conical.

Last whorl with 3 frondose varices and 1-3 low to high intervaricial axial ridges. Spiral sculpture consisting of finely crenulated cords and intermediate fine threads in each interspace. Major or minor foliated spines present where cords cross varices. Varices generally bearing 5 frondose spines, shoulder spine and two abapical spines longest ; second spine shortest, middle spine of intermediate length ; with small intermediate spinelets. Shoulder spine sometimes obsolete. Abapical spines connected by webbing in some South African specimens (*kilburni* form).

Aperture broadly ovate to rounded. Columellar lip adherent, usually smooth, occasionally bearing 2 or 3 small denticles abapically ; 1 or 2 denticles adapically in some juvenile specimens, strong callus, delineating large, deep anal notch. Outer lip crenulate, partially lirate within. Siphonal canal moderately short, narrowly open, distally recurved, with 3 or 4 frondose spines.

Usually brown with darker spiral cords and threads. Occasional specimens deep mauve fronds ; rarely pure orange or cream. The Southern African *kilburni* form has 2 pale bands on last whorl. Aperture white, columellar lip light peach to dark orange.

Radula (Figs 94-97).

REMARKS. A neotype for *C. torrefactus* was designated by HOUART & PAIN (1982 : 51) to stabilise the concept of the species which had been erroneously synonymised with *C. microphyllus* by RADWIN & D'ATTILIO (1976 : 39). There are constant differences between these two, often extremely similar species (Table 1).

The type material for *Murex rubiginosus* consists of 7 specimens, one of which is not conspecific but could not be identified. Of the 5 paralectotypes (labelled " Philippines ") only one shows the long fronds, typical of most Australian specimens (HOUART & PAIN, 1983a : 16, fig. 1).

SOWERBY (1879 : 14) noted " at first it was thought that *M. rubiginosus* might be *M. torrefactus* with more fully developed fronds, but the latter has a much longer spine and shorter canal ". Nevertheless, his fig. 61 is without any doubt *C. torrefactus* localised " China ". POIRIER (1883 : 252) had already synonymised *C. rubiginosus* with *C. torrefactus*. VAL SWEETMAN (in litt.) personally observed that " biological habits " for both forms are widely different whilst the shells of both species are much alike. He observed, both in aquaria and in the " natural environment ", that *C. torrefactus* lays its eggs in " early summer " and the egg capsules are hatched before the form called *C. rubiginosus* commences laying eggs in " late February to early March ". Both share the same habitat but he never found any evidence of interbreeding (no more details were obtained).

An explanation for this is perhaps that the *rubiginosus* form is a long established local population of *C. torrefactus*, and that " typical " *C. torrefactus* with shorter varicial fronds is a more recent arrival in that area (Broome, West Australia), what could explain these differences. I examined many shells of the Australian form including one with preserved protoconch and I could not separate it from *C. torrefactus* on protoconch and teleoconch characters.

Murex affinis is another subjective synonym of *M. torrefactus*. Moreover, this name is preoccupied by *M. affinis* Gmelin, 1791, *M. affinis* Risso, 1826 and *M. affinis* Eichwald, 1830.

A letter of inquiry with respect to type material led to the rediscovery (BOSCHEINEN, 1983 : 16) of Löbbecke's long lost material, including the holotype of *Murex benedictinus*. The shell of this specimen is 88.7 mm long, not 90 mm as originally described. BOSCHEINEN (1983) considered that the shell was probably damaged after it was originally measured, which would explain the minor discrepancy. The name *C. benedictinus* was used by FAIR (1976 : 25) for a Red Sea muricid later named as *C. corrugatus ethiopius* Vokes, 1978, and also by RADWIN & D'ATTILIO (1976 : 34), who used the name for another Red Sea species later named *C. peledi* Vokes, 1978. In fact, *M. benedictinus* is merely the form of *C. torrefactus* with pale violet fronds and strong intervaricial axial nodes, that is characteristic of eastern African coasts. This is most probably the *Chicoreus kilburni* of authors (not HOUART & PAIN, 1982), recently offered by dealers.

Murex rochebruni is based on a shell with somewhat slender form and short shoulder spines, but it is undoubtedly *C. torrefactus* since absence or reduction of shoulder spine is not uncommon in *C. torrefactus* or, for that matter, in most other *Chicoreus (Triplex)* species.

C. kilburni was separated on basis of its more ovate aperture, appressed suture, single and shallow axial node, among other differences. Unfortunately, at that time, the authors had not seen a specimen retaining the protoconch. Intensive searching led to discovery of a specimen from southern Mozambique with a partially intact protoconch (2 last whorls and terminal varix). The nuclear whorls and terminal varix are identical to these of " typical " *C. torrefactus*, which suggests that *C. kilburni* is only a variant of this morphologically variable *Chicoreus* species. Again, this confirms the importance of the protoconch for species recognition.

GROUP 2

Chicoreus (Triplex) microphyllus (Lamarck, 1822)
Figs 13-19, 87-91, 133, 164, 299-301, 309-323

Murex microphyllus Lamarck, 1816 : pl. 145, fig. 5 ; 1822 : 163.

Murex poirieri Jousseaume, 1881 : 349.
Murex jousseaumei Poirier, 1883 : 58, pl. 6, fig. 1.
Chicoreus akritos Radwin & D'Attilio, 1976 : 228, pl. 4, fig. 1

ADDITIONAL REFERENCES

Murex (Chicoreus) microphyllus. — SMITH, 1953 : 5, pl. 8, fig. 10.
Chicoreus (Triplex) microphyllus. — KIRA, 1965 : 64, pl. 24, fig. 17.
Chicoreus microphyllus. — ZHANG, 1965 : 21,pl. 2, fig. 6 ; CERNOHORSKY, 1967a : 120, fig. 4, pl. 4, fig. 10 ; CERNOHORSKY, 1967b : 122, pl. 25, fig. 151 ; HINTON, 1972 : 36, pl. 18, fig. 8 ; KAICHER, 1974 : card 576 ; RADWIN & D'ATTILIO, 1976 : 39 (in part), pl. 4, fig. 7 ; HINTON, 1979 : 26, fig. 7 ; MUHLHAUSSER & ALF, 1983 : fig. 2 (in part) ; FREEMAN, 1986 : 9, text fig. ; RIPPINGALE, 1987 : 7, fig. 18.
Chicoreus (Chicoreus) microphyllus. — FAIR, 1976 : 58, pl. 7, fig. 87 ; HOUART & PAIN, 1983b : 3, text figs (syntype of *Murex poirieri* Jousseaume and *Murex jousseaumei* Poirier), text fig., p. 4 ; HOUART, 1984b : pl. 1, fig. 5 ; SPRINGSTEEN & LEOBRERA, 1986 : 135, pl. 36, fig. 16a.
Chicoreus penchinati. — CERNOHORSKY, 1971 : 187 (in part), fig. 7 ; CERNOHORSKY, 1972 : 122, pl. 34, fig. 5 (not *Murex penchinati* Crosse, 1861).
Chicoreus huttoniae. — WILSON & GILLETT, 1971 : 86, pl. 58, fig. 10 (not *Murex huttoniae* Wright, 1878).
Chicoreus (Chicoreus) huttoniae. — FAIR, 1976 : 49 (in part), pl. 6, fig. 76A (unnumbered on plate) (not *Murex huttoniae* Wright, 1878).
Chicoreus akritos. — RIPPINGALE, 1987 : 3, fig. 1 ; SHORT & POTTER, 1987 : 56, pl. 27, fig. 9.
Chicoreus (Chicoreus) rubescens. — FAIR, 1976 : 73, pl. 8, fig. 110 (not *Murex rubescens* Broderip, 1833).
NOT *Chicoreus microphyllus*. — RADWIN & D'ATTILIO, 1976 : 39 (in part), pl. 5, fig. 7 ; WELLS & BRYCE, 1985 : 86, pl. 25, fig. 282 [= *Chicoreus (Triplex) torrefactus* (Sowerby, 1841)] ; ABBOTT & DANCE, 1982 : 136, text fig. [= *Chicoreus (Triplex) strigatus* (Reeve, 1845)].
NOT *Murex (Chicoreus) microphyllus*. — EISENBERG, 1981 : 91, pl. 73, fig. 9 [= *Chicoreus (Triplex) torrefactus* (Sowerby, 1841)], figs 9A, B [= *Chicoreus (Triplex) strigatus* (Reeve, 1845)].
not *Chicoreus (Chicoreus) microphyllus*. — SPRINGSTEEN & LEOBRERA, 1986 : 135 (in part), pl. 36, fig. 16b [= *Chicoreus (Triplex) torrefactus* (Sowerby, 1841)].

TYPE LOCALITIES. *M. microphyllus* : none ; *M. poirieri* : New Caledonia ; *M. jousseaumei* : probably New Caledonia ; *C. akritos* : off Keppel Island, Keppel Bay (Queensland), Australia.

TYPE MATERIAL. *M. microphyllus* : probable syntype MHNG 1099/22, here designated as lectotype ; *M. poirieri* : lectotype MNHN, here selected from 3 syntypes ; *M. jousseaumei* : lectotype MNHN, here selected from 3 syntypes ; *C. akritos* : holotype SDSNH 53173.

OTHER MATERIAL EXAMINED : more than 200 specimens from throughout the geographical range.

DISTRIBUTION. (Fig. 164). South-western Indian Ocean : Madagascar (Nossi-bé and Tuléar) ; West Thailand ; the Philippine Is (Mindanao) ; South China Sea ; Ryukyu Is ; Australia : North West Australia to southern Queensland ; New Guinea ; Solomon Is ; New Caledonia to the Fiji Is (East limit) ; on coral reefs.

DESCRIPTION. Shell up to 90 mm in length, stout. Spire high, acute, with 1 3/4-3 protoconch whorls and up to 9 roundly teleoconch whorls. Suture impressed to slightly appressed. Protoconch whorls glossy, rounded.
Last whorl with 3 frondose varices, each with 5 or 6 short, open, frondose spines, intermediate spinelets short. Adapical spines usually short to obsolete but specimens with long shoulder spine not uncommon. Other axial sculpture consisting of 3 or occasionally 2 low intervaricial nodulose ridges. Spiral sculpture of 6 or 7 squamous cords flanked by squamous cords and threads. Small nodules formed where major cords cross axial sculpture.

Aperture ovate. Columellar lip fully adherent, usually bearing series of weak denticles along edge. Moderately strong callus delineating large, deep anal sulcus. Outer lip crenulate, strongly lirate for short distance within. Siphonal canal moderately short, narrowly open, bent abaperturally at tip, bearing 2 or 3 short, frondose, open spines.
Colour highly variable, ranging from uniformly pale brown to bright orange, or pale brown mottled with dark brown and orange. Typically pale brown with darker spiral cords and dark brown siphonal canal. Aperture glossy white or bluish-white. Columellar lip peach, pale orange or bluish-white.

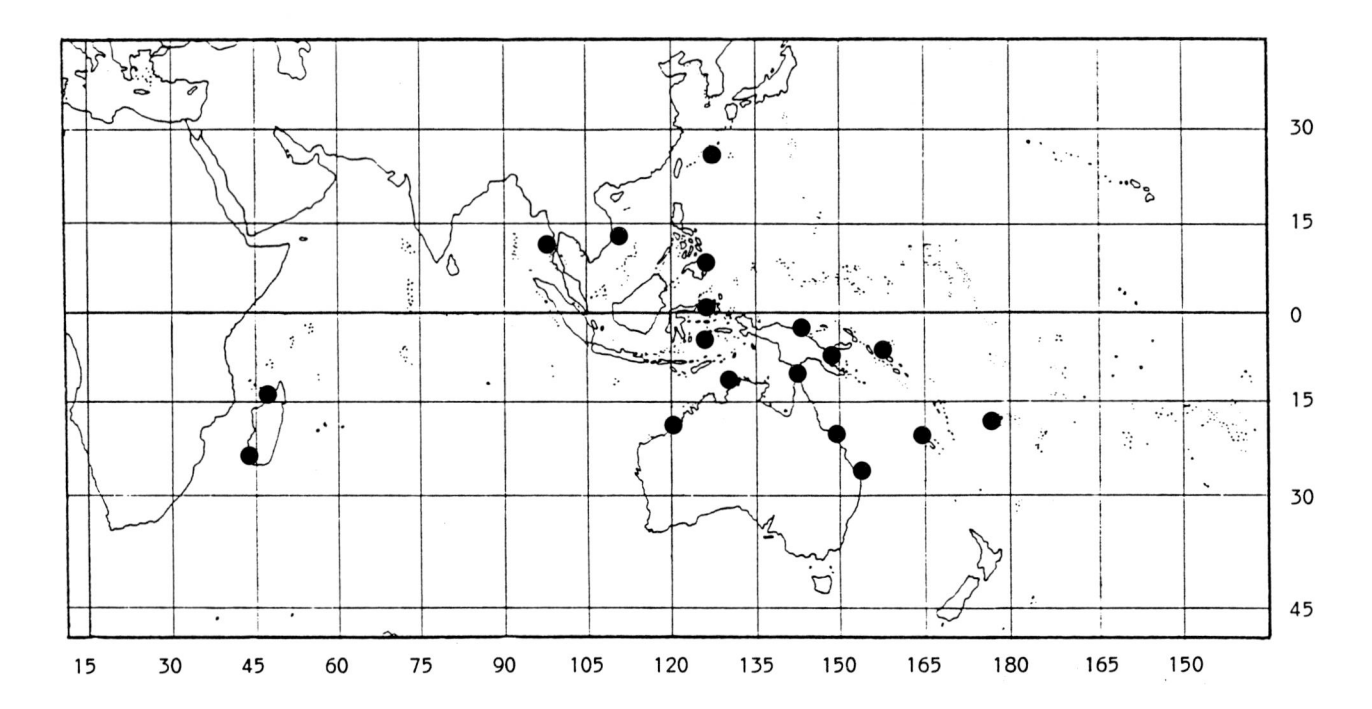

FIG. 164. — Distribution of *Chicoreus microphyllus* (Lamarck).

REMARKS. The so called " typical " *akritos* form differs from *C. microphyllus* in being golden brown and smaller, but examination of numerous specimens reveals complete intergradation between the two. The operculum is identical in both forms, while the protoconch, comprising 1 3/4-3 whorls is identical in the two forms.

C. microphyllus is extremely variable. Most shells have short frondose spines with generally a short shoulder spine. In some specimens the shoulder spine may be long, as in the syntypes of *M. jousseaumei*. The shell is usually lightly built and narrowly elongate but may be heavier and more broadly fusiform. The colour is typically light brown with darker spiral cords but others may be very pale orange with darker spiral bands or entirely pale to deep orange. Specimens from the Philippines are consistantly orange with brown blotches (Fig. 323).

Chicoreus (Triplex) paini Houart, 1983
Figs 21, 98-99, 165, 326-328

Chicoreus paini Houart, 1983 (30 June) : 28, figs 1-2, pl. 1, figs 3-4.

Chicoreus kengaluae Mühlhäusser & Alf, 1983 (1 July) : 101, text figs.

ADDITIONAL REFERENCES

Chicoreus sp. — HINTON, 1979 : 26, fig. 11.
Chicoreus paini. — HOUART, 1985 : 12, text fig. (paratype) ; CERNOHORSKY, 1985 : 49 (in part), fig. 9 (paratype) ; RIPPINGALE, 1987 : 9, fig. 20.

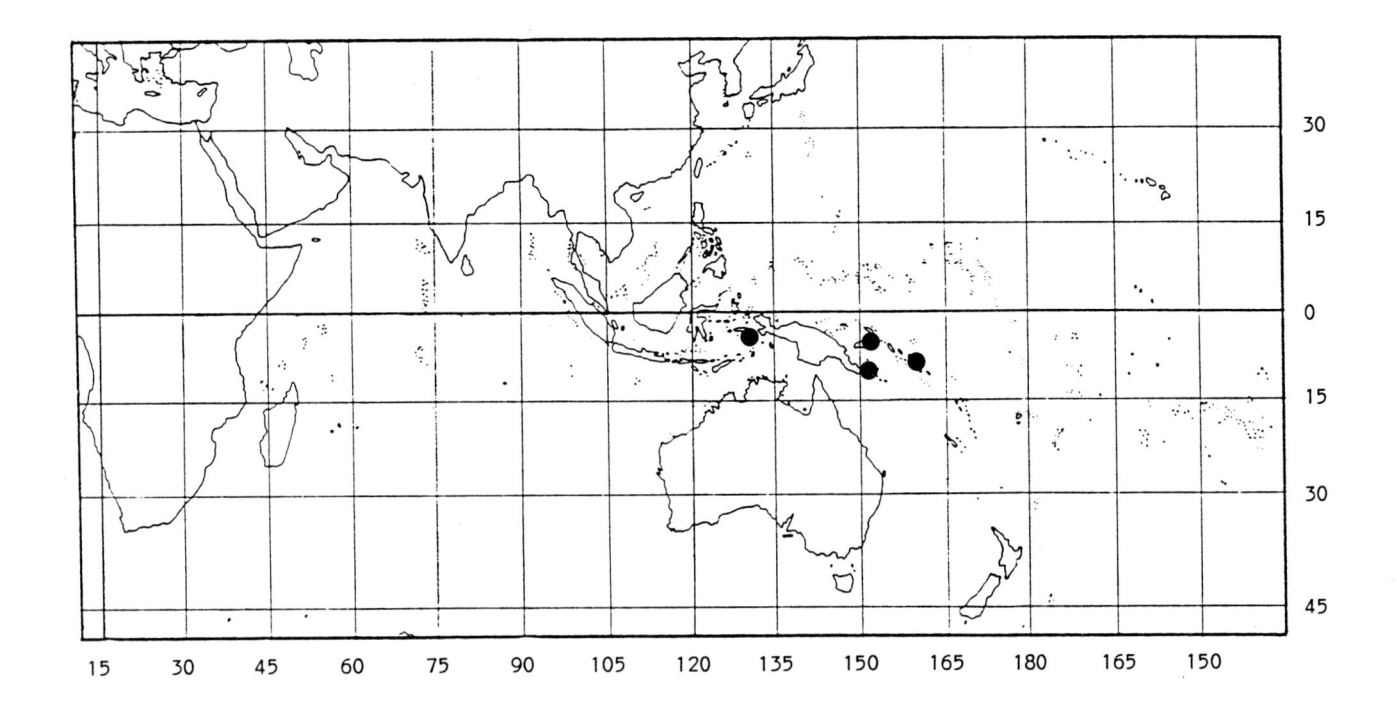

FIG. 165. — Distribution of *Chicoreus paini* Houart.

TYPE LOCALITY. *C. paini* and *C. kengaluae* : Honiara, Guadalcanal I., Solomon Is.

TYPE MATERIAL. *C. paini* : Holotype IRSNB 26554/396 ; *C. kengaluae* : Holotype ZSM 1742.

OTHER MATERIAL EXAMINED : c. 30 specimens from throughout the geographical range.

DISTRIBUTION. (Fig. 165). Banda Sea (Moluccas) to the Solomon Is.

DESCRIPTION. Shell up to 71 mm in length, fusiform. Spine high, 1 1/2 protoconch whorls and up to 8 convex teleoconch whorls. Suture impressed. Protoconch whorls smooth, weakly shouldered.
Last whorl with 3 frondose varices and 2 or 3 relatively low intervaricial ridges. Varices of last whorl each with 5 frondose aperture spines and small intermediate spinelets. Spiral sculpture consisting of 5 or 6 strong cords, with small, finely squamous threads.
Aperture ovate. Columellar lip smooth throughout or weakly denticulate adapically ; adherent abapically, rim detached and erect abapically. Anal notch small, deep, delineated by strong callus. Outer lip weakly erect, strongly denticulate, lirate for short distance within. Siphonal canal moderately long, open, distally recurved, with 3 or 4 frondose spines.
Spiral ridges dark brown, interspaces pale brown. Aperture bordered with pink or peach, occasionally uniformly white. Columellar lip pink or peach. Orange coloured specimens have been collected from the Solomon Islands.
Radula (Figs 98-99).

REMARKS. This species was originally compared with *C. brunneus, C. trivialis* and *C. microphyllus*. The shell of *C. paini* has 2 or 3 intervaricial ridges, while adult specimens of *C. brunneus* always have only a single strong intervaricial node. The anal notch is shallower though larger and the protoconch is more narrowly conical than in *C. brunneus*. For differences between *C. microphyllus* and *C. trivialis* see table 2.

TABLE 2. — Comparisons of *Chicoreus* species of group 2.

Character	*C. microphyllus*	*C. strigatus*	*C. paini*	*C. trivialis*	*C. rubescens*
Protoconch	2-3 rounded protoconch whorls. Terminal varix weakly curved	2 rounded protoconch whorls. First whorl globular. Curved terminal varix	1.5 weakly shouldered protoconch whorls. Terminal varix unknown	2-2 1/4 rounded weakly flattened protoconch whorls. Terminal varix undulate, heavy	Probably multispiral. Last whorl shouldered
Number of teleoconch whorls	7-9	6-7	7-8	8-9	7
Spines on last whorl	5-6 short frondose spines with intermediate spinelets	6 short frondose spines with intermediate spinelets	5 spines with intermediate spinelets	6-7 short and straight frondose spines with numerous spinelets	5 spines. 2-3 adapical spines obsolete
Intervaricial axial sculpture	2-4 ridges (usually 3)	1-3 low ridges (usually 2-3)	2-3 ridges	1 strong, elongate node	1 strong ridge
Adult shell length	40-90 mm	23-40 mm	43-53 mm	42-58 mm	25-48 mm

CERNOHORSKY (1985 : 50, fig. 9) illustrated one of the paratypes of *C. paini*, presently in my collection. This paratype, according to CERNOHORSKY (1985), is a specimen of *C. trivialis*, but it is obvious that the shell of *C. paini* consistantly has 2 or 3 intervaricial ridges while *C. trivialis* has only 1 strong axial node. Although CERNOHORSKY also noted that the paralectotype of *C. trivialis* has 2 intervaricial nodes, that specimen is subadult, while adult shells of *C. trivialis* has 1 single node on the 2 or 3 last whorls (adults have 8 or 9 teleoconch whorls). Another distinctive feature of *C. paini* are the long varicial shoulder fronds on the last adult whorls, and sometimes on earlier whorls, as in the illustrated paratype.

C. paini has one day priority over *C. kengaluae* (1 July), having been introduced in June of the same year [i.e. 30 June following ICZN art 21 (c)(i)].

Chicoreus (Triplex) rubescens (Broderip, 1833)
Figs 38, 166, 252, 329

Murex rubescens Broderip, 1833 : 174

ADDITIONAL REFERENCES

Murex rubescens. — SOWERBY, 1834 : pl. 58, fig. 7.
Murex (Chicoreus) rubescens. — SMITH, 1953 : 7, pl. 23, fig. 10.
Chicoreus rubescens. — RADWIN & D'ATTILIO, 1976 : 41, pl. 6, fig. 5; KAICHER, 1979 : card 1993; ABBOTT & DANCE, 1982 : 137, text fig.
NOT *Chicoreus (Chicoreus) rubescens*. — FAIR, 1976 : 73, pl. 8, fig. 110 [= *Chicoreus (Triplex) microphyllus* (Lamarck, 1816)].

TYPE LOCALITY. Tahiti.

TYPE MATERIAL. Lectotype BMNH 197480, here selected from 3 syntypes.

OTHER MATERIAL EXAMINED : Tuamotus, RH (3 dd), MNHN (1 dd), IRSNB (4 dd); locality unknown (Marquesas, Tahiti or New Caledonia), RH (1 lv); Tahiti, MNHN (8 dd); New Caledonia, MNHN (2 dd); Wallis, coll. J. COLOMB (1 dd).

DISTRIBUTION. (Fig. 166). Only known from the Tuamotu and the Society archipelagoes, Wallis, and New Caledonia.

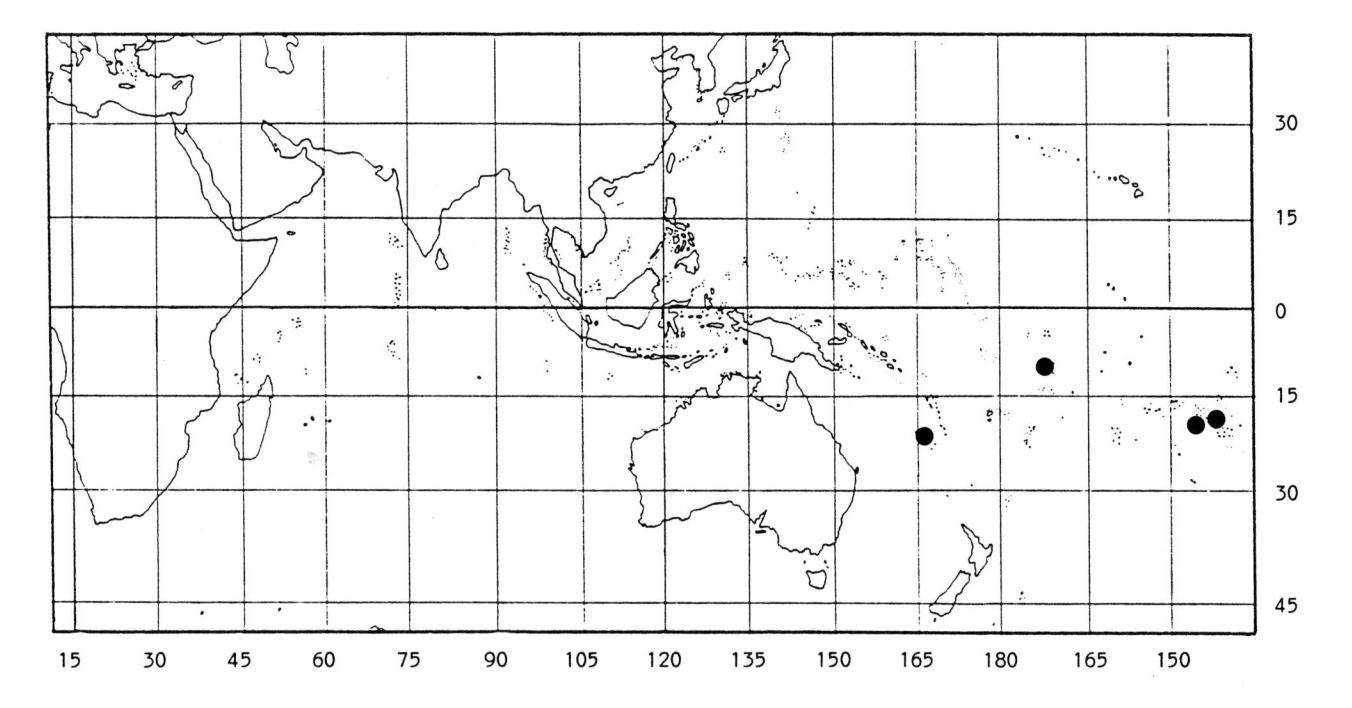

FIG. 166. — Distribution of *Chicoreus rubescens* (Broderip).

DESCRIPTION. Shell up to 48 mm in length, stout. Spire high, protoconch unknown (probably multispiral), up to 7 weakly shouldered teleoconch whorls. Suture appressed.
Last whorl bearing 3 rounded varices, each ornamented with 5 spines, 2 or 3 adapical spines obsolete. Most abapical spine longest. Other axial sculpture after first or second teleoconch whorl, consisting of single, strong intervaricial ridge, most prominent on shoulder. Spiral sculpture of 5 strong cords and 8-10 squamous threads in each interspace.
Aperture rounded. Columellar lip adherent, narrow adapically, with a small elongate node abapically. Anal notch rather deep, delineated by strong callus. Outer lip denticulate, strongly lirate for short distance within. Siphonal canal medium-sized, narrowly open, ornamented with 2 or 3 abapically bent open spines.
Pinkish-orange with darker spiral cords. Aperture white. Edge of columellar lip and outer lip stained pale violet or pink.

REMARKS. The only live collected specimen seen comes from a mixed lot of shells collected at the Marquesas, Tahiti and New Caledonia, unfortunately without any other information. Apparently it has never been collected alive before. The species is apparently not rare as several specimens are present in European museum collections. It is possible that it lives sublittorally but the shell does not present any of the features characteristic of deep-living species, such as palmate or frondose spines, lightly built shell, long siphonal canal etc...

In the original description, Broderip mentioned that the shells were found on coral reefs, but his shells were quite definitely collected dead.

A small juvenile shell measuring 15 mm in length, was recently collected in Wallis. Although the tip of the protoconch is missing, the remaining whorls indicate a multispiral protoconch of 3-3 1/2 whorls (Fig. 38).

Chicoreus (Triplex) strigatus (Reeve, 1849)
Figs 20, 86, 167-170, 171, 304-305, 324-325

Murex strigatus Reeve, 1849 : pl. 1, fig. 189.

Murex penchinati Crosse, 1861 : 351, pl. 16, fig. 6.
Murex multifrondosus Sowerby, 1879 : 16, fig. 192.

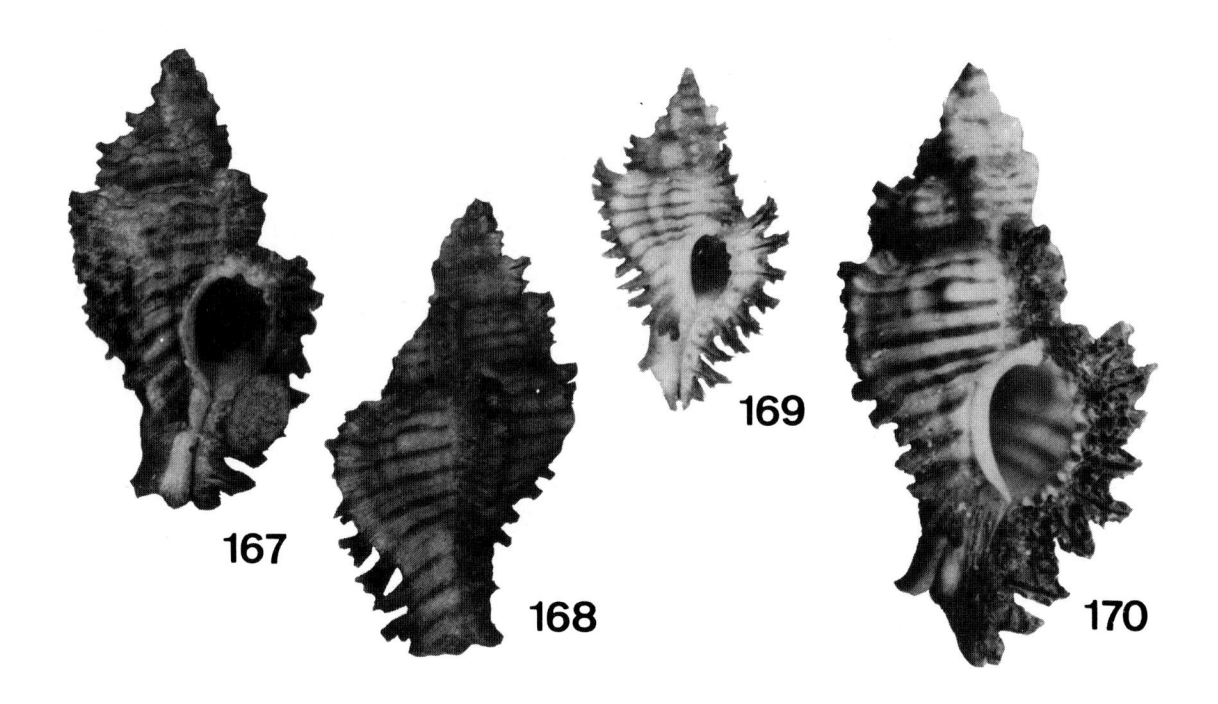

FIGS 167-170. — *Chicoreus (Triplex) strigatus* (Reeve).
167-168, locality unknown, 40 mm (paralectotype, BMNH 1980132).
169, locality unknown, 35.5 mm (lectotype, BMNH 1980132).
170, Philippines, 44 mm (RH)

ADDITIONAL REFERENCES

Murex (Chicoreus) penchinati. — SMITH, 1953 : 6, pl. 5, fig. 10.
Chicoreus penchinati. — ARAKAWA, 1964 : 361, pl. 21, figs 15-16 (radula); CERNOHORSKY, 1971 : 187 (in part), text fig. 12, fig. 6 (syntype); KURODA, HABE & OYAMA, 1971 : 140, pl. 40, fig. 2; RADWIN & D'ATTILIO, 1976 : 40, pl. 6, fig. 4; KAICHER, 1980, card 2543 (holotype); ABBOTT & DANCE, 1982 : 137, text fig.
Chicoreus (Chicoreus) penchinati. — FAIR, 1976 : 66, pl. 7, fig. 90; SPRINGSTEEN & LEOBRERA, 1986 : 131, pl. 35, fig. 15.
Chicoreus (Chicoreus) strigatus. — FAIR, 1976 : 79, fig. 61.
Chicoreus (Chicoreus) trivialis. — FAIR, 1976 : 83, pl. 7, fig. 96 (not *Murex trivialis* A. Adams, 1854).
Murex (Chicoreus) microphyllus. — EISENBERG, 1981 : 91, pl. 73, fig. 9A-B (not *Murex microphyllus* Lamarck, 1816).
Chicoreus microphyllus. — ABBOTT & DANCE, 1982 : 136, text fig. (not *Murex microphyllus* Lamarck, 1816).
Chicoreus multifrondosus. — RIPPINGALE, 1987 : 9, fig. 19.
Chicoreus brunnea. — RIPPINGALE, 1987 : 11, fig. 27 (not *Purpurea brunnea* Link, 1807).
NOT *Chicoreus penchinati*. — CERNOHORSKY, 1971 : 187 (in part), fig. 7; CERNOHORSKY, 1972 : 122, pl. 34, fig. 5 [= *Chicoreus (Triplex) microphyllus* (Lamarck, 1816)].

TYPE LOCALITIES. *M. strigatus* : unknown; *M. penchinati* : Liou-Tcheou (= Ryukyu Is); *M. multifrondosus* : unknown.

TYPE MATERIAL. *M. strigatus* : lectotype BMNH 1980132, here selected from 2 syntypes; *M. penchinati* : holotype BMNH 1896, 12.1.5; *M. multifrondosus* : holotype MNHN.

OTHER MATERIAL EXAMINED : c. 150 specimens from throughout the geographical range.

DISTRIBUTION. (Fig. 171). From the Moluccas to the Philippine Is; Taiwan Strait; Okinawa (as north limit). Two specimens were dead collected in French Polynesia (Tuamotu and Society archipelagoes) (RH and M. BOUTET coll.).

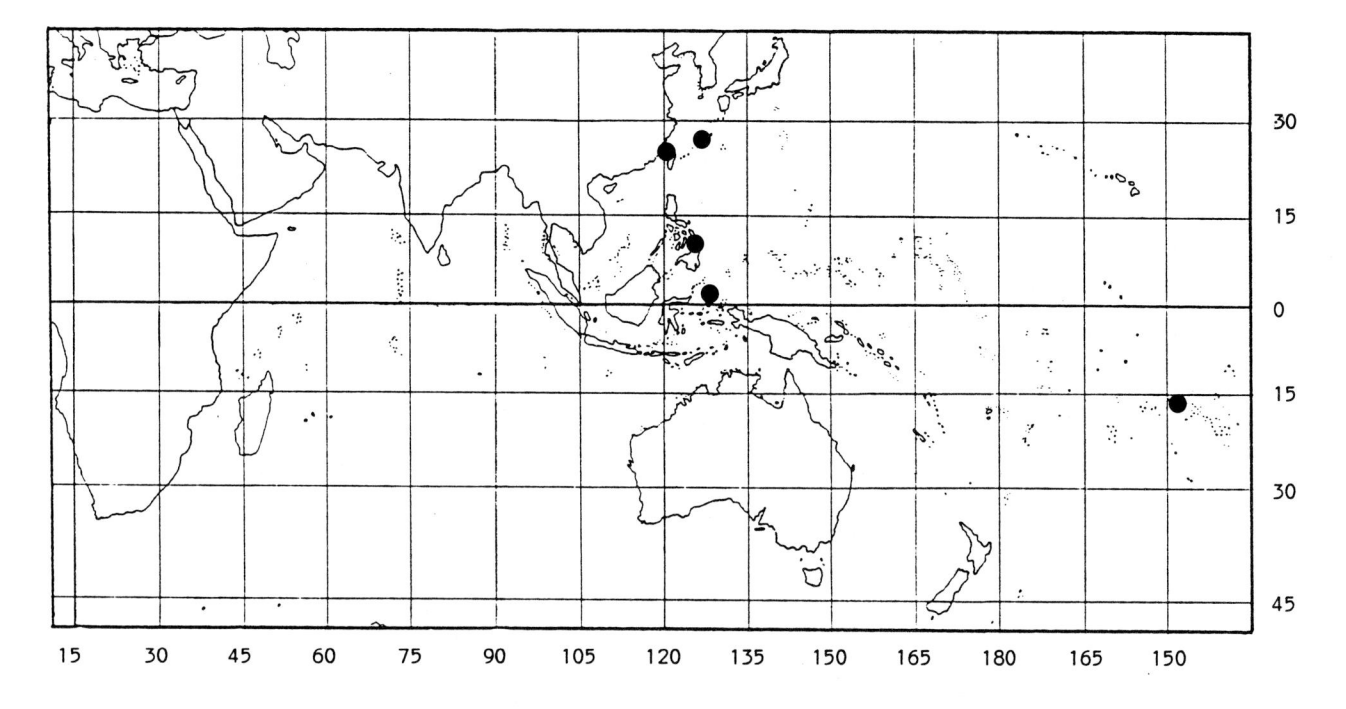

FIG. 171. — Distribution of *Chicoreus strigatus* (Reeve).

DESCRIPTION. Shell up to 47 mm in length. Spire high with 2 protoconch whorls and up to 7 ovate teleoconch whorls. Suture slightly appressed. Protoconch whorls rounded and glossy.

Last whorl with 3 frondose varices, each ornamented with 6 short frondose spines and short intermediate spinelets. Other axial sculpture consisting of 1-3 low intervaricial ridges (generally 2 or 3), that extend from suture to siphonal canal. Spiral sculpture of 6 or 7 low, lightly squamous cords, flanked by scabrous spiral threads; 1 small intermediate cord between each pair of cords. Small knobs nodules where spiral cords cross axial ribs.

Aperture ovate. Columellar lip smooth, adherent, rim weakly erect abapically. Anal notch deep, narrow. Outer lip crenulate, lirate for short distance within. Siphonal canal moderately short, narrowly open, slightly bent abaperturally at tip, with 3 or 4 short open, frondose, spines.

Shell ranging from pure white to uniformly orange or pink, or light brown with darker or almost black spiral cords. Aperture whitish, columellar lip white, suffused with pink, or uniformy pink or pale orange.

REMARKS. Comparison of type material and numerous specimens reveals that *M. strigatus*, *M. penchinati* and *M. multifrondosus* are based on forms of a single variable species. *M. strigatus* has a brown shell with darker spiral cords; *M. penchinati* has an orange, red or white shell. *M. multifrondosus*, also with orange coloured shell, has long shoulder spines. Apart from colour and length of spines there are no significant differences between them.

Chicoreus (Triplex) trivialis (A. Adams, 1854)
Figs 22, 172, 330-335

Murex trivialis A. Adams, 1854 : 71.

ADDITIONAL REFERENCES

Murex trivialis. — SOWERBY, 1879, fig. 80.
Chicoreus trivialis. — KAICHER, 1974 : card 568; RADWIN & D'ATTILIO, 1976 : 43, pl. 6, fig. 12; HINTON, 1979 : 26, fig. 12; ABBOTT & DANCE, 1982 : 137, text fig.; CERNOHORSKY, 1985 : 49 (in part), figs 4-8 (lectotype and paralectotype); RIPPINGALE, 1987 : 5, fig. 7.
NOT *Chicoreus (Chicoreus) trivialis*. — FAIR, 1976 : 83, pl. 7, fig. 96 [= *Chicoreus (Triplex) strigatus* (Reeve, 1845)].
NOT *Chicoreus (Chicoreus) trivialis*. — VOKES, 1978 : 386 (in part), pl. 2, fig. 6 [?= *Chicoreus (Triplex) groschi* Vokes, 1978].

TYPE LOCALITY. Unknown.

TYPE MATERIAL. Lectotype and paralectotypes BMNH 1980136, designated by CERNOHORSKY (1985).
OTHER MATERIAL EXAMINED : c. 100 specimens from throughout the geographical range.

DISTRIBUTION. (Fig. 172). North-western Australia : from Broome to Cockatoo Islands, Buccaneer Archipelago, on rocky reefs and on coral rubble. Other localities such as Japan (FAIR, 1976 : 83) are based on misidentifications.

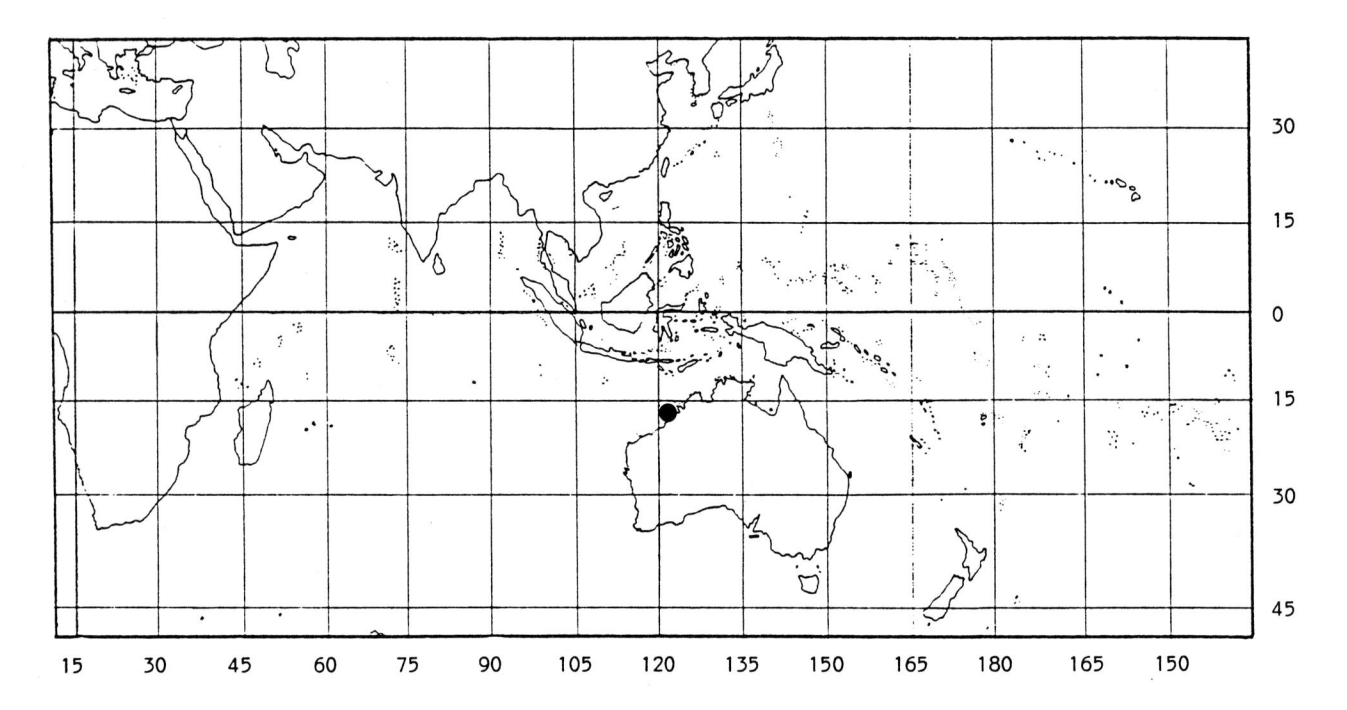

FIG. 172. — Distribution of *Chicoreus trivialis* (A. Adams).

DESCRIPTION. Shell up to 58 mm in length, stout, fusiform. Spire high, with 2-2 1/4 protoconch whorls and up to 9 angulate teleoconch whorls. Suture strongly appressed. Protoconch whorls rounded and glossy, weakly flattened.
Last whorl with 3 frondose varices and 1 large and heavy intervaricial node. Each varice with 6 or 7 short, straight, frondose spines and numerous short frondose spinelets. Spiral sculpture consisting of 6 or 7 cords that interconnect the spines, flanked by small squamous threads.
Aperture ovate. Columellar lip smooth, rim weakly detached abapically and adherent to the shell adapically. Anal notch small, of moderate depth, delineated by small callus. Outer lip crenulate, lirate for short distance within. Siphonal canal relatively short and broad, narrowly open, bent abapically at tip, with 3 or occasionally 4 crowded, straight, frondose spines.
Light brown to pale orange with darker spiral cords and spines. Intervaricial node and siphonal canal yet more darkly pigmented. Shoulder area most lightly pigmented. Edge of aperture and entire columellar lip dark pink.

REMARKS. CERNOHORSKY (1985) incorrectly interpreted the Australian shell identified as *C. trivialis* by RADWIN & D'ATTILIO (1976 : pl. 6, fig. 12) as a form of *C. brunneus*, but the shell morphology and colour are in fact perfectly accordant with *C. trivialis*.

GROUP 3

Chicoreus (Triplex) axicornis (Lamarck, 1822)
Figs 23-25, 173-175, 176, 336-337, 360

Murex axicornis Lamarck, 1822 : 163.

Murex kawamurai Shikama, 1964 : 116, pl. 65, fig. 4.

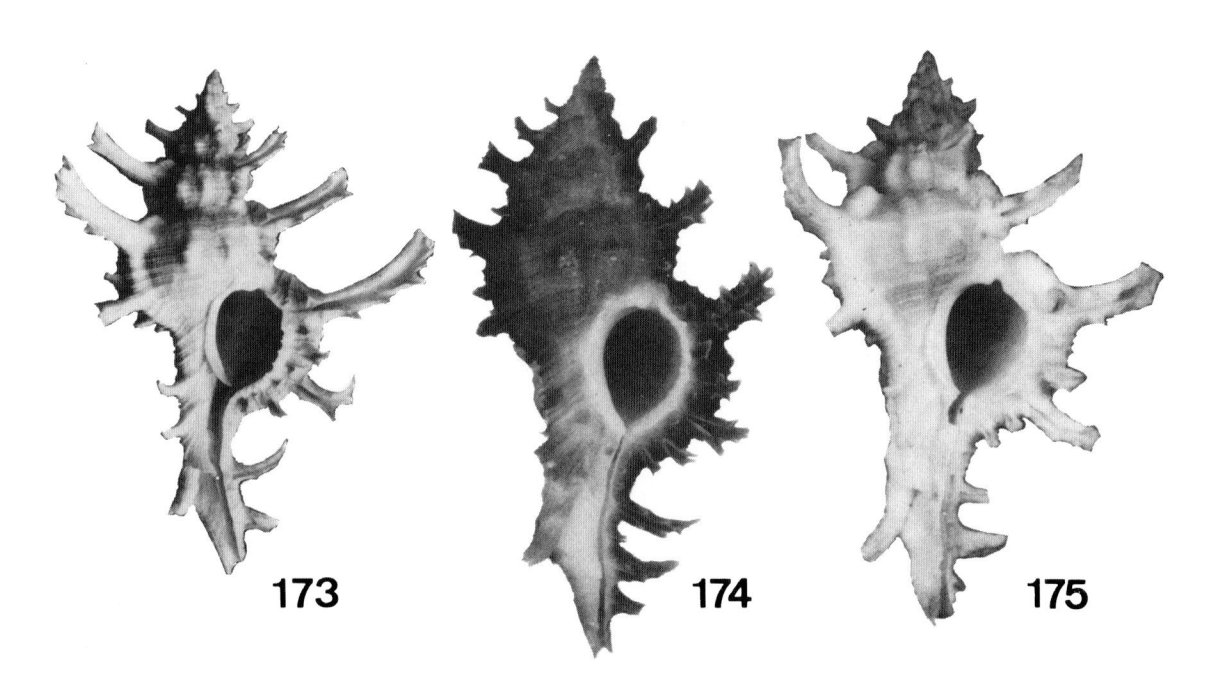

FIGS 173-175. — *Chicoreus (Triplex) axicornis* (Lamarck).
173, Indian Ocean, 59 mm (lectotype, MNHN).
174, Papua New Guinea, 38.5 mm (MNHN).
175, Taiwan, 65.5 mm (MNHN).

ADDITIONAL REFERENCES

Murex axicornis. — KIENER, 1842 : pl. 42, fig. 2 ; STOTT, 1983 : 4, text fig.
Murex (Chicoreus) axicornis. — SMITH, 1953 : 7, pl. 5, fig. 14 ; EISENBERG, 1981 : 87, pl. 69, fig. 12.
Chicoreus axicornis. — ZHANG, 1965 : 21, pl. 1, fig. 6 ; RADWIN & D'ATTILIO, 1976 : 32, pl. 4, fig. 2 ; KAICHER, 1979 ; card 2071 ; ABBOTT & DANCE, 1982 : 138, text fig. ; LAI, 1987 : 57, pl. 27, fig. 4.
Chicoreus (Chicoreus) axicornis. — FAIR, 1976 : 23, pl. 6, fig. 75.
Chicoreus kawamurai. — RIPPINGALE, 1987 : 7, fig. 14.
NOT *Murex axicornis*. — BARNARD, 1959 : 196 [= *Chicoreus (Chicoreus) litos* Vokes, 1978].
NOT *Murex (Chicoreus) axicornis*. — SHIKAMA, 1963 : 70, pl. 53, fig. 1 ; EISENBERG, 1981 : 87, pl. 69, fig. 12A [= *Chicoreus (Triplex) banksii* (Sowerby, 1841)].
NOT *Euphyllon axicornis*. — CERNOHORSKY, 1967b : 124, pl. 26, fig. 54 ; HINTON, 1972 : 36, pl. 18, fig. 11 [= *Chicoreus (Triplex) banksii* (Sowerby, 1841)].
NOT *Murex axicornis*. — KENSLEY, 1973 : 140, fig. 472 [= *Chicoreus (Chicoreus) litos* Vokes, 1978].
NOT *Chicoreus (Chicoreus) axicornis*. — FAIR, 1976 : 23 (in part), pl. 8, fig. 100 ; SPRINGSTEEN & LEOBRERA, 1986 : 132, pl. 36, fig. 5b [= *Chicoreus (Triplex) banksii* (Sowerby, 1841)].
NOT *Chicoreus axicornis*. — HINTON, 1979 : 26, fig. 10 [= *Chicoreus (Triplex) banksii* (Sowerby, 1841)].

TYPE LOCALITIES. *M. axicornis* : Indian Ocean and Moluccas ; *M. kawamurai* : Southwest Taiwan.

TYPE MATERIAL. *M. axicornis* : lectotype MNHN, here selected from 2 syntypes ; *M. kawamurai* : holotype NSMT 61245.

OTHER MATERIAL EXAMINED : c. 400 specimens from throughout the geographical range.

DISTRIBUTION. (Fig. 176). From the eastern African coast (North Mozambique) and the Seychelles to Sri Lanka ; the Andaman Sea ; South China Sea to Taiwan ; the Philippine Is ; North-East Australia ; Papua New Guinea ; New Caledonia (MNHN). Depth range : 40-80 m.

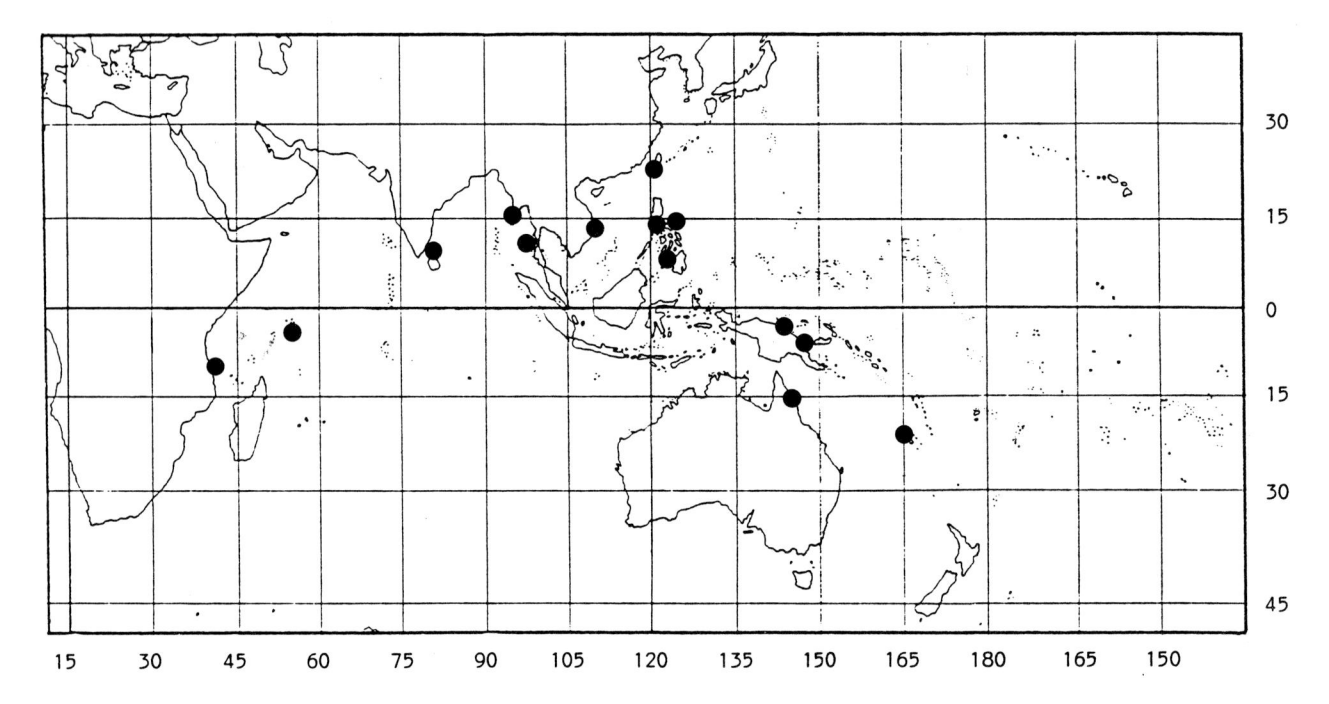

FIG. 176. — Distribution of *Chicoreus axicornis* (Lamarck).

DESCRIPTION. Shell up to 86.5 mm in length. Spire high, acutely conical, with 1 3/4-2 1/4 protoconch whorls and up to 8 rounded teleoconch whorls. Suture impressed to slightly appressed. Protoconch whorls rounded, sometimes weakly subcarinate.

Last whorl with 3 frondose varices, each with long shoulder spine, followed abapically by short spinelet, 1 or 2 medium-sized spines, and 1 or 2 spinelets. Second and third spines may be reduced into strong spinelets. Intervaricial area with 2 or occasionally 3 moderately low ridges. Spiral sculpture of 6 or 7 cords with smaller intermediate cords and squamous threads in each interspace.

Aperture ovate. Columellar lip smooth and erect. Anal notch deep, narrow. Outer lip crenulate, weakly erect, lirate for short distance within. Siphonal canal relatively long, narrowly open and recurved abaperturally, with 2 or 3 acute or frondose spines.

White to dark brown. Spiral cords sometimes darker. Aperture white. Specimens from the Eastern African coasts are uniformly pale orange or orange with darker spiral cords to my knowledge.

REMARKS. A variable species with a more extensive geographical range than usually credited by recent authors. It has often been confused with an Australian form of *C. banksii* (e.g. SOWERBY, 1841 : fig. 66 ; HINTON, 1972 : pl. 18, fig. 11 ; FAIR, 1976 : pl. 8, fig. 100).

Considering its variability, it is surprising to see that only one, relatively recently named taxon (*Murex kawamurai*), could be synonymised with this species. In some localities, such as Papua New Guinea, the shell exhibits only small spines and does not exceed 40 mm in length (subfossil specimens from Rabaul, New Britain, attain larger size), while most specimens from elsewhere range betweeen 40 and 65 mm in length. In some cases the breadth of the shell (spines included) exceed its length.

Chicoreus (Triplex) banksii (Sowerby, 1841)
Figs 26-28, 100-101, 178, 338-343, 358

Murex banksii Sowerby, 1841 : pl. 191, fig. 82 ; 1841b : 140.

Triplex cornucervi Perry, 1811 : pl. 7, fig. 4 (non *Purpura cornucervi* Röding, 1798).
Murex crocatus Reeve, 1845 : pl. 33, fig. 168.

ADDITIONAL REFERENCES

Murex (Chicoreus) banksii. — SMITH, 1953 : 6, pl. 23, fig. 6.
Chicoreus banksii. — WILSON & GILLETT, 1971 : 86, pl. 58, fig. 2 ; RADWIN & D'ATTILIO, 1976 : 33 (in part) ; HINTON, 1979 : 26, fig. 9 ; WELLS & BRYCE, 1985 : 86, pl. 25, fig. 279 ; RIPPINGALE, 1987 : 3, fig. 4 ; SHORT & POTTER, 1987 : 56, pl. 27, fig. 14 ; VAUGHT, 1989 : 3, figs 1, 3-9.
Chicoreus banksi (sic). — KAICHER, 1973 : card 157.
Chicoreus (Chicoreus) banksii. — FAIR, 1976 : 23, pl. 8, fig. 109 ; VOKES, 1978 : 382 (in part), SPRINGSTEEN & LEOBRERA, 1986 : 134, pl. 36, fig. 8.
Murex (Chicoreus) axicornis. — SHIKAMA, 1963 : 70, pl. 53, fig. 1 ; EISENBERG, 1981 : 87, pl. 69, fig. 12A (not *Murex axicornis* Lamarck, 1822).
Euphyllon axicornis. — CERNOHORSKY, 1967b ; 124, pl. 26, fig. 54 ; HINTON, 1972 : 36, pl. 18, fig. 11 (not *Murex axicornis* Lamarck, 1822).
Chicoreus (Chicoreus) axicornis. — FAIR, 1976 : 23 (in part), pl. 8, fig. 100 ; SPRINGSTEEN & LEOBRERA, 1986 : 136, pl. 36, fig. 5b (not *Murex axicornis* Lamarck, 1822).
Chicoreus axicornis. — HINTON, 1979 : 26, fig. 10.
Murex (Chicoreus) crocatus. — FAIR, 1974b : 12, text fig.
Chicoreus (Chicoreus) crocatus. — FAIR, 1976 : 35, pl. 7, fig. 92 ; LEEHMAN, 1976B : 5, TEXT G. ; SPRINGSTEEN & LEOBRERA, 1986 : 148, pl. 41, fig. 1 ; 153, text fig. (lectotype and type figure) ; VAUGHT, 1989 : 3 (in part).
Chicoreus crocatus. — ABBOTT & DANCE, 1982 : 137, text fig. ; RIPPINGALE, 1987 : 3, fig. 6.
Chicoreus ryosukei. — VAUGHT, 1989 : 3 (in part), fig. 6 [not *Chicoreus (Triplex) ryosukei* SHIKAMA, 1978].
NOT *Chicoreus banksii.* — SPRY, 1961 : 19, pl. 4, fig. 133 ; RADWIN & D'ATTILIO, 1976 : 33 (in part), pl. 4, fig. 12 ; RIPPINGALE, 1987 : 3, fig. 5 ; VAUGHT, 1989 : 3 (in part), fig. 2 [= *Chicoreus (Triplex) bourguignati* (Poirier, 1883)].
NOT *Chicoreus (Chicoreus) banksii.* — VOKES, 1978 : 382 (in part), pl. 3, figs 2-3 [= *Chicoreus (Triplex) bourguignati* (Poirier, 1883)].
NOT *Chicoreus banksii.* — ABBOTT & DANCE, 1982 : 136, text fig. [= *Chicoreus (Rhizophorimurex) capucinus* (Lamarck, 1822)].

TYPE LOCALITIES. *M. banksii* : Moluccas ; *T. cornucervi* : South Seas ; *M. crocatus* : unknown.

TYPE MATERIAL. *M. cornucervi* : no material located in BMNH ; *M. banksii* : lectotype BMNH 197478, here selected from 3 syntypes ; *M. crocatus* : 1 probable syntype BMNH 1984076, here designated as lectotype.

OTHER MATERIAL EXAMINED : c. 300 specimens from throughout the geographical range.

DISTRIBUTION. (Fig. 177). Malaysia ; the Philippine Is ; West and North Eastern Australia ; Papua New Guinea ; the Solomons Is ; New Caledonia. Sand with algae. Depth range : 10-60 m.

DESCRIPTION. Shell up to 82 mm in length. Spire high, with 2 1/4-2 1/2 protoconch whorls and up to 8 weakly angulate teleoconch whorls. Suture slightly appressed. Protoconch whorls rounded, sometimes slightly carinate abapically.

Last whorl with 3 rounded varices, each with 5 frondose spines. Shoulder spine long or very long, followed abapically by small spinelet that may be as the short spines in some specimens, and by 4 short frondose spines ; sometimes a fifth spine immediately adapical the siphonal canal. Other axial sculpture consisting of 1, 2 or occasionally 3 elongate intervaricial nodes or ridges. Spiral sculpture consisting of 5 primary cords and 5 or 6 intermediate secondary cords and threads in each interspace.

Aperture roundly-ovate. Columellar lip smooth, rim slightly erect abapically and adherent adapically. Anal notch relatively shallow, delineated by low callus. Outer lip crenulate, lirate for short distance within.

Siphonal canal medium-sized, narrowly open and abaperturally recurved at tip, with 3 or 4 long, frondose, adapically curved spines ; second abapical spine usually longest.

Usually light brown with darker spiral cords, varices and spines. Occasionally ochre or orange or darker with orange fronds.

Radula. (Fig. 100-101).

REMARKS. *C. banksii* is a highly variable and widely distributed species. The Australian form (Fig. 343), frequently misidentified as *C. axicornis*, has a more slender shell and longer shoulder spines.

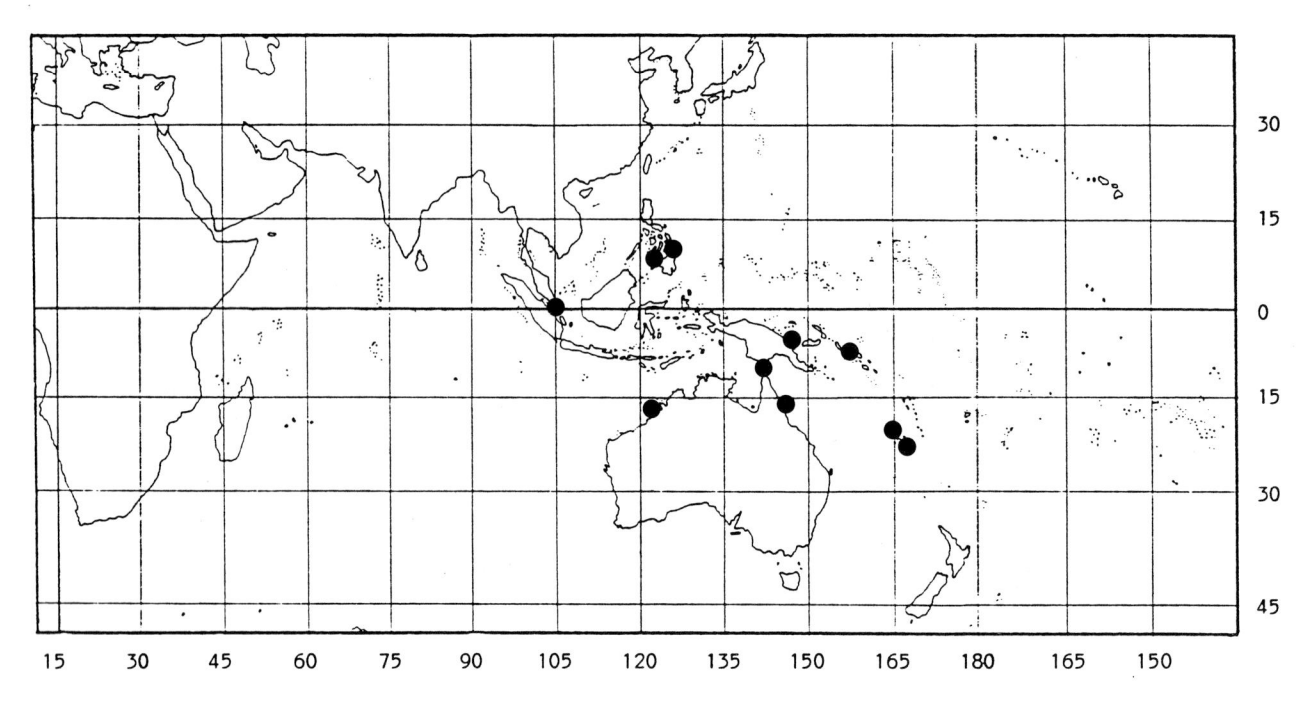

FIG. 177. — Distribution of *Chicoreus banksii* (Sowerby).

C. crocatus has long been regarded as a valid species by several authors, but this orange to dark brown form from the Philippines Islands and Australia is merely a variant of *C. banksii*, since there is complete intergradation between the extremes.

C. crocatus was based on a shell from the Norris collection, much of which was incorporated into the Tomlin collection (NMW). Unfortunately this shell cannot now be located in the Tomlin collection.

Nevertheless, a specimen in the BMNH (SPRINGSTEEN, 1982 : fig. 9) agrees very well with the original description and illustration and is probably the specimen examined by REEVE. It is here designated as lectotype of *Murex crocatus* Reeve, 1845;

Another form of *C. banksii* occurs on the reef at 8-10 m in the vicinity of Anna Plains, Western Australia. These shells are white with brown fronds or rather uniform light brown or nearly white. Although this form is particularly distinctive, the presence of intermediate specimens in the area, and the protoconch morphology confirmed its identity.

The East African shell identified as *C. banksii* by recent authors is *C. bourguignati*, a distinct though closely similar species.

Chicoreus (Triplex) bourguignati (Poirier, 1883)
Figs 29-30, 181, 261, 344-345

Murex bourguignati Poirier, 1883 : 48, pl. 5, fig. 2.

ADDITIONAL REFERENCES

Chicoreus banksii. — SPRY, 1961 : 19, pl. 4, fig. 133 ; RADWIN & D'ATTILIO, 1976 : 33 (in part), pl. 4, fig. 12 ; RIPPINGALE, 1987, p. 3, fig. 5 (not *Murex banksii* Sowerby, 1841).
Chicoreus (Chicoreus) banksii. — VOKES, 1978 : 382 (in part), pl. 3, figs 2-3 (not *Murex banksii* Sowerby, 1841).
Chicoreus groschi. — DRIVAS & JAY, 1988 : 70, pl. 20, fig. 2 [not *Chicoreus (Triplex) groschi* Vokes, 1978].

FIGS 178-180. — Subgenus *C. (Triplex)*.
178, *C. banksii* (Sowerby). Moluccas, 77.5 mm (lectotype, BMNH 197478).
179-180, *C. ryosukei* Shikama. 179, Arafura Sea, 53.1 mm (holotype, NSMT 60928). 180, locality unknown, 41 mm (RH).

TYPE LOCALITY. Nossi-Bé (Madagascar).

TYPE MATERIAL. Lectotype MNHN, here selected from 2 syntypes.

OTHER MATERIAL EXAMINED : c. 50 specimens from throughout the geographical range.

DISTRIBUTION. (Fig. 181). Durban, Southern Africa ; North Mozambique to Assab (Southern Ethiopia, Red Sea) ; Madagascar ; Réunion and Mauritius ; the Seychelles ; Trincomalee (Sri Lanka) ; on muddy sand and gravel or on coral, on coral reefs. Depth range : 0.50-80 m.

DESCRIPTION. Shell up to 95 mm in length, stout, fusiform. Spire high, with 2 3/4-3 protoconch whorl and up to 10 weakly angulate teleoconch whorls. Suture slightly appressed. Protoconch conical, whorls rounded, smooth.
Last whorl with 3 rounded varices, each with 5 short frondose spines and intermediate spinelets. Shoulder spine longest ; relatively strong spinelet between shoulder and abapical next spine. Other axial sculpture consisting of 2 or 3 low to moderately heavy intervaricial ridges. Spiral sculpture consisting of 5 or 6 squamous primary cords with 8-10 intermediate secondary cords and threads in each interspace.

Aperture ovate. Columellar lip smooth, adherent, detached at abapical extremity. Anal notch small, relatively shallow, delineated by low callus. Outer lip crenulate, lirate for short distance within. Siphonal canal medium-sized, narrowly open, lightly bent abaperturally at tip, with 3 or 4 relatively long, fronded spines, the abapical second of which is longest.
Light brown with darker spiral sculpture and varicial spines.

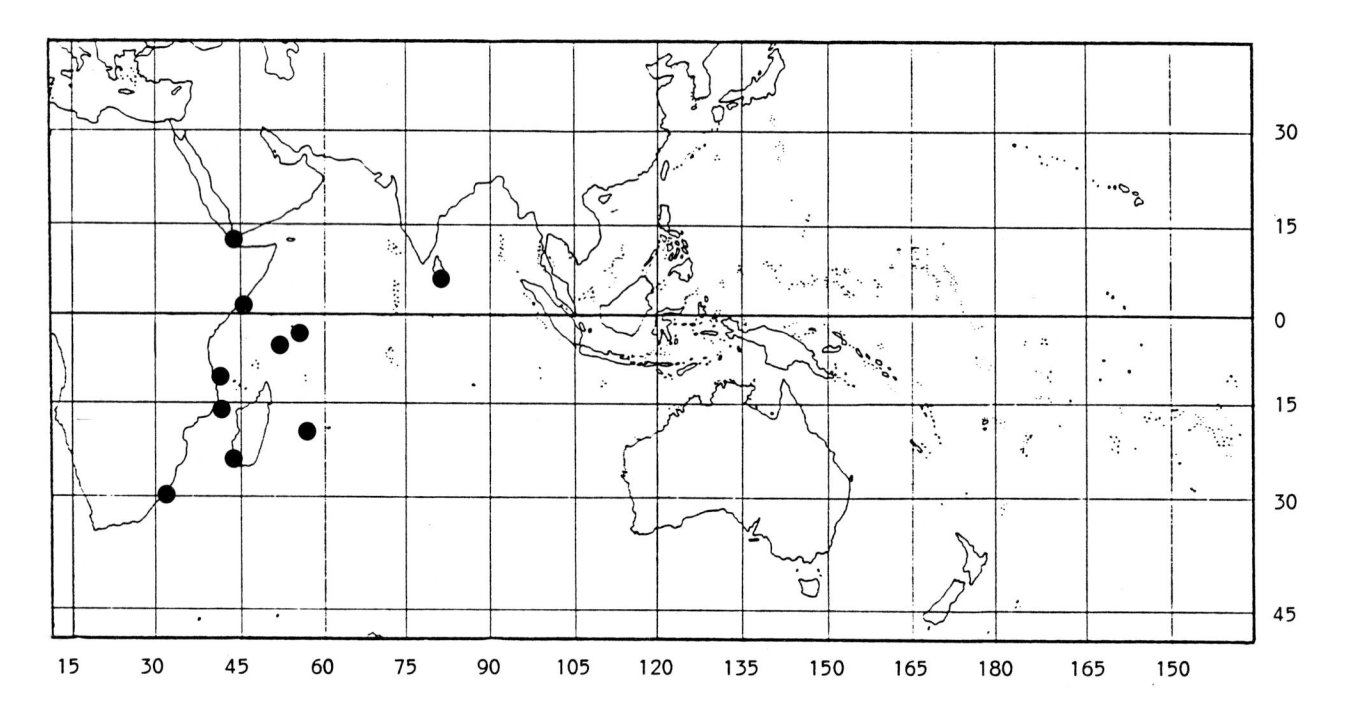

FIG. 181. — Distribution of *Chicoreus bourguignati* (Poirier).

REMARKS. *C. bourguignati* is the species regularly misidentified as *C. banksii* from Eastern Africa. Examination of an adult shell with intact protoconch indicates planktotrophic larval development for *C. bourguignati* and non planktotrophic larval development for *C. banksii*, which separates them surely. Other minor differences include more numerous whorls, comparatively larger size, and more uniformly light brown shell in *C. bourguignati*.

Chicoreus (Triplex) brunneus (Link, 1807)
Figs 31-33, 102-104, 182, 346-355

Purpura brunnea Link, 1807 : 121 (reference to MARTINI, 1777 : figs 990, 991, 993, 994).

Murex versicolor Gmelin, 1791 : 3530 (not *M. versicolor* Gmelin, 1791 : 3531).
Triplex rubicunda Perry, 1810 : pl. 25.
Triplex flavicunda Perry, 1810 : pl. 25 ; 1811, pl. 6, fig. 2.
Murex adustus Lamarck, 1822 : 162.
Murex erithrostomus Dufo, 1840 : 56.
Purpura scabra Mörch, 1852 : 97.
Murex australiensis A. Adams, 1854 : 72.
Murex despectus A. Adams, 1854 : 72.
Murex huttoniae Wright, 1878 : 85, pl. 9, figs 1-2.
Murex oligacanthus Euthyme, 1889 : 269, pl. 7, figs 2-3.

ADDITIONAL REFERENCES

Murex (Chicoreus) rubicundus. — SMITH, 1953 : 6, pl. 3, fig. 11.
Chicoreus rubicundus. — ARAKAWA, 1964 : 361, pl. 21, figs 13-14 (radula).
Chicoreus adustus. — SPRY, 1961 : 19.
Murex (Chicoreus) brunneus. — SHIKAMA, 1963 : 70, pl. 54, fig. 10 ; EISENBERG, 1981 : 88, pl. 70, fig. 7.
Chicoreus (Triplex) brunneus. — KIRA, 1965 : 64, pl. 24, fig. 18.

Chicoreus brunneus. — ZHANG, 1965 : 19, pl. 2, figs 5, 8 ; CERNOHORSKY, 1967a : 117, fig. 2, pl. 14, fig. 6 ; CERNOHORSKY, 1967b : 120, pl. 25, fig. 148 ; HINTON, 1972 : 36, pl. 18, fig. 1 ; KAICHER, 1973, card 130 ; RADWIN & D'ATTILIO, 1976 : 35, pl. 4, fig. 9 ; HINTON, 1979 : 26, fig. 8 ; ABBOTT & DANCE,1982 : 137, text fig. , MUHLHAUSSER & ALF, 1983, fig. 2 (in part) ; LAI, 1987 : 61, pl. 27, fig. 3 ; SHORT & POTTER, 1987 : 56, pl. 27, fig. 10 ; DRIVAS & JAY, 1988 : 70, pl. 20, fig. 1.
Chicoreus (Chicoreus) brunneus. — FAIR, 1976 : 27, pl. 7, fig. 9 ; SPRINGSTEEN & LEOBRERA, 1986 : 134, pl. 36, fig. 10.
Chicoreus (Chicoreus) brunneus brunneus. — VOKES, 1978 : pl. 2, figs 2-3.
Chicoreus brunneus var. *huttoniae*. — KAICHER : 1974, card 500.
Chicoreus (Chicoreus) huttoniae. — FAIR, 1976 : 49, pl. 6, fig. 76.
Chicoreus huttoniae. — RIPPINGALE, 1987 : 7, fig. 13.
Chicoreus (Chicoreus) brunneus flavicundus. — VOKES, 1978 : 381,pl. 2, fig. 1.
NOT *Chicoreus huttoniae*. — WILSON & GILLETT, 1971 : 86, pl. 58, fig. 10 [= *Chicoreus (Triplex) microphyllus* (Lamarck, 1816)].
NOT *Chicoreus (Chicoreus) huttoniae*. — FAIR, 1976 : 49 (in part), pl. 6, fig. 76A [= *Chicoreus (Triplex) microphyllus* (Lamarck, 1816)].
NOT *Chicoreus brunnea*. — RIPPINGALE, 1987 : 11, fig. 27 [= *Chicoreus (Triplex) strigatus* (Reeve, 1845)].

TYPE LOCALITIES. *P. brunnea* : unknown ; *M. versicolor* : unknown ; *T. rubicunda* : Amboyna and the Persian Gulf ; *T. flavicunda* : Botany Bay (New South Wales, Australia) ; *M. adustus* : Indian Ocean ; *M. erithrostomus* : Seychelles and Amirantes ; *P. scabra* : China (*fide* MARTYN, 1784) ; *M. australiensis* : Australia ; *M. despectus* : West Indies (error) ; *M. huttoniae* : New Caledonia ; *M. oligacanthus* : New Caledonia.

TYPE MATERIAL. *M. adustus* : Uncertain type MHNG 1099/18, here designated as lectotype ; *M. australiensis* : Lectotype BMNH 1980130, here selected from 4 syntypes ; *M. despectus* : Lectotype BMNH 1980135, here selected from 2 syntypes. No material found for the other names.

OTHER MATERIAL EXAMINED. Hundreds of specimens from throughout the distributional range.

DISTRIBUTION. (Fig. 182). Throughout the Indo-West Pacific. From Mozambique to the Amirantes and Seychelles Is ; Maldives ; Sri Lanka ; Andaman Sea ; throughout Indonesia and Malaysia ; the Philippine Is ; South China Sea ; Hong Kong ; Taiwan ; southern Japan ; Guam ; Marshall Is ; Papua New Guinea ; north-eastern and north-western Australia ; New Caledonia to Fiji and Samoa ; in mud, sand or on coral reefs. Depth range : 0-20 m.

DESCRIPTION. Shell up to 115.5 mm in length, usually stout. Spire high, with 2 or 3 protoconch whorls and up to 9 teleoconch whorls. Suture slightly appressed. Protoconch whorls glossy and rounded.
Last whorl with 3 frondose varices, each with 5-7 foliated spines (usually 6). Shoulder spine usually rather long ; second spine slightly abaperturally recurved, third and fourth moderately to very strongly abaperturally recurved, fifth and sixth spine straight. When varices have 5 spines instead of 6, second and third bent abaperturally, fourth and fifth straight. In rare cases specimens with 7 spines on each varix on last whorl, first, sixth and seventh straight, second, third and fourth more or less recurved abaperturally ; small intermediate spinelets are also present. Other axial sculpture consisting of a single, heavy, more or less elongate intervaricial node. Spiral sculpture consisting of 5-7 cords, that interconnect the spines and with numerous tuberculated intermediate threads in each interspace.
Aperture ovate to roundly-ovate. Columellar lip smooth, rim weakly erect abapically, adherent adapically, or fully adherent. Anal notch deep, narrow. Outer lip crenulate, lirate for short distance within. Siphonal canal moderately long, narrowly open abaperturally bent, with 3 or 4 straight spines of which first always bent abaperturally.
Usually dark brown with darker spiral cords, but (particularly off New Caledonia), it may be orange, orange stained with black, beige with black spines, etc... Orange specimens are also known from Guam.
Aperture bluish-white or pinkish. Columellar and outer lips bright pink (yellow to orange in Indian Ocean specimens).
Radula. (Figs 102-104).

REMARKS. Perhaps connected with its extended geographical distribution and to the different ecological conditions in which it is encountered, this species has possibly the most variable shell of all *Chicoreus* species from the Indo-West Pacific, together with *C. microphyllus, C. banksii* and *C. torrefactus*.
The typical form occurs throughout the range of species, but form and colour variants may occur together with them, particularly off New Caledonia where the typical form is found with the extreme *M. huttoniae* and *M. oligacanthus* forms.

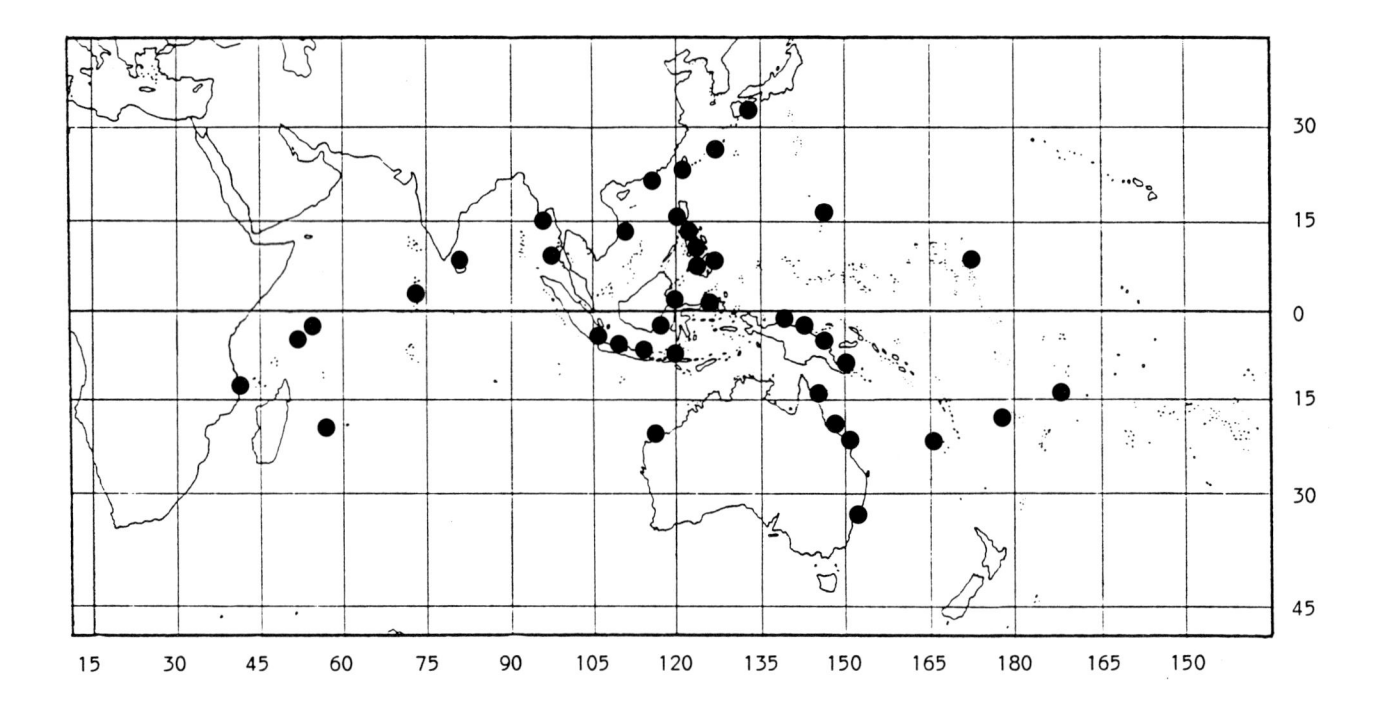

FIG. 182. — Distribution of *Chicoreus brunneus* (Link).

Colour, size and spine length are variable but the arrangement of the spines, anal notch, protoconch and operculum are identical, as is, to a lesser extent, the form of the aperture.

Triplex flavicunda is based on a form with yellow aperture occuring in the Indian Ocean, although similarly coloured specimens also occur in Hong Kong Harbour, together with typical colour form.

Murex despectus is based on a perfectly typical specimen of *C. brunneus* and is a subjective synonym. The type-locality is erroneous, which unfortunately happened with many specimens described from the Cuming collection.

M. huttoniae is based on an orange shell, such as occur off New Caledonia and Guam. Bryce Wright, who described the species from a specimen in her own collection, was a dealer in the latter half of the nineteenth century (K. M. WAY, *in litt.*).

Murex oligacanthus is a form with high spire and short spines. Some of Euthyme's shells were found at the Université Catholique of Lyon (France), but not any Muricidae (P. BOUCHET, *in litt.*); nevertheless, the original illustration clearly represents *C. brunneus*, a similar form of which is illustrated here (Fig. 350).

Although *Murex versicolor* Gmelin, 1791 (= *C. brunneus*) has page priority over the second *M. versicolor* (a *Fusinus* species), VOKES (1971 : 115) as first revisor, selected the second name so that the first name became the homonym. This was done in the interest of stability, to preserve the widely accepted *C. brunneus*.

Chicoreus (Triplex) elisae Bozzetti, 1991
Figs 183, 356-357

Chicoreus elisae Bozzetti, 1991 : 43, figs 1-3.

TYPE LOCALITY. Off Capo Ras Hafun, 150 kms south of Capo Guardafui, Somalia, 100-150 m.

TYPE MATERIAL. Holotype MNHN, and 3 paratypes.

MATERIAL EXAMINED. Holotype and 1 paratype ; Somalia, RH (2 dd).

DISTRIBUTION. (Fig. 183). Off Capo Ras Hafun, Somalia.

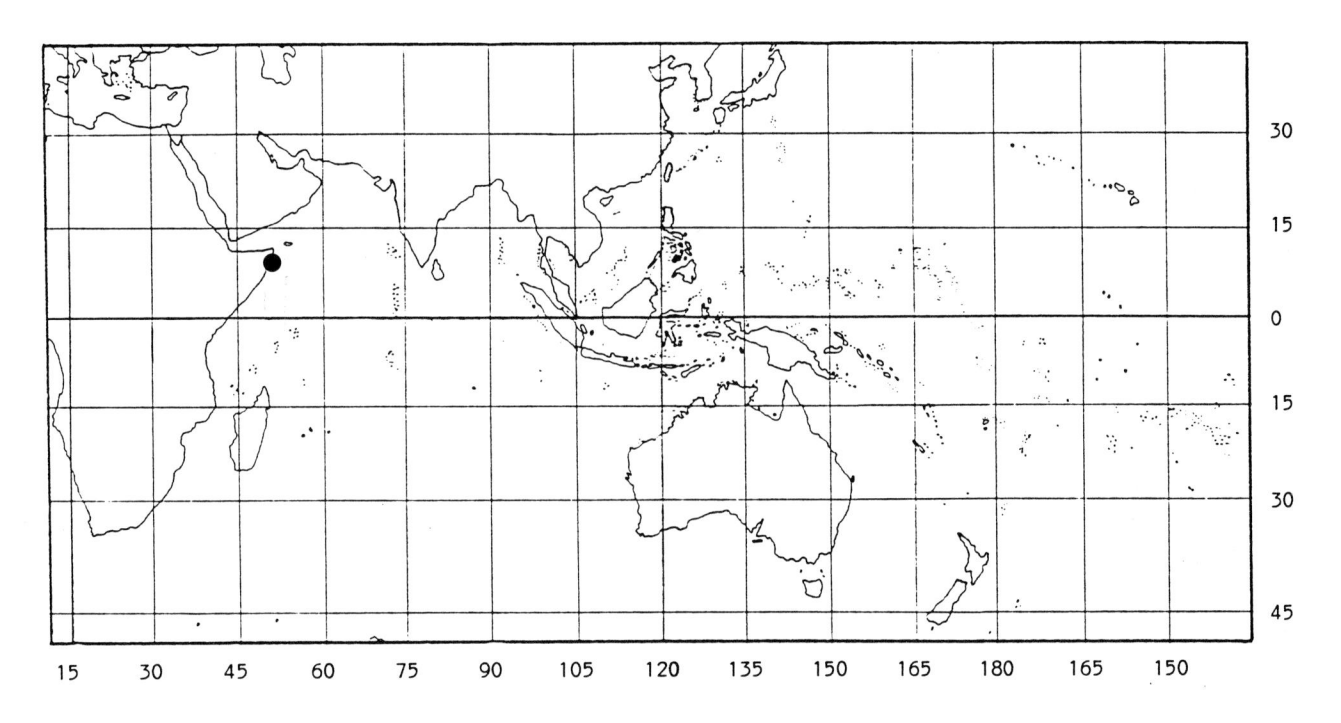

FIG. 183. — Distribution of *Chicoreus elisae* Bozzetti.

DESCRIPTION. Shell up to 36.1 mm in length. Spire high, up to 6 teleoconch whorls. Protoconch unknow. Suture lightly appressed.
Last whorl with 3 frondose varices, each with very small open spines. Shoulder spine longest. Intervaricial axial sculpture consisting of one strong, rounded, prominent node. Spiral sculpture of 6 major cords and numerous, lightly squamous threads in each interspace.

Aperture roundly-ovate. Columellar lip smooth, adherent, rim slightly erect abapically. Anal notch deep, narrow. Outer lip crenulate, lirate for short distance within. Siphonal canal moderately long, straight, with one or two small spines.
Pale brown to reddish brown with dark brown blotches on the intervaricial node and on the spiral sculpture. Aperture and columellar lip light pink.

REMARKS. Originally compared with *Chicoreus rubescens* (Broderip, 1833), *C. trivialis* (A. Adams, 1854) and *C. groschi* Vokes, 1978.
It is not very related to *C. rubescens* and differs from *C. trivalis* and *C. groschi* in having much shorter varical spines, a smaller size for a same number of teleoconch whorls and a lighter coloured shell. *C. elisae* also has a shorter spire compared to *C. trivialis*.

Chicoreus (Triplex) groschi (Vokes, 1978)
Figs 34, 184, 362-364

Chicoreus (Chicoreus) groschi Vokes, 1978 : 386, pl. 2, figs 4, 5.

ADDITIONAL REFERENCES

Chicoreus (Chicoreus) trivialis. — VOKES, 1978 : 386 (in part), pl. 2, fig. 6 (not *Murex trivialis* A. Adams, 1854).
Chicoreus groschi. — KAICHER, 1979 : card 1971 (holotype) ; HOUART, 1981c : 16, text fig. ; HOUART, 1985c : 240 ; RIPPINGALE, 1987 : 5, fig. 11.
NOT *Chicoreus groschi.* — DRIVAS & JAY, 1988 : 70, pl. 20, fig. 2 [= *Chicoreus (Triplex) bourguignati* (Poirier, 1883)].

TYPE LOCALITY. South-west Conducia Bay, north-west of Chocas, Mozambique.

TYPE MATERIAL. Holotype NM H192/T2134.

OTHER MATERIAL EXAMINED, c. 80 specimens from throughout the distributional range.

DISTRIBUTION. (Fig. 184). Aliwal Shoal, Natal, southeastern South Africa ; Mozambique coast, in gravel with sparse *Thalassodendron* (VOKES, 1978 : 386) ; Tulear and Ifaty Lagoon, Madagascar.

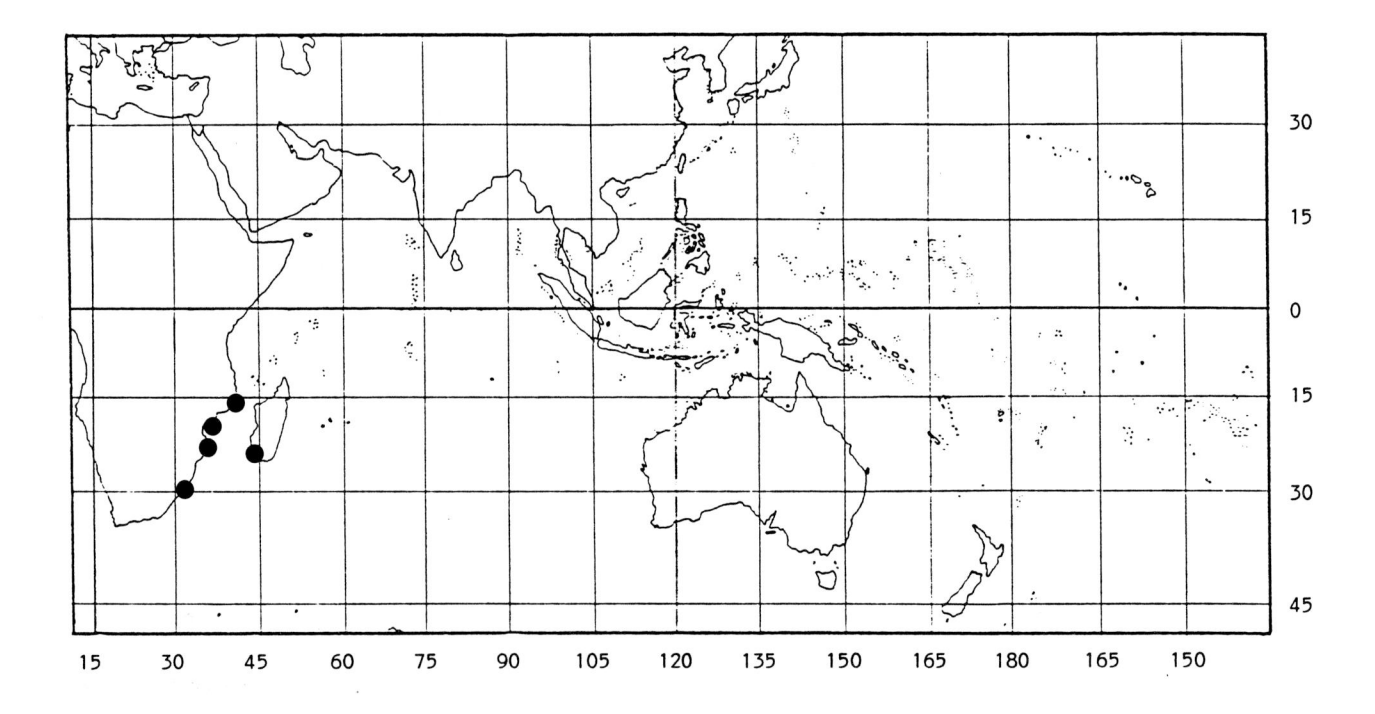

FIG. 184. — Distribution of *Chicoreus groschi* Vokes.

DESCRIPTION. Shell up to 65 mm in length, stout. Spire moderately high, with 2 protoconch whorls and up to 8 rounded teleoconch whorls. Suture slightly appressed. Protoconch whorls rounded and glossy.

Last whorl with 3 frondose varices, each with 6 or 7 short, straight, frondose spines of whith third always bent abaperturally ; numerous short intermediate spinelets. Other axial sculpture consisting of a single, strong intervaricial node. Spiral sculpture of 6 or occasionally 7 primary cords flanked by small squamous threads, 1 secondary cord, 3 or 4 threads between each pair of primary cords.

Aperture ovate. Columellar lip rim erect, smooth. Anal

notch narrow, small, moderately deep. Outer lip crenulate, lirate for short distance within. Siphonal canal moderately short, narrowly open, tip bent abaperturally, with 3 abapically bent frondose spines.

Pale brown with darker spiral sculpture, varices and siphonal canal. Columellar lip pale yellow to peach or pinkish, occasionally bordered with darker line.

REMARKS. *C. groschi*, previously confused with *C. brunneus*, differs from that species in the shape of the terminal varix on the protoconch (Fig. 34), and by the position of the varical spines. While *C. brunneus* has 5-7 spines, of which the second, third, and sometimes fourth are bent abaperturally, *C. groschi* has 6 or 7 spines of which only the third is abaperturally bent. Other differences, as indicated by VOKES (1978), include the coloration of the aperture (yellow or orange in the east African form of *C. brunneus* vs pale apricot or light pink in *C. groschi*) and the spinose rather than ramose ornamentation as in *C. brunneus*.

Elongate shells, somewhat resembling *C. trivialis* or with a wide varical flange, occur off Mozambique and Madagascar. An elongate specimen of *C. groschi* was identified as *C. trivialis* by VOKES (1978 : pl. 2, fig. 6).

Chicoreus (Triplex) ryosukei Shikama, 1978
Figs 179-180, 185

Chicoreus (Triplex) ryosukei Shikama, 1978 : 36, pl. 7, figs 5-8 (not 3-4 as originally stated).

ADDITIONAL REFERENCES

Chicoreus ryosukei. — OKUTANI, 1983 : 8, pl. 24, fig. 5 (holotype); HOUART, 1984a : 12, text fig.
NOT *Chicoreus ryosukei*. — VAUGHT, 1989 : fig. 6 [= *Chicoreus (Triplex) banksii* (Sowerby, 1841)].
not *Chicoreus ryosukei*. — SHIKAMA, 1978 : figs 3, 4 [= *Chicoreus (Triplex) ryukyuensis* Shikama, 1978].

TYPE LOCALITY. Arafura Sea.

TYPE MATERIAL. Holotype NSMT 60928.

OTHER MATERIAL EXAMINED. Unknown locality, RH (2 dd).

DISTRIBUTION. (Fig. 185). Only known from the type locality.

DESCRIPTION. Shell up to 53,1 mm in length, stout. Spire high, up to 8 teleoconch whorls, protoconch unknown. Suture appressed.
Last whorl with 3 frondose varices, each with 5 ramose spines. Shoulder spine short, followed abapically by a fairly long spinelet; second, third and fourth abapical spines each longer than the other, last abapical spine shorter, intermediate spinelets present. Intervarical axial sculpture consisting of 1 relatively strong ridge or 2 weak ridges. Spiral sculpture of numerous cords and threads.
Aperture roundly-ovate. Columellar lip smooth, adherent. Anal notch shallow, narrow, delineated by small callus. Outer lip crenulate, lirate for short distance within. Siphonal canal short or of moderate length, with 3 adapically bent, open, frondose spines.
First 2 or 3 teleoconch whorls pinkish. Following whorls whitish to pinkish with darker varical fronds, usually some more deeply pigmented spiral cords.

REMARKS. Without knowledge of the protoconch it is difficult to be certain if this species is really distinct from *C. banksii*. It is distinguishable by the position and length of the varical spines on the last whorl and by the consistently fully adherent columellar lip. The colour, if constant, seems to be characteristic also. A forms of *C. banksii* occuring off Anna Plains, Western Australia, sometimes exhibits similar shell coloration but has an erect columellar lip rim, and varical spines positioned as in typical *C. banksii*. Two specimens illustrated by SHIKAMA (1978 : figs 3, 4) as *C. ryosukei* actually represent *C. ryukyuensis* Shikama, 1978.

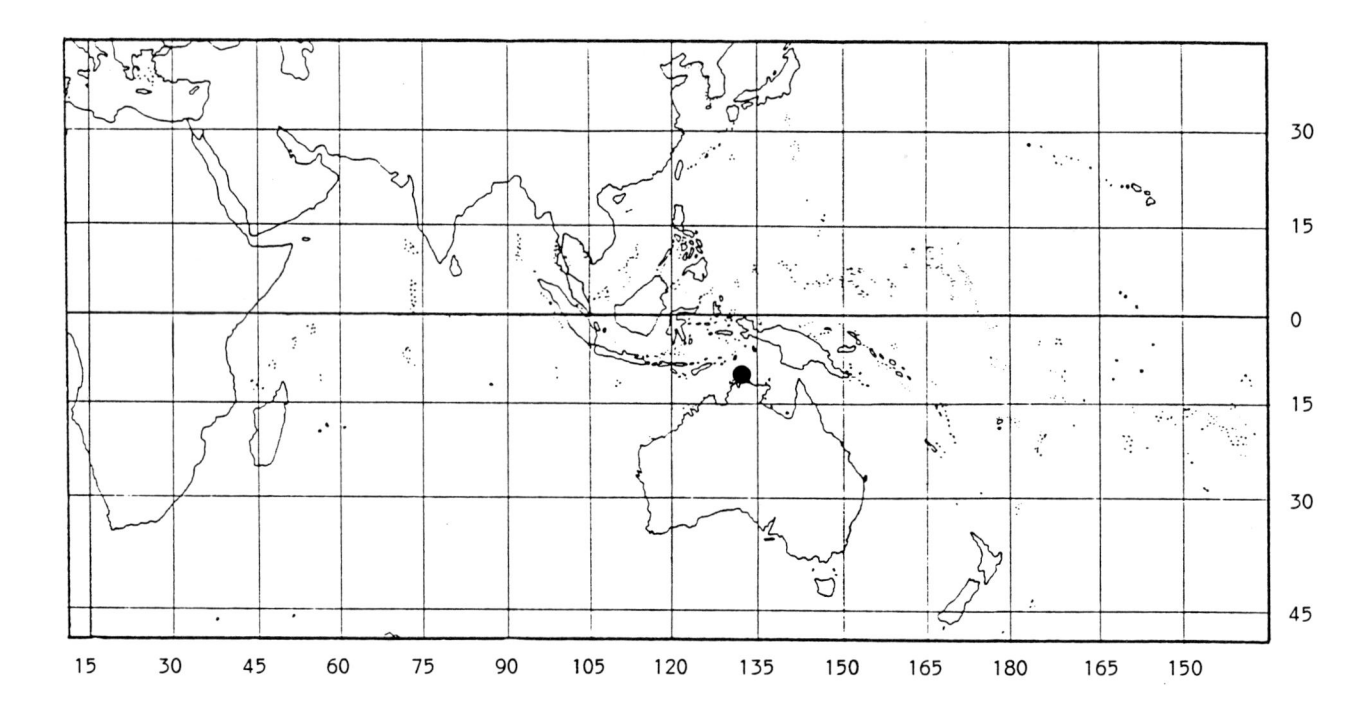

FIG. 185. — Distribution of *Chicoreus ryosukei* Shikama.

GROUP 4

Chicoreus (Triplex) cnissodus (Euthyme, 1889)
Figs 35, 186, 365-367

Murex cnissodus Euthyme, 1889 : 263, pl. 6, figs 1, 2.

ADDITIONAL REFERENCES

Chicoreus aculeatus. — ZHANG, 1965 : 21, pl. 2, fig. 7 (not *Murex aculeatus* Lamarck, 1822).
Chicoreus cnissodus. — KAICHER, 1973 : card 154; RADWIN & D'ATTILIO, 1976 : 35, pl. 5, fig. 3; ABBOTT & DANCE, 1982 : 136, text fig.; LAI, 1987 : 61, pl. 29, fig. 5.
Chicoreus (Chicoreus) cnissodus. — FAIR, 1976 : 32, pl. 6, fig. 74 : SPRINGSTEEN & LEOBRERA, 1986 : 134, pl. 36, fig. 9.
Murex (Chicoreus) cnissodus. — EISENBERG, 1981 : 89, pl. 71, fig. 2.

TYPE LOCALITY. " New Caledonia " (doubtful — see remarks).

TYPE MATERIAL. None.

OTHER MATERIAL EXAMINED : c. 80 specimens from throughout the geographical range.

DISTRIBUTION. (Fig. 186). Sri Lanka; the Philippine Is; Taiwan; South China Sea, and middle Japan (Wakayama, Kii Channel). Depth range : 30-70 m.

DESCRIPTION. Shell up to 88 mm in length, spire high, with 3 protoconch whorls and up to 10 rounded teleoconch whorls. Suture impressed. Protoconch conical, of 3 rounded and glossy whorls.
Last whorl with 3 rounded varices, each with 5 or occasionally 6 short to moderately long, approximately equal-sized, frondose spines. Shoulder spine and abapical spine sometimes weakly longer than the others. Other axial sculpture consisting of 2 or occasionally 3 strong intervaricial nodes. Spiral sculpture of 9-12 spiral cords with 2-4 fine intermediate squamous threads in each interspace.
Aperture rounded. Columellar lip smooth, rim erect aba-

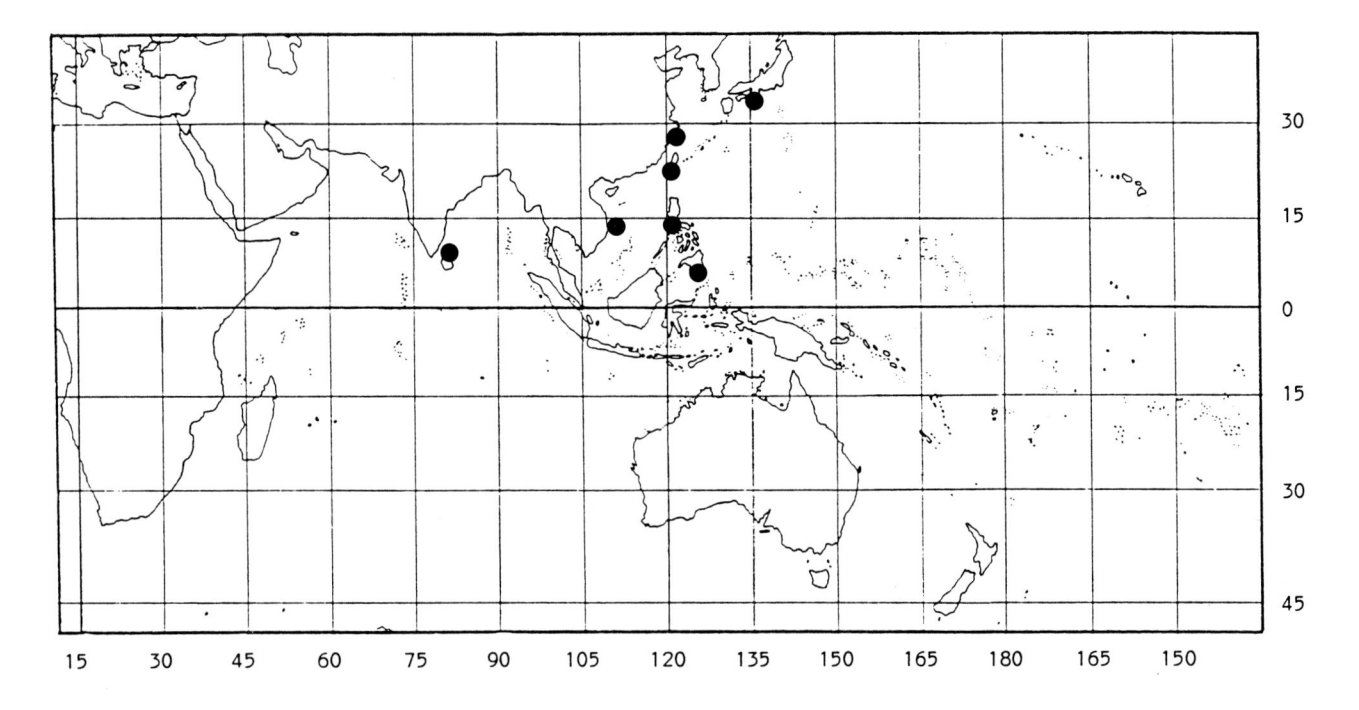

FIG. 186. — Distribution of *Chicoreus cnissodus* (Euthyme).

pically and lightly adherent adapically. Anal notch deep, small, sometimes obsolete. Outer lip crenulate, lirate for short distance within. Siphonal canal short or of moderate length, weakly bent abaperturally, with 2 or 3 straight, open, frondose spines.

Usually milky white with brown spiral sculpture and brown spots. Specimens from Sri Lanka light brown with slightly darker spiral sculpture and dark spiral bands.

REMARKS. Although described from New Caledonia, this species has never been subsequently recorded from there, and the type locality seems very doubtful.

This species is highly distinctive and is very stable in form and colour. Brown shells with short spines recently found off Sri Lanka are otherwise typical in sculpture, apertural characters and shape. The Sri Lankan form somewhat resembles elongate forms of *C. peledi*, and indeed, shape and colour of the shells are sometimes identical, including the darkly pigmented spiral banks. Although similar, *C. cnissodus* has a shell with spiral sculpture of 9-12 cords and intermediate threads, while *C. peledi* has numerous spiral threads of similar size. *C. cnissodus* has a more globose shell with impressed suture and varices with 5 or 6 spines, while *C. peledi* has only 4 spines.

Chicoreus (Triplex) peledi Vokes, 1978
Figs 187, 250, 368

Chicoreus (Chicoreus) peledi Vokes, 1978 : 391, pl. 3, figs 5, 6.

ADDITIONAL REFERENCES

Chicoreus (Chicoreus) sp. — FAIR, 1976 : 88, pl. 9, fig. 118.
Chicoreus benedictinus. — RADWIN & D'ATTILIO, 1976 : 34, fig. 10 (not *Murex benedictinus* Löbbecke, 1879).
Chicoreus peledi. — KAICHER, 1979 : card 2011 (holotype and paratype); HOUART, 1981a : 10, text fig.; RIPPINGALE, 1987 :
   9, fig. 21.

TYPE LOCALITY. Eilat, northern end of the Gulf of Aqaba, Red Sea.
TYPE MATERIAL. Holotype HUJ 10129.

OTHER MATERIAL EXAMINED. Eilat, 40 m, on mud, RH (1 lv); Eilat, coll. Rippingale (1 dd); Nuweiba, Sinai, Egypt (AMS C138494, C133601, C133604) (4 dd).

DISTRIBUTION. (Fig. 187). Gulf of Aqaba, Red Sea, from Eilat to Nuweiba, on mud. Depth range : c. 40 m.

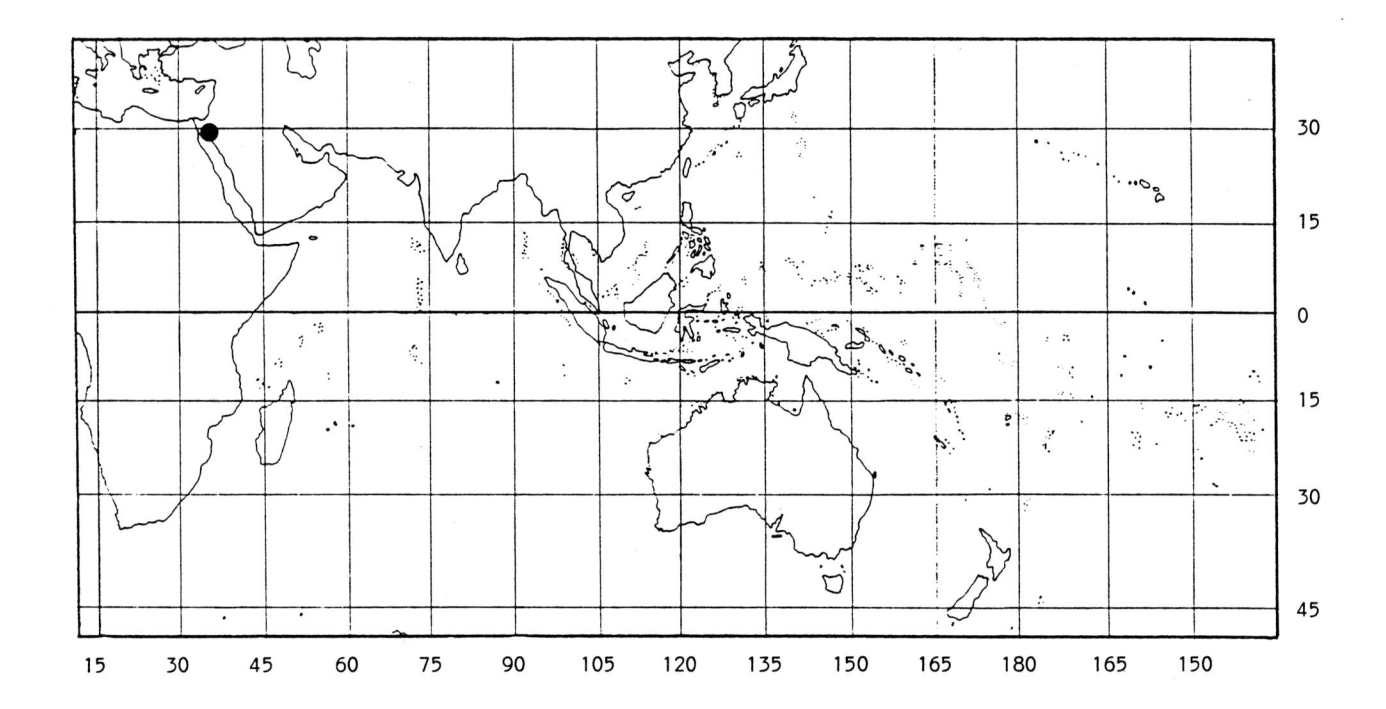

FIG. 187. — Distribution of *Chicoreus peledi* Vokes.

DESCRIPTION. Shell up to 81 mm in length, stout. Spire moderately high, up to 8 teleoconch whorls, protoconch unknown. Suture slightly appressed.
Last whorl with 3 frondose varices, each with long shoulder spine, followed abapically by 3 smaller ones; numerous intermediate spinelets. Intervaricial sculpture generally consisting of 2 strong nodes, occasionally with a third weaker node. Spiral sculpture of numerous threads of approximately similar size.

Aperture rounded, outer lip angulate. Columellar lip smooth, rim erect, partially adherent adapically. Anal notch deep, narrow, delineated by small callus. Outer lip crenulate, lirate for short distance within. Siphonal canal of moderate length, bent abaperturally, with 2 long open spines.
Light brown with 3 bands of darker spiral threads, one on shoulder, one the middle of last whorl, one on base of siphonal canal.

REMARKS. This rare species was illustrated both by FAIR (1976 : pl. 9, fig. 118) and RADWIN & D'ATTILIO (1976 : pl. 23, fig. 10). FAIR referred to it as an unnamed species from the Red Sea, while RADWIN & D'ATTILIO believed it to be *Murex benedictinus*, which is based on a form of *Chicoreus torrefactus*. The species is in danger of extermination in the area between Eilat and Taba due to the extreme pollution from an oil terminal, electric power station and port (VOKES, 1978).

GROUP 5

Chicoreus (Triplex) corrugatus

Two subspecies are recognized :

Chicoreus (Triplex) corrugatus corrugatus (Sowerby, 1841)
Figs 39-40, 188, 376

Murex corrugatus Sowerby, 1841 : pl. 189, fig. 72 ; 184b : 142.

ADDITIONAL REFERENCES

Chicoreus corrugatus. — KAICHER, 1973 : card 137 ; SHARABATI, 1984 : pl. 18, fig. 5 ; RIPPINGALE, 1987 : 5, fig. 8.
Murex corrugatus. — LEEHMAN, 1973b : 3, fig. 1.
Chicoreus (Chicoreus) corrugatus. — FAIR, 1976 : 34, pl. 7, fig. 89 ; VOKES, 1978 : 392, pl. 4, fig. 4.
Chicoreus denudatus. — RADWIN & D'ATTILIO, 1976 : 36 (in part) (not *Triplex denudata* Perry, 1811).
NOT *Chicoreus palmiferus*. — HINTON, 1972 : 36 (in part), pl. 18, figs 2-3 [= *Chicoreus (Triplex) torrefactus* (Sowerby, 1841)].

TYPE LOCALITY. *M. corrugatus* : Suez [subsequent designation by VOKES (1978)].

TYPE MATERIAL. No type material could be located in BMNH.

MATERIAL EXAMINED : c. 30 specimens from throughout the geographical range.

DISTRIBUTION. (Fig. 188). Apparently restricted to the northern end of the Red Sea, Gulf of Suez and Gulf of Aqaba.

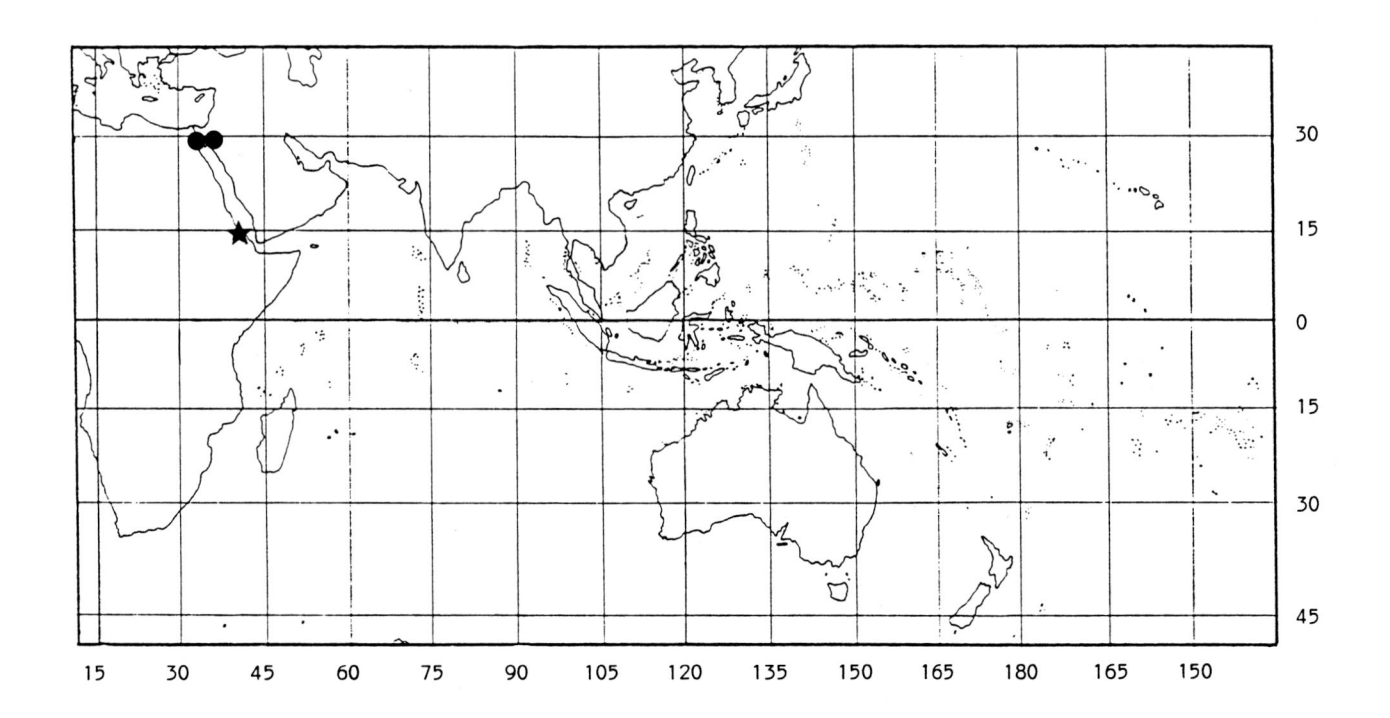

FIG. 188. — Distribution of *Chicoreus corrugatus corrugatus* and of *Chicoreus corrugatus ethiopius*.
● *C. corrugatus corrugatus*
★ *C. corrugatus ethiopius*

DESCRIPTION. Shell up to 50 mm in length, stout. Spire high, with 1 1/2-3/4 protoconch whorls and up to 8 tuberculate teleoconch whorls. Suture slightly appressed. Protoconch whorls rounded, smooth.

Last whorl with 3 rounded varices, each with 6 or 7 open frondose spines; the 2, or occasionally 3 adapical spines generally interconnected by a low varicial flange. Intermediate spinelets either absent or very weak. Other axial sculpture consisting of 2 strong intervaricial ridges that cross the spiral cords and impart a nodose appearance. Spiral sculpture of 6-8 strong cords flanked by squamous threads, 3 or 4 intermediate finely squamous threads in each interspace.

Aperture roundly-ovate. Columellar lip smooth, rim weakly erect abapically, elsewhere adherent. Anal notch broad, shallow. Outer lip crenulate, strongly lirate for short distance within. Siphonal canal short, narrowly open, bearing 2-4 anteriorly bent, frondose spines.

Whitish or orange to reddish-brown. Aperture white or light purple.

REMARKS. Although *C. corrugatus* was synonymised with *C. denudatus* by RADWIN & D'ATTILIO (1976 : 36), the shell differs in having spinose instead of palmate varicial fronds. Moreover, *C. denudatus* has a narrower and deeper anal notch, a proportionally shorter siphonal canal and one strong intervaricial node on the last whorl whereas *C. corrugatus* has 2 low axial ridges in this position. VOKES (1978 : 108) considered *Murex tirondus* de Gregorio, 1885 as a possible synonym. Unfortunately that species was never illustrated and it is impossible to locate type material in the de Gregorio "collection" stored in the IMG (RUGGIERI, in litt.). Accordingly *M. tirondus* must be regarded as a *nomen dubium*.

It is difficult to confuse *C. corrugatus* with any other species. Rather small for the genus, the shell is usually whitish although populations from the Gulf of Aqaba include pinkish, brown-red or orange shells.

Chicoreus (Triplex) corrugatus ethiopius (Vokes, 1978)
Figs 188, 189

Chicoreus (C.) corrugatus ethiopius Vokes, 1978 : 393, pl. 4, fig. 5.

ADDITIONAL REFERENCES

Chicoreus (Chicoreus) benedictinus. — FAIR, 1976 : 25, pl. 9, fig. 116 (not *Murex benedictinus* Löbbecke, 1879).
Chicoreus corrugatus ethiopius. — KAICHER, 1979 : card 2046; HOUART, 1981c : 16, text fig.
Chicoreus ethiopius. — RIPPINGALE, 1987 : 5, fig. 9.

TYPE LOCALITY. Mandat Island, Dahlak Archipelago, off the coast of Ethiopia, near Massawa.

TYPE MATERIAL. Holotype HUJ 10.131.

OTHER MATERIAL EXAMINED. Dahlak Archipelago, Red Sea, RH (3 lv).

DISTRIBUTION. (Fig. 188). Apparently restricted to the southern end of the Red Sea.

DESCRIPTION. Shell up to 40 mm in length. Spire high, with 1 1/4 protoconch whorls and up to 6 elongate teleoconch whorls. Suture appressed. Protoconch whorl smooth (VOKES, 1978 : 393).

Last whorl with 3 rounded varices, each with 5 or 6 short frondose spinelets. Intervaricial axial sculpture consisting of 2 or 3 strong ridges. Spiral sculpture of 5 or 6 cords and intermediate threads between each cords.

Aperture ovate. Columellar lip adherent, smooth, or with series of small denticles abapically. Anal notch broad, shallow. Outer lip crenulate, strongly lirate for short distance within. Siphonal canal short, narrowly open, with 2 short, abapically bent spines.

Brown with darker coloured spiral cords and axial ridges. Aperture white.

REMARKS. The shell is smaller and narrower than *C. (T.) corrugatus corrugatus*, with a more ovate, whitish aperture, and is dark brown instead of whitish, orange or reddish brown. Actually, lack of specimens from localities between the geographical range of *C. (T.) corrugatus ethiopius* and its nominate subspecies does not allow to do more research on the specific or subspecific status of this taxon.

FIGS 189-191. — Subgenus *C. (Triplex)*.
189, *C. corrugatus ethiopius* Vokes. Dahlak Archipelago, Red Sea, 35 mm (holotype, HUJ 10131. Courtesy of E.H. Vokes).
190-191, *C. denudatus* Perry. 190, east of Sydney, 26 mm (holotype of *Murex immunitus* Iredale, AMS C60671). 191, Port Western, Australia, 34.9 mm (lectotype of *Murex australis* Quoy & Gaimard, MNHN).

Chicoreus (Triplex) damicornis (Hedley, 1903)
Figs 41, 192, 377-378, 381

Murex damicornis Hedley, 1903 : 378, fig. 92.

Chicoreus falsinii Nicolay, 1976 : 18.

ADDITIONAL REFERENCES

Torvamurex damicornis. — MCPHERSON & GABRIEL, 1962 : 168, fig. 200.
Euphyllon damicornis. — CERNOHORSKY, 1967b : 124, pl. 26, fig. 157 ; HINTON, 1972 : 38, pl. 19, figs 14-15.
Chicoreus damicornis. — WILSON & GILLETT, 1971 : 86, pl. 58, fig. 6 ; RADWIN & D'ATTILIO, 36, pl. 4, fig. 3 ; ABBOTT & DANCE, 1982 : 138, text fig. ; SHORT & POTTER, 1987 : 56, pl. 27, fig. 8.
Chicoreus (Chicoreus) damicornis. — FAIR, 1976 : 35, pl. 7, fig. 83.
Murex (Chicoreus) damicornis. — EISENBERG, 1981 : 89, pl. 71, figs 5, 8.

TYPE LOCALITY. Shoalhaven Bight, New South Wales (Australia), in 19-20 fms (35-37 m).

TYPE MATERIAL. — Holotype AMS C16416.

OTHER MATERIAL EXAMINED : c. 50 specimens from throughout the geographical range.

DISTRIBUTION. (Fig. 192). Port Macquarie, New South Wales, to Victoria, Australia, Depth range : 40-100 m.

DESCRIPTION. Shell up to 70 mm in length, lightly built. Spire high, acutely conical, with 1 3/4-2 1/4 protoconch whorls and up to 7 roundly-fusiform teleoconch whorls. Suture impressed. Protoconch whorls smooth, rounded.

Last whorl with 3 frondose varices, each with (usually) 5 spinelike fronds, shoulder frond very long and broad, bifurcated at tip. Other spines somewhat smaller, straight. Abapical spines often interconnected by webbing. Intervari

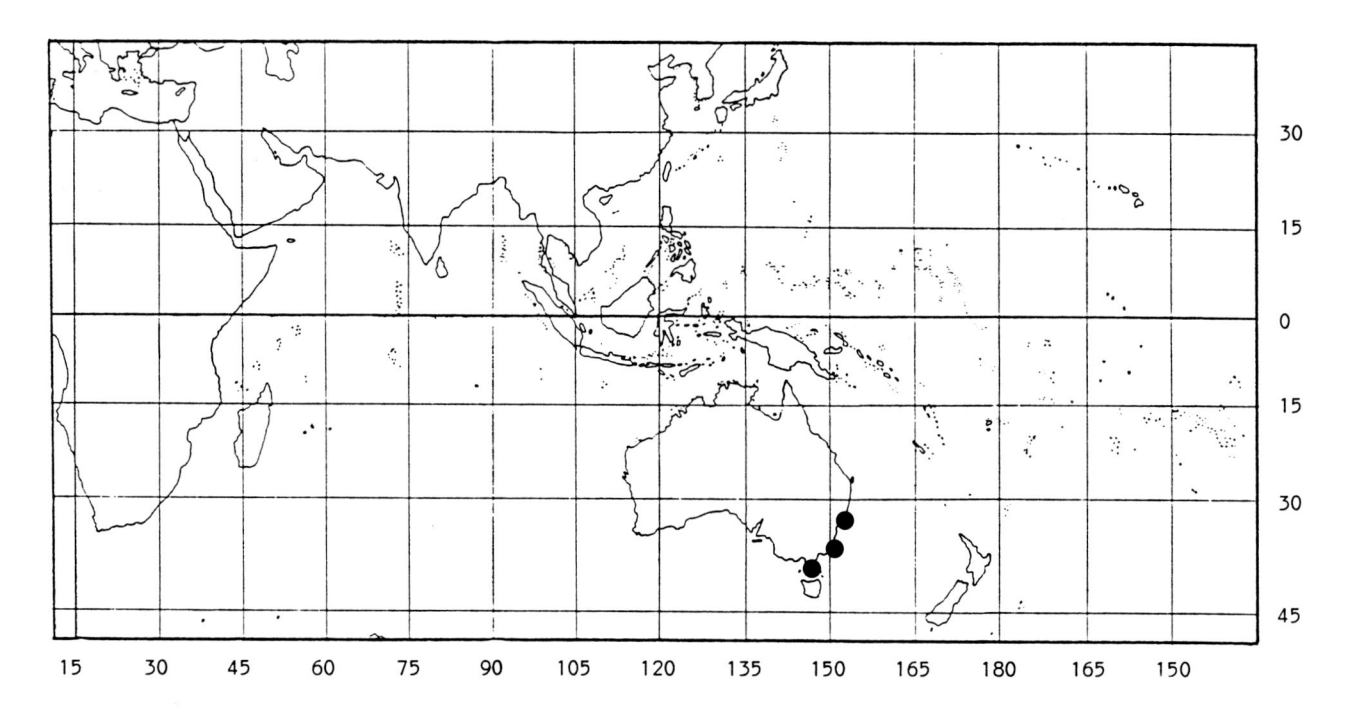

FIG. 192. — Distribution of *Chicoreus damicornis* (Hedley).

cial axial sculpture consisting of 1-3 low, shallow ridges. Spiral sculpture of numerous, nodose, low cords and finer threads.

Aperture roundly-ovate. Columellar lip smooth, rim detached abapically, elsewhere adherent. Anal notch deep, narrow. Outer lip weakly crenulate, lightly lirate for short distance within. Siphonal canal short, straight, narrowly open, slightly bent abaperturally at tip, 2 or 3 open, straight spines.

Usually uniform light cream, occasionally tinged with brown.

REMARKS. This is distinguishable from all other Australian and Indo-West Pacific species by the long bifid shoulder spines. It has been included in the genera *Torvamurex* and *Euphyllon* by previous authors but *Torvamurex* is here considered to be a synonym of *Triplex*, and *Euphyllon* of *Chicoreus s. s.* The name *Chicoreus falsinii* conditionally proposed by NICOLAY (1976 : 18) is unavailable (ICZN, article 15) and is based on a specimen of *C. damicornis*.

Chicoreus (Triplex) denudatus (Perry, 1811)
Figs 42, 190-191, 193, 379-380, 384

Triplex denudata Perry, 1811 : pl. 7, fig. 2.

Triplex frondosa Perry, 1811 : pl. 6, fig. 11.
Murex australis Quoy & Gaimard, 1833 : 536.
Torvamurex extraneus Iredale, 1936 : 324, pl. 23, fig. 12.
Torvamurex denudatus immunitus Iredale, 1936 : 324, pl. 23, fig. 14.
Murex palmiferus Sowerby, 1841 : pl. 195, fig. 104 ; 1841b : 142.

ADDITIONAL REFERENCES

Murex (Chicoreus) palmiferus. — SMITH, 1953 : 6, pl. 7, fig. 7.
Torvamurex denudatus. — MCPHERSON & GABRIEL, 1962 : 167, fig. 199.
Chicoreus denudatus. — WILSON & GILLETT, 1971 : 86, pl. 58, fig. 7 ; HINTON, 1972 : 36, pl. 18, fig. 15 ; RADWIN & D'ATTILIO, 1976 : 37, fig. 17, pl. 4, fig. 5 ; ABBOTT & DANCE, 1982 : 136, text fig. ; SHORT & POTTER, 1987 : 56, pl. 27, fig. 12.

Chicoreus (Chicoreus) denudatus. — FAIR, 1976 : 36, pl. 7, fig. 88 ; VOKES, 1978 : pl. 4, fig. 6.
Chicoreus extraneus. — CERNOHORSKY, 1967b : 120, pl. 25, fig. 150.
Murex sp. — COCHRANE, 1980 : 7, fig. 2.
NOT *Chicoreus denudatus.* — CERNOHORSKY, 1972 : 122, pl. 34, fig. 4 [= *Chicoreus (Triplex) territus* (Reeve, 1845)].
NOT *Murex (Chicoreus) denudatus.* — EISENBERG, 1981 : 89, pl. 71, fig. 7 [= *Chicoreus (Triplex) territus* (Reeve, 1845)].

TYPE LOCALITIES. *T. denudata* : " Van Diemen's Land " (Tasmania), see below ; *T. frondosa* : coasts of New Holland ; *M. australis* : Port Western (Australia) ; *T. extraneus* : Sydney Harbour ; *T. immunitus* : 70 fathoms east of Sydney ; *M. palmiferus* : Red Sea (error).

TYPE MATERIAL. *T. denudata, T. frondosa* and *M. palmiferus* : none ; *M. australis* : lectotype MNHN, here selected from 2 syntypes ; *T. extraneus* : holotype AMS C60673 ; *T. denudatus immunitus* : holotype AMS C60671.

OTHER MATERIAL EXAMINED. c. 100 specimens from throughout the geographical range.

DISTRIBUTION. (Fig. 193). Australia : Queensland (Cape Moreton) ; New South Wales ; Bass Strait ; Victoria and (?)Tasmania.

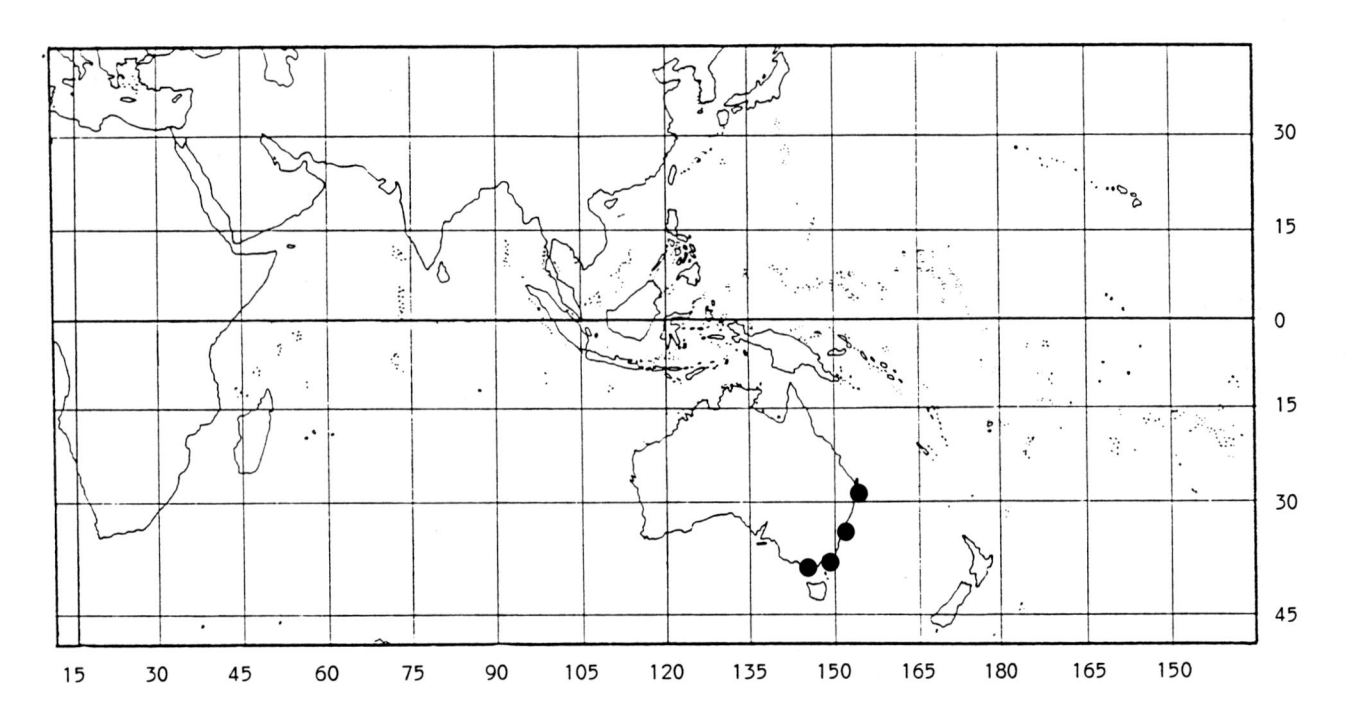

FIG. 193. — Distribution of *Chicoreus denudatus* (Perry).

DESCRIPTION. Shell up to 55 mm in length, rather stout. Spire high, acutely conical, with 2 protoconch whorls and up to 7 rounded, weakly angulate teleoconch whorls. Suture impressed. Protoconch whorls rounded ; last nuclear whorl axially sculptured.

Last whorl with 3 frondose varices, each with 6 short, open, palmate fronds, of which 2 adapical and 2 abapical spines often interconnected. Intervaricial axial sculpture consisting generally of 2 strong nodes. Spiral sculpture of numerous coarse cords.

Aperture rounded. Columellar lip smooth, adherent. Anal notch narrow, relatively deep. Outer lip crenulate, lirate for short distance within. Siphonal canal short, narrowly open, with generally 3 widely open, short spines.

Colour varying from yellowish-brown to rosy-pink, occasionally with brown spiral bands.

REMARKS. Type-species of the genus *Torvamurex* Iredale, 1936 (= *Triplex*). *Torvamurex immunitus*, described as a subspecies, has a smaller, narrower shell, but all other characters are accordant with the typical form. So-called "dwarf" specimens (25-35 mm long) are juveniles with fewer teleoconch whorls, numbering generally 6 or 7 in adults.

SHORT & POTTER (1987 : 56) erroneously included the distinct *Murex australiensis* A. Adams, 1854 as a synonym, suggesting that they confused that name with *Murex australis* Quoy & Gaimard, 1833.

Tasmania is the type locality for the species but no other material has ever been seen by me from that locality.

Chicoreus (Triplex) territus (Reeve, 1845)
Figs 43, 194, 382-383, 385-387

Murex territus Reeve, 1845 : pl. 33, fig. 167.

Murex nubilis Sowerby, 1860 : 428, pl. 49, fig. 4.

ADDITIONAL REFERENCES

Chicoreus territus. — WILSON & GILLETT, 1971 : 86, pl. 58, fig. 11 ; KAICHER, 1973 : card 158 ; RADWIN & D'ATTILIO, 1976 : 43, pl. 6, fig. 9.
Chicoreus (Chicoreus) territus. — FAIR, 1976 : 80, pl. 7, fig. 95.
Chicoreus denudatus. — CERNOHORSKY, 1972 : 122, pl. 34, fig. 4 (not *Triplex denudata* Perry, 1811).
Murex (Chicoreus) denudatus. — EISENBERG, 1981 : 89, pl. 71, fig. 7 (not *Triplex denudata* Perry, 1811).

TYPE LOCALITIES. None.

TYPE MATERIAL. *M. territus* : lectotype BMNH 198239, here selected from 2 syntypes ; *M. nubilis* : not located.

OTHER MATERIAL EXAMINED. c. 100 specimens from throughout the geographical range.

DISTRIBUTION. (Fig. 194). Common at low tide on mud flats, Queensland, Australia ; off southern New Caledonia, 50-80 m (MNHN). Reputedly from south-east Malaya and Broome (Western Australia) (RH), but confirmation required.

DESCRIPTION. Shell up to 70 mm in length, triangular shape. Spire high, with 2 teleoconch whorls and up to 8 rounded teleoconch whorls. Suture slightly appressed. Protoconch whorls smooth, rounded.

Last whorl with 3 sharp, frondose varices, each with 5 or 6 short, open abaperturally recurved spines, inter connected on each other by a varicial flange. Shoulder spine longest. When varices are strongly rounded spines may be obsolete. Other axial sculpture consisting of 1 or 2 more or less strong intervaricial nodes. Spiral sculpture of 5 or 6 major cords, flanked by finer threads ; 1 intermediate minor cord in each interspace.

Aperture roundly-ovate. Columellar lip smooth, rim weakly detached abapically, elsewhere adherent. Anal notch small. Outer lip crenulate, lirate for short distance within. Siphonal canal relatively long, narrowly open, slightly bent abaperturally at tip, generally bearing 3 rather long spines.

Varying from pure white to grey or pinkish brown with darker blotches. Aperture white to dark brown or grey.

REMARKS. Although previous authors have synonymised *M. nubilis* with *C. denudatus*, the original illustration agrees more closely with short spined specimens of *C. territus*.

Chicoreus (Triplex) thomasi (Crosse, 1872)
Figs 37, 195, 388-389

Murex thomasi Crosse, 1872 : 212 ; 1873 : pl. 11, fig. 4.

ADDITIONAL REFERENCES

Chicoreus thomasi. — SALVAT & RIVES, 1975 : 312, fig. 194 ; CERNOHORSKY, 1978a : 74, fig. 20 (syntype), 21 ; CERNOHORSKY, 1978b : 65, pl. 18, fig. 4 ; KAICHER, 1980 : card 2529 (lectotype) ; ABBOTT & DANCE, 1982 : 138, text fig. (paralectotype).
Chicoreus (Chicoreus) thomasi. — FAIR, 1976 : 81, fig. 62.
Chicoreus maurus. — RADWIN & D'ATTILIO, 1976 : 39 (in part) (not *Murex maurus* Broderip, 1833).

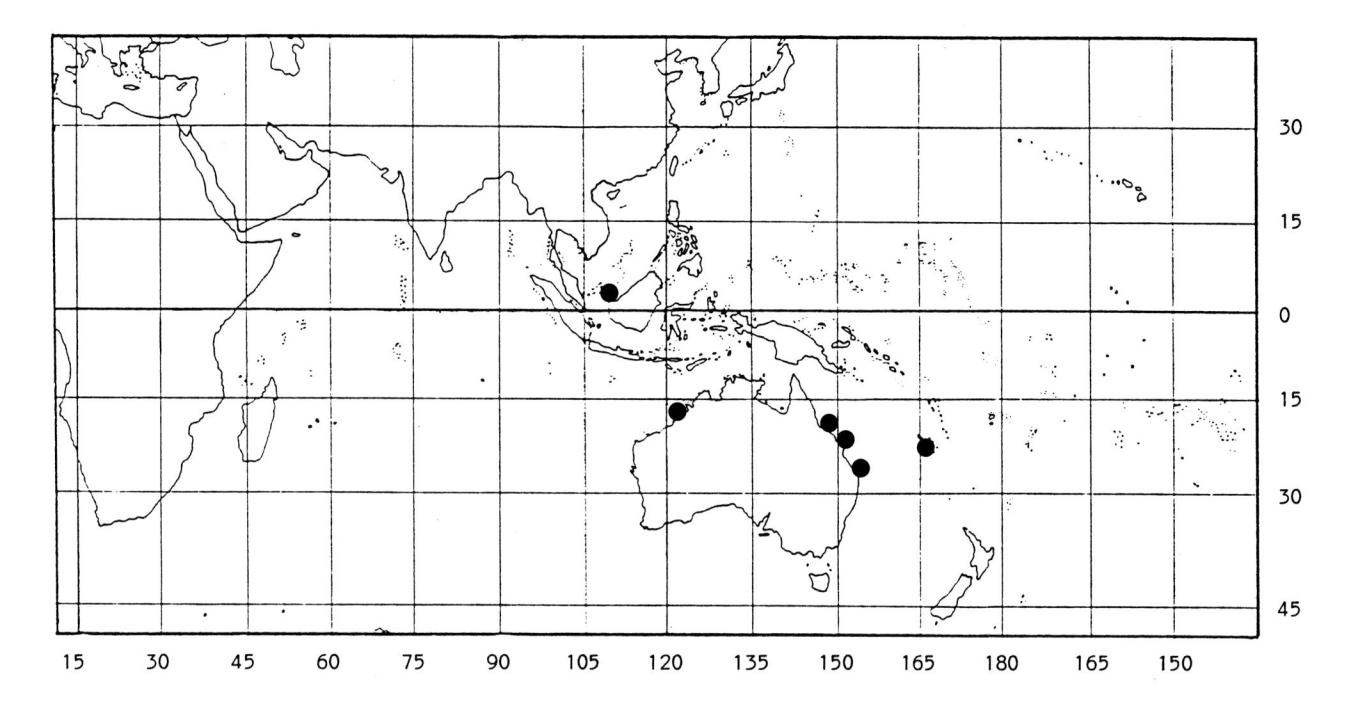

FIG. 194. — Distribution of *Chicoreus territus* (Reeve).

TYPE LOCALITY. Marquesas Islands.

TYPE MATERIAL. 1 syntype BMNH 1902.5.28.53, here designated as lectotype; 1 syntype NMW 1955.158.12, here designated as paralectotype.

MATERIAL EXAMINED. Marquesas Is, coll. RH (2 lv); Tahiti, MNHN (1 lv); Nuku Hiva, Marquesas Is, MNHN (1 lv); Nuka Hiva, Marquesas Is, IRSNB (2 lv), Tahuata, Marquesas Is, 09°54.32′ S, 139°6.51′ W, 48 m, MNHN (7 lv).

DISTRIBUTION. (Fig. 195). Only known from the Marquesas and doubtfully from Tahiti.

DESCRIPTION. Shell up to 60 mm in length. Spire high, up to 8 teleoconch whorls. Suture appressed. Protoconch conical weakly subcarinate, partially broken in examined specimens.
Last whorl with 3 rounded varices, each with short open and straight spine, 1 or 2 obsolete spines on shoulder, followed by 1 or 2 small spinelets and 2 or 3 somewhat longer spines abapically. Other axial sculpture consisting of 1 strong intervaricial node. Spiral sculpture of 8 or 9 cords surmounted by nodose threads and separated by other numerous nodose spiral threads in each interspace.
Aperture ovate, small. Columellar lip edge with small nodes, weakly erect, adapically adherent. Anal notch narrow, moderately deep. Outer lip crenulate, strongly lirate for short distance within. Siphonal canal moderately long, straight, narrowly open, with 2 or 3 adapically bent spines.
Pale orange to pinkish; aperture light pink; edge of the columellar lip glossy white.

REMARKS. *C. thomasi* was incorrectly synonymised with *Chicoreus maurus* (Broderip, 1833) by RADWIN & D'ATTILIO (1976 : 39) although the two species differ in numerous characters.

C. thomasi is a rare species, at least in collections, and most specimens examined are from nineteen century collections. Recent dredgings in the Marquesas (J. Poupin-SMCB coll. in MNHN) however finally resulted in the discovery of several specimens in 48 m.

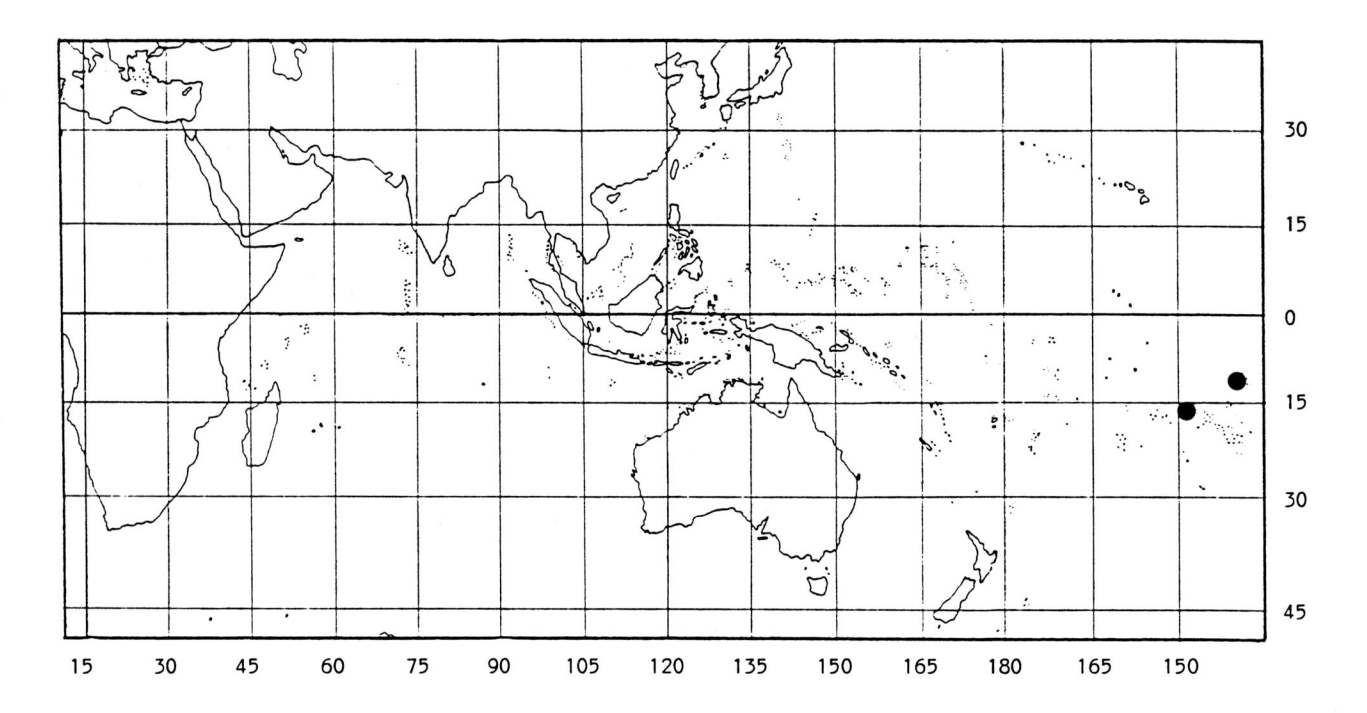

FIG. 195. — Distribution of *Chicoreus thomasi* (Crosse).

GROUP 6

Chicoreus (Triplex) boucheti Houart, 1983
Figs 44, 106-107, 196, 255, 399-400

Chicoreus boucheti Houart, 1983 : 27, pl. 1, figs 1-2, text figs 3-4.

ADDITIONAL REFERENCE

Chicoreus boucheti HOUART, 1984a : 12, text fig. (holotype).

TYPE LOCALITY. Dredged south of New Caledonia, 22°08′ S, 167°04′ E, in 230-260 m.

TYPE MATERIAL. Holotype MNHN.

OTHER MATERIAL EXAMINED. 22 specimens from throughout the geographical range.

DISTRIBUTION. (Fig. 196). Southern New Caledonia. Depth range : 170-345 m.

DESCRIPTION. Shell up to 38 mm in length, lightly built. Spire high, with 1 1/2 protoconch whorls and up to 5 rounded, weakly angulate teleoconch whorls. Suture appressed. Protoconch whorls rounded, smooth, last whorl finely carinate.

Last whorl with 3 varices, each with 3 open, adapical recurved, weakly frondose spines, additional spinelets present between second and third spine, shoulder spine longest. Other axial sculpture consisting of 2 moderately strong intervaricial costae. Spiral sculpture of numerous, fine, almost smooth threads.

Aperture ovate or roundly-ovate. Columellar lip smooth, rim adherent over most of adapical part, weakly erect abapically. Anal notch broad, moderately deep, delineated by small callus. Outer lip erect, denticulate, lirate for short distance within. Siphonal canal long, open, bent abaperturally, with one adapically curved, open spine.

Cream with light brown blotches. The first 3 whorls, axial costae, subsutural area and space between each shoulder spine paler cream ; spines and other surfaces light brown. Occasional uniform light yellow specimens.

Radula (Figs 106-107).

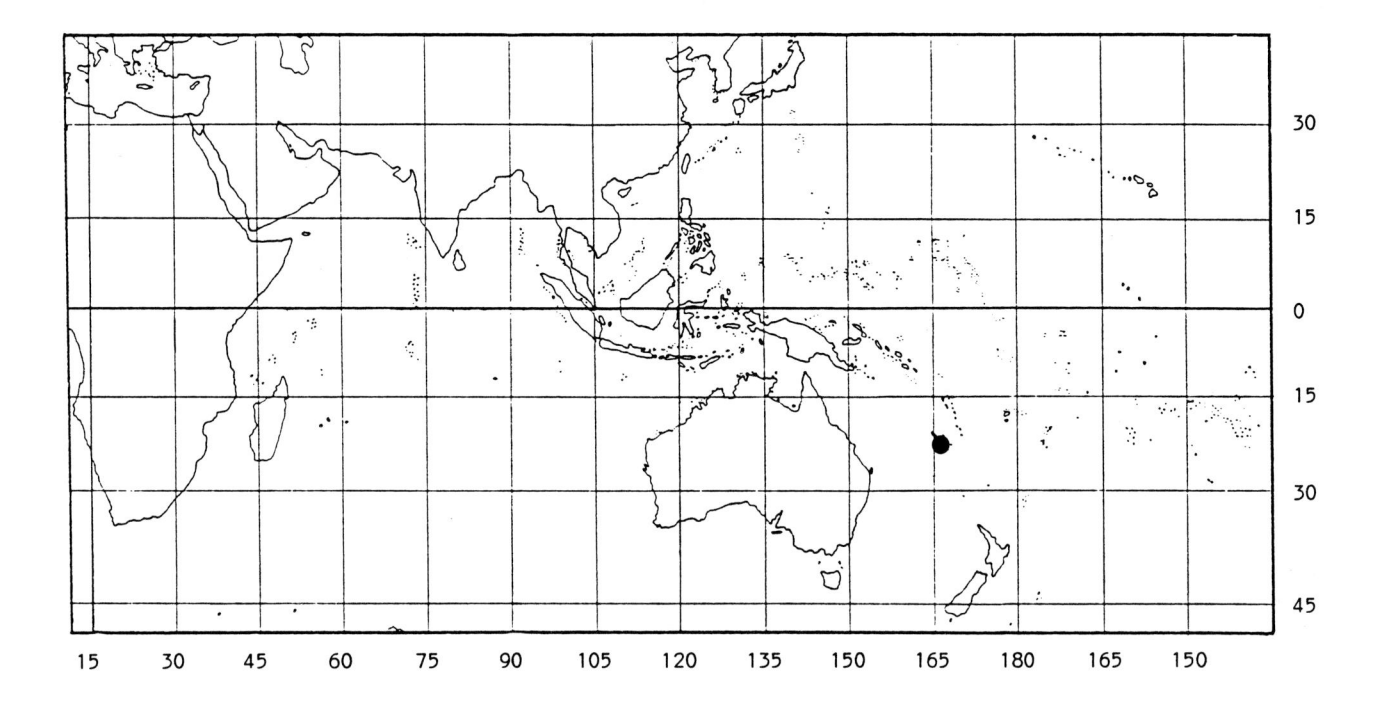

FIG. 196. — Distribution of *Chicoreus boucheti* Houart.

REMARKS. The species was originally compared with *C. longicornis* from which it differs in having 3 open, slightly frondose instead of closed, acute varicial spines. *C. boucheti* also has a more strongly appressed suture, while the aperture is longitudinally lirate instead of smooth within.

Although the protoconchs of the two species are superficially similar (Figs 44 and 48), that of *C. boucheti* differs in having a fine carina on the last whorl. Moreover, *C. boucheti* has 4 spiral threads and 12 sharp axial costae on the first teleoconch whorl, whereas *C. longicornis* has 3 spiral threads and 12-15 rounded axial costae. The second whorl of *C. longicornis* has 3 obsolete axial varices, each with 1 sharp, closed, shoulder spine, and 2 or 3 intervaricial axial costae, while *C. boucheti* has 3 varices, each with 2 open spines, and 3 or 4 intervaricial axial costae.

Chicoreus (Triplex) cervicornis (Lamarck, 1822)
Figs 47, 197, 361

Murex cervicornis Lamarck, 1822 : 163.

ADDITIONAL REFERENCES

Murex cervicornis. — KIENER, 1842 : pl. 20, fig. 2; RADWIN & D'ATTILIO, 1976 : 63, pl. 11, fig. 5; WELLS & BRYCE, 1985 : 86, pl. 25, fig. 280.
Murex (Euphyllon) cervicornis. — SHIKAMA, 1963 : 70, pl. 53, fig. 2.
Chicoreus cervicornis. — WILSON & GILLETT, 1971 : 86, pl. 58, fig. 9; KAICHER, 1973 : card 140; HINTON, 1979 : 26, fig. 13; ABBOTT & DANCE, 1982 : 138, text fig.; SHORT & POTTER, 1987 : 56, pl. 27, fig. 7.
Euphyllon cervicornis. — HINTON, 1972 : 38, pl. 19, figs 11-13.
Chicoreus (Chicoreus) cervicornis. — FAIR, 1976 : 30, pl. 7, fig. 98.
Murex (Murex) cervicornis. — EISENBERG, 1981 : 88, pl. 70, fig. 17.
Murex (Chicoreus) longicornis. — SMITH, 1953 : 7 (in part), pl. 5, fig. 9 (not *Murex longicornis* Dunker, 1864).

TYPE LOCALITY. New Holland (= Australia).

TYPE MATERIAL. Holotype MNHN.

OTHER MATERIAL EXAMINED. c. 40 specimens from throughout the geographical range.

DISTRIBUTION. (Fig. 197). Australia : Rottnest Island, North Queensland ; Western Australia ; north-western Australia ; southern coast of Papua New Guinea ; Gulf of Papua ; Aru Islands (Moluccas) ; sandy bottom. Depth range : 1-180 m.

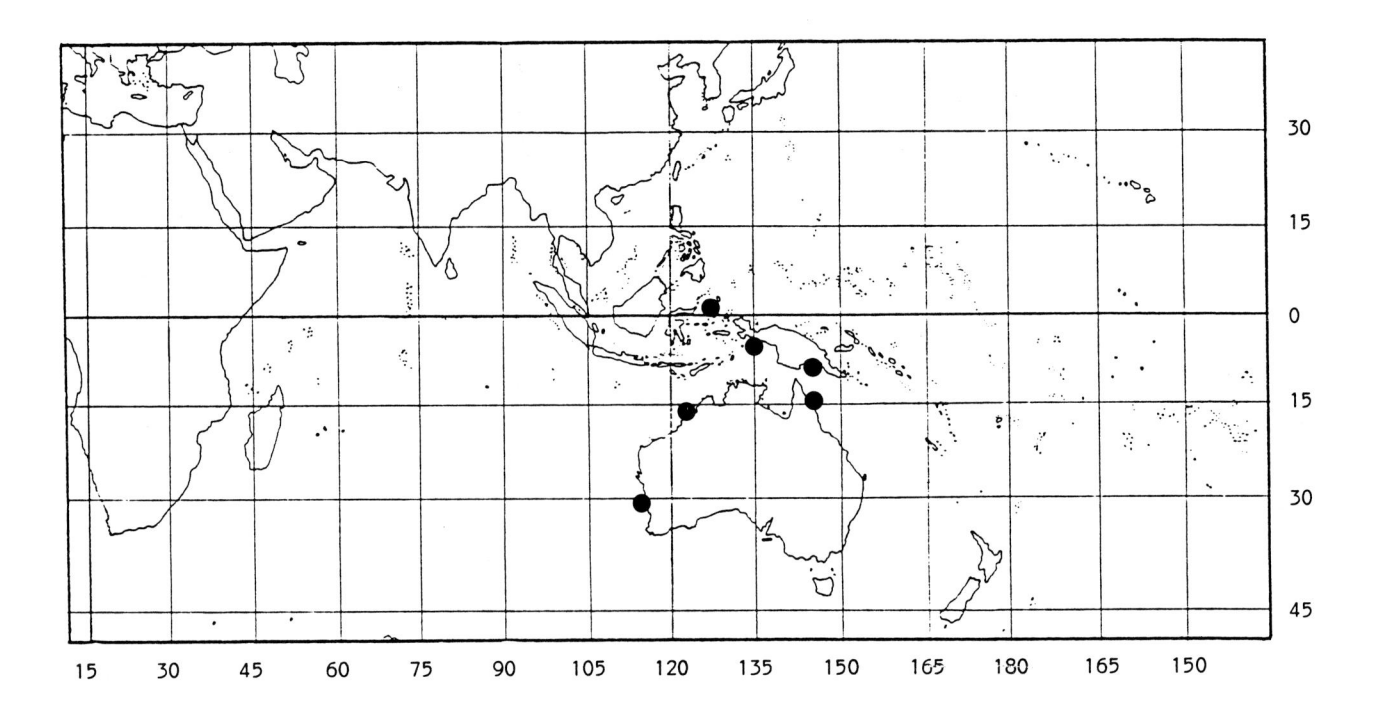

FIG. 197. — Distribution of *Chicoreus cervicornis* (Lamarck).

DESCRIPTION. Shell up to 75 mm in length. Spire high, with 1 3/4-2 protoconch whorls and up to 7 rounded teleoconch whorls. Suture impressed. Protoconch whorls rounded and smooth.
Last whorl with 3 spinose varices, each with 2 long, bifurcated, closed spines. Shoulder spine longest ; short spinelets present between the two varical spines and below abapical spine. Other axial sculpture consisting of 3 or 4 weak ridges, more strongly developed on spire than on last whorl. Spiral sculpture of 4 major cords, situated adapically and abapically on last whorl, two of which interconnect varical spines ; numerous smaller cords and threads on entire shell.
Aperture rounded. Columellar lip smooth, rim either adherent or completely but very weakly erect. Anal notch shallow, small, delineated by small callus. Outer lip smooth, slightly undulating. Siphonal canal long, narrowly open, slightly abaperturally bent, with 1 long closed spine adapically and occasionally a smaller spine abapically.
Creamy white to light brown, aperture of same colour.

REMARKS. The divided, closed spines and the rather long siphonal canal with its 1 or 2 closed, straight, acute spines render this species extremely distinctive.

Chicoreus (Triplex) longicornis (Dunker, 1864)
Figs 48, 198, 390-393

Murex longicornis Dunker, 1864 : 99, pl. 22, figs 5-6.

Murex recticornis von Martens *in* Löbbecke & Kobelt, 1880 : 81, pl. 3, fig. 3.
Poirieria kurranula Garrard, 1961 : 27, pl. 2, fig. 4.

ADDITIONAL REFERENCES

Murex (Chicoreus) recticornis. — SMITH, 1953 : 7, pl. 5, fig. 19.
Chicoreus recticornis. — WILSON & GILLETT, 1971 : 86, pl. 58, fig. 8 ; KAICHER, 1973 : card 138.
Euphyllon longicornis. — HINTON, 1972 : 38, pl. 19, figs 16-17.
Chicoreus (Chicoreus) longicornis. — FAIR, 1976 : 55, pl. 6, fig. 78.
Murex longicornis. — RADWIN & D'ATTILIO, 1976 : 67, pl. 11, fig. 4 ; SHORT & POTTER, 1987 : 56, pl. 27, fig. 6.
Murex (Murex) longicornis. — EISENBERG, 1981 : 90, pl. 72, fig. 19.
NOT *Murex (Chicoreus) longicornis.* — SMITH, 1953 : 7, pl. 5, fig. 9 [= *Chicoreus (Triplex) cervicornis* (Lamarck, 1822)].

TYPE LOCALITIES. *M. longicornis* : Amboina Island, Indonesia ; *M. recticornis* : East Australia, 26°05′ S, 76 fms (139 m) ; *P. kurranula* : off Cape Moreton, Queensland.

TYPE MATERIAL. *M. longicornis* : not located ; *M. recticornis* : syntype LM, here designated as lectotype, 2 syntypes ZMB, here designated as paralectotypes ; *P. kurranula* : holotype MV F21118.

OTHER MATERIAL EXAMINED. c. 40 specimens from throughout the geographical range.

DISTRIBUTION. (Fig. 198). South-eastern Queensland, Australia, from Cape Moreton to Gladstone. Depth range : 130-180 m.

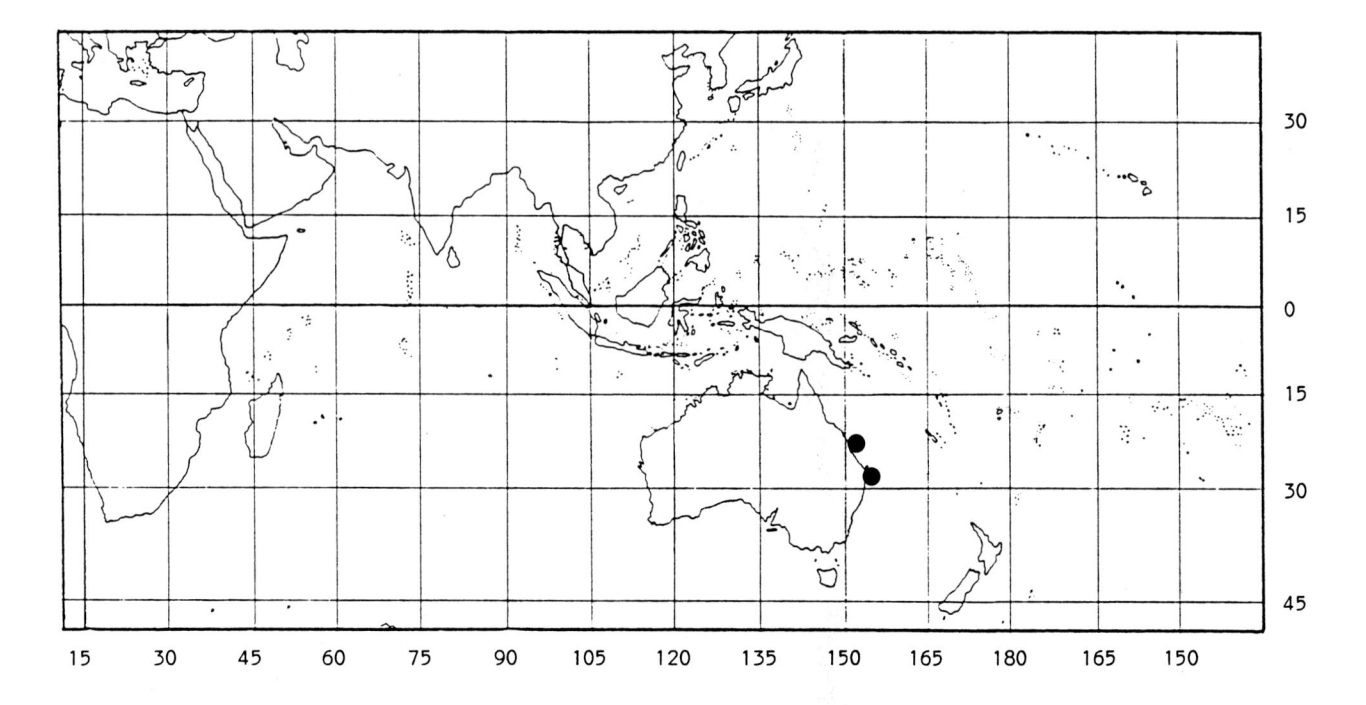

FIG. 198. — Distribution of *Chicoreus longicornis* (Dunker).

DESCRIPTION. Shell up to 52 mm in length. Spire high, with 1 1/2-1 3/4 teleoconch whorls and up to 5 rounded teleoconch whorls. Suture deeply impressed. Protoconch whorls rounded, smooth.

Last whorl with 3 rounded varices, each with 2 long, sharp spines, one on shoulder, the other situated abapically, 2 or 3 intermediate open spinelets between the spines. Other axial sculpture consisting of 2 or 3 low intervaricial nodes. Spiral sculpture of 6 or 7 low cords and numerous intermediate threads in each interspace.

Aperture rounded. Columellar lip smooth, rim weakly erect abapically. Anal notch shallow, rather deep, delineated by small callus. Outer lip smooth, slightly undulating. Siphonal canal long, slender, open, very weakly bent abaperturally, with 1 small adapically recurved acute spine.

Whitish to light cream, sometimes with light brown traces.

REMARKS. The type specimen of *M. longicornis*, thought to be in the Hessisches Landesmuseum, Darmstadt, could not be located, but from the original illustration, there can be little doubt about the identity of the species.

It was included in the genus *Murex (s.s.)* by RADWIN & D'ATTILIO (1976 : 67) together with *C. cervicornis*, on the basis of the lack of spine foliations and the overall *Murex*-like form of the last whorl. In agreement with PONDER & VOKES (1988), I prefer to treat both as *Chicoreus* species. Although the shell morphology is unusual for the genus, the siphonal canal in both species is shorter than in *Murex (s.s.)* and bears only 1 or 2 small spine, and they lack the labral tooth that is characteristic of all species of *Murex (s.s.)*.

No material has ever been seen by me from Amboina, the type locality for *C. longicornis*.

Chicoreus (Triplex) paucifrondosus Houart, 1988
Figs 46, 199, 259

Chicoreus paucifrondosus Houart, 1988 : 186, figs 3, 16.

TYPE LOCALITY. New Caledonia, off Grand Récif Sud, 22°39' S, 167°07' E, 222 m.

TYPE MATERIAL. Holotype MNHN, and 7 paratypes.

MATERIAL EXAMINED. Type material.

DISTRIBUTION. (Fig. 199). Off southern New Caledonia. Depth range : 105-222 m.

DESCRIPTION. Shell up to 29 mm in length, narrowly elongate. Spire high, with 2 protoconch whorls and up to 6 elongate, convex teleoconch whorls. Suture impressed. Protoconch whorls rounded, glossy.

Last whorl with 3 weakly frondose varices, each with 3 major, frondose, short, open spines, occasionally with a small intermediate spinelet between each spine. Shoulder and abapical spine largest, median spine narrower and shorter. Intervaricial axial sculpture consisting of 2 or 3 elongate nodes. Spiral sculpture of numerous scabrous cords and threads.

Aperture roundly-ovate. Columellar lip narrow, smooth, rim adherent but weakly detached abapically. Anal notch broad, shallow. Outer lip slightly erect and crenulate, lirate for short distance within. Siphonal canal long, slender, narrowly open, slightly abaperturally bent, with short, acute spine on its middle, 2 or 3 short spinelets adapically and occasionally with additional short, open spinelet almost at tip of siphonal canal.

Bluish-white to light brown with darker coloured spines.

REMARKS. *C. paucifrondosus* differs from the superficially similar *C. boucheti* in being more elongate with lower axial sculpture, and shorter, weakly palmate fronds. The protoconch is more elongate, not subcarinate and has a rounded rather than angulate terminal varix (Fig. 46), while the columellar lip is narrower and the shell is smaller, relative to the number of whorls.

Chicoreus (Triplex) subpalmatus Houart, 1988
Figs 45, 108-109, 200, 256

Chicoreus subpalmatus Houart, 1988 : 188, figs 1, 2, 15.

TYPE LOCALITY. South of New Caledonia, 23°42' S, 167°59' E, 338 m.

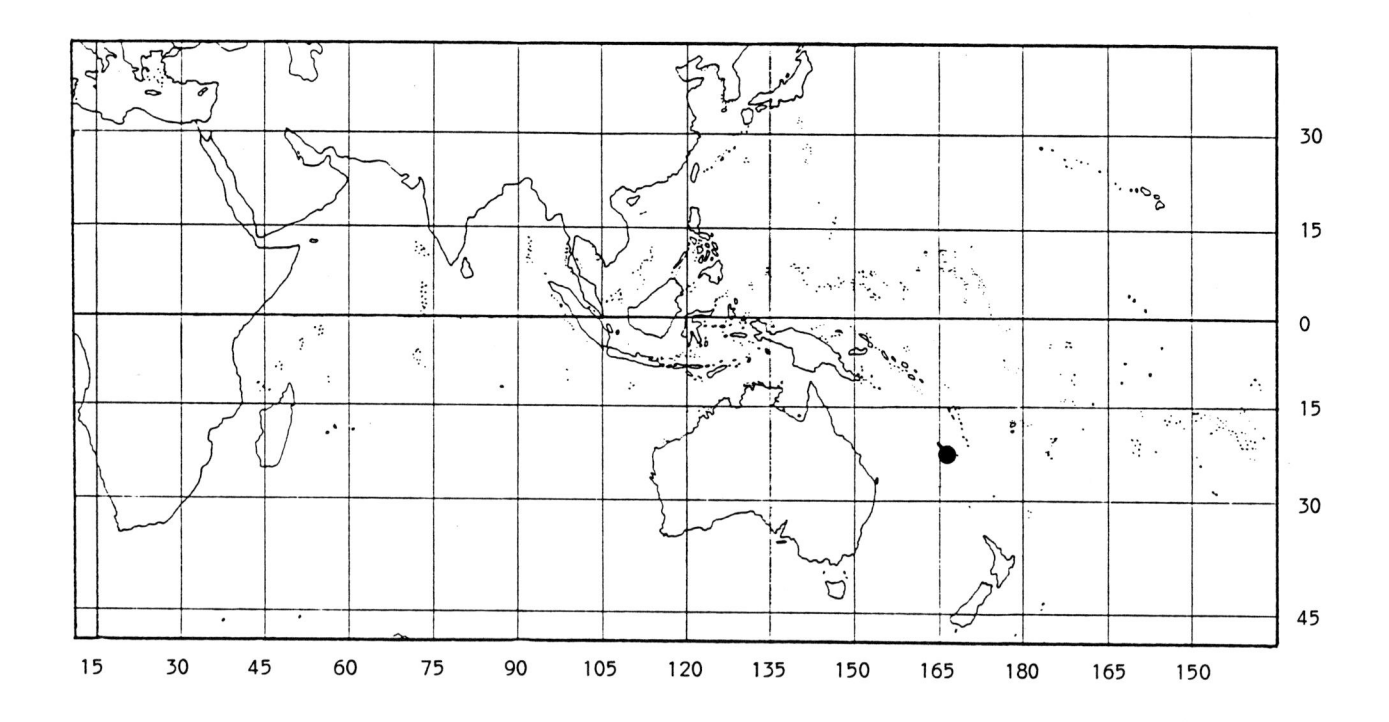

FIG. 199. — Distribution of *Chicoreus paucifrondosus* Houart.

TYPE MATERIAL. Holotype MNHN, and 14 paratypes.

OTHER MATERIAL EXAMINED. 33 specimens from throughout the geographical range.

DISTRIBUTION. (Fig. 200). Off southern New Caledonia. Depth range : 238-338 m.

DESCRIPTION. Shell up to 30 mm in length. Spire high, with 1 3/4-2 protoconch whorls and up to 5 rounded teleoconch whorls. Suture impressed. Protoconch large, smooth, ending with a thin, erect, bladelike terminal varix, slightly carinate.
Last whorl with 3 rounded varices, each with short palmate fronds ; adapical and abapical fronds largest. Intermediate small spinelets broadly open. Other axial sculpture consisting of 2 strong intervaricial nodes. Spiral sculpture of numerous equal sized cords and occasionally one small supplementary intermediate thread in each interspace.

Aperture roundly-ovate. Columellar lip smooth, rim adherent, weakly erect abapically. Anal notch small. Outer lip strongly erect, crenulate, lirate for short distance within. Siphonal canal long, slender, abaperturally bent, strongly curved abaxially, with 1 or 2 small spinelets on it adapically.
Whitish, yellow or pink with paler coloured axial nodes and varices, and darker coloured fronds and siphonal canal.
Radula (Figs 108-109).

REMARKS. Compared with the superficially similar *C. boucheti*, *C. subpalmatus* differs by its less spinose and smaller shell, its more deeply impressed suture, its more regular sculpture, and its narrower columellar lip, more erect outer lip and heavier, more regular interior sculpture.

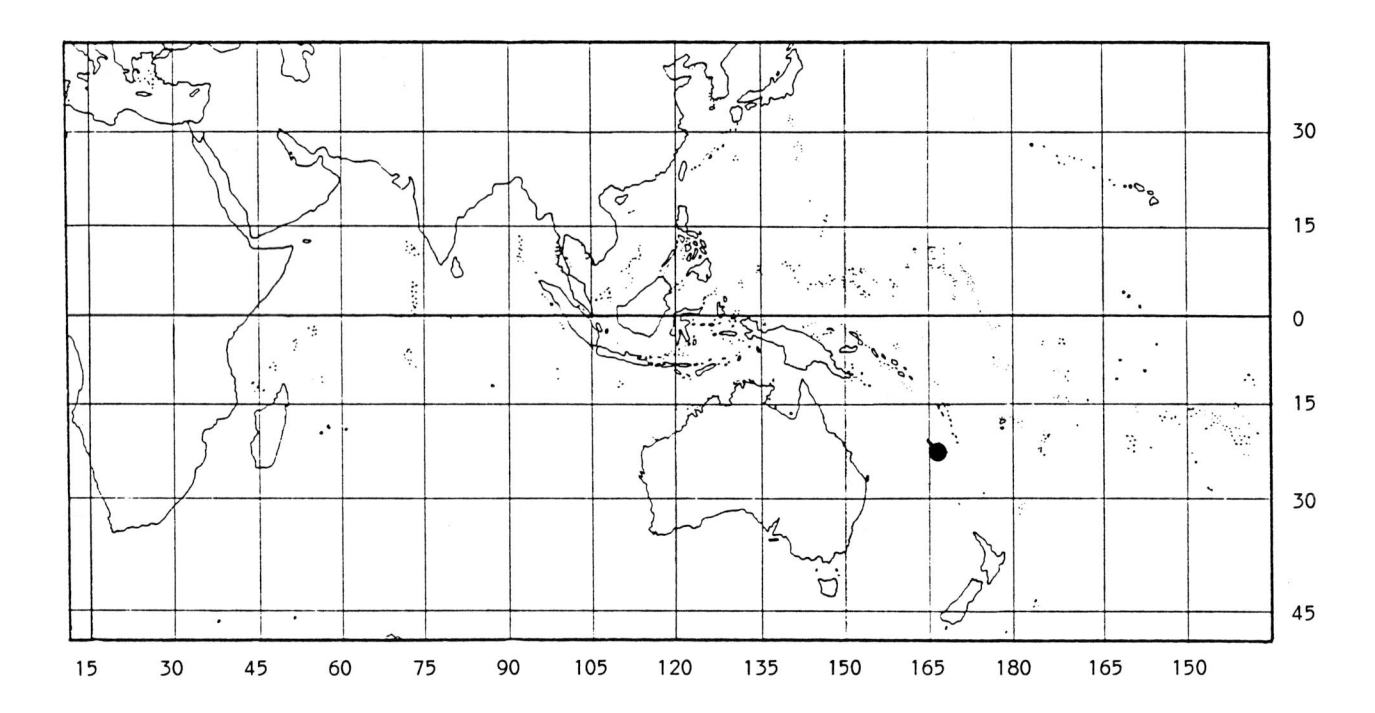

FIG. 200. — Distribution of *Chicoreus subpalmatus* Houart.

FIGS 201-203. — Subgenus *C. (Triplex)*.
201, *C. aculeatus* (Lamarck). Zululand, South Africa, 39.9 mm (NM D6517).
202, *C. nobilis* Shikama. New Caledonia, 51.1 mm (coll. GLASS & FOSTER).
203, *C. rossiteri* (Crosse). Japan, 59.5 mm (RH).

GROUP 7

Chicoreus (Triplex) aculeatus (Lamarck, 1822)
Figs 51, 139, 201, 204, 394, 397

Murex aculeatus Lamarck, 1822 : 163.

Chicoreus artemis Radwin & D'Attilio, 1976 : 32, pl. 4, fig. 4 (unnecessary n.n. for *Murex aculeatus* Lamarck, 1822).

ADDITIONAL REFERENCES

Murex aculeatus. — KIENER, 1842 : pl. 39, fig. 3.
Murex (Chicoreus) aculeatus. — SHIKAMA, 1963 : 70, pl. 53, fig. 6.
Chicoreus aculeatus. — D'ATTILIO, 1966 : 4, fig. 5 ; CERNOHORSKY, 1971 : 188 ; KAICHER, 1973 : card 160 ; CERNOHORSKY, 1985 : 47 (in part), figs 1-2 ; SPRINGSTEEN & LEOBRERA, 1986 : 154, text fig. C ; LAI, 1987 : 63, pl. 30, fig. 30 ; RIPPINGALE, 1987 : 9, fig. 22.
Murex aculeatus. — LEEHMAN, 1978b : 1, text fig., second row.
Chicoreus artemis. — HOUART, 1981a : 8, text fig. ; HOUART, 1981b : 7, fig. 2.
Chicoreus rossiteri. — DE COUET & MUHLHAUSSER, 1983 : 3, bottow row of text fig. (not *Murex rossiteri* Crosse, 1872).
NOT *Murex aculeatus*. — REEVE, 1845 : pl. 15, fig. 60 [= *Chicoreus (Triplex) nobilis* Shikama, 1977].
NOT *Murex (Chicoreus) axicornis aculeatus*. — SMITH, 1953 : 7, pl. 22, fig. 3 [= *Chicoreus (Triplex) nobilis* Shikama, 1977].
NOT *Chicoreus aculeatus*. — ZHANG, 1965 : 21, pl. 2, fig. 7 [= *Chicoreus (Triplex) cnissodus* (Euthyme, 1883)].
NOT *Chicoreus aculeatus*. — CERNOHORSKY, 1967a : 117, fig. 1, pl. 14, fig. 5 ; CERNOHORSKY, 1967b : 118, pl. 25, fig. 147 ; ABBOTT & DANCE, 1982 : 137, text fig. ; DE COUET & MUHLHAUSSER, 1983 : upper row of text fig. ; CERNOHORSKY, 1985 : 47 (in part), fig. 3 ; SPRINGSTEEN & LEOBRERA, 1986 : 154, figs a & b [= *Chicoreus (Triplex) nobilis* Shikama, 1977)].
NOT *Chicoreus (Chicoreus) aculeatus*. — FAIR, 1976 : 18, pl. 7, fig. 86 [= *Chicoreus (Triplex) nobilis* Shikama, 1977].
NOT *Murex artemis*. — LEEHMAN, 1978b : 1, text fig. upper row [= *Chicoreus (Triplex) nobilis* Shikama, 1977].
NOT *Murex aculeatus*. — LEEHMAN, 1978b : 1, text fig., bottom row, shell on the left [= *Chicoreus (Triplex) rossiteri* (Crosse, 1872)], shell on the right [= *Chicoreus (Triplex) nobilis* Shikama, 1977)].
NOT *Murex (Chicoreus) artemis*. — EISENBERG, 1981 : 87, pl. 69, fig. 10 [= *Chicoreus (Triplex) rossiteri* (Crosse, 1872)].
NOT *Chicoreus (Chicoreus) artemis*. — SPRINGSTEEN & LEOBRERA, 1986 : 131, pl. 35, fig. 16 [= *Chicoreus (Triplex) nobilis* Shikama, 1977)].

TYPE LOCALITY. Unknown. Punta Engano, Mactan, Philippine Is, here designated.

TYPE MATERIAL. None, the presumed syntype in the MNHN is *Chicoreus banksii* (Sowerby, 1834) ; no specimen was located in the MHNG. Neotype (MNHN) here designated.

OTHER MATERIAL EXAMINED. C. 200 specimens from throughout the geographical range.

DISTRIBUTION. (Fig. 204). Southern Africa, north Zululand, 27°32'8" S, 32°42'6" E (NM D6517) ; Malaysia (Phuket I.) ; the Philippine Is ; Taiwan and southern Japan (Tosa Bay). Usually caught in tangle nets or in lobster nets. Habitat unknown, the south African specimen was taken with sponges and coral rubble.

DESCRIPTION. Shell up to 67 mm in length, stout. Spire high, with 3-3 1/4 protoconch whorls and up 9 teleoconch whorls. Suture appressed. Protoconch conical, whorls weakly subcarinate.

Last whorl with 3 rounded varices, each with 3 or 4 major, straight, slightly frondose spines, abapical spine shortest. A short to medium sized intermediate spine between shoulder and second spine rarely present ; short intermediate spinelets present. Intervaricial axial sculpture consisting of one prominent node, rarely with an additional ridge. Spiral sculpture of 8 or 9 strong cords, 1 or 2 finer cords and 3 or 4 scabrous threads in each interspace.

Aperture roundly-ovate. Columellar lip smooth, rim partially erect abapically. Anal notch broad, moderately deep, delineated by shallow callus. Outer lip crenulate, lirate for short distance within. Siphonal canal moderately long, narrowly open and bent abaperturally at tip, with 2 or 3 abapically curved spines.

Light brown, orange or pink, with darker coloured cords.

REMARKS. KIENER (1842 : pl. 39, fig. 3) was the first to illustrate *Murex aculeatus* Lamarck, 1822, probably figuring the specimen used by Lamarck for its description. Three years later, REEVE (1845 : pl. 15, fig. 60) illustrated another species as *Murex aculeatus*, his illustration clearly represents the species now known as *C. nobilis* Shikama, 1977.

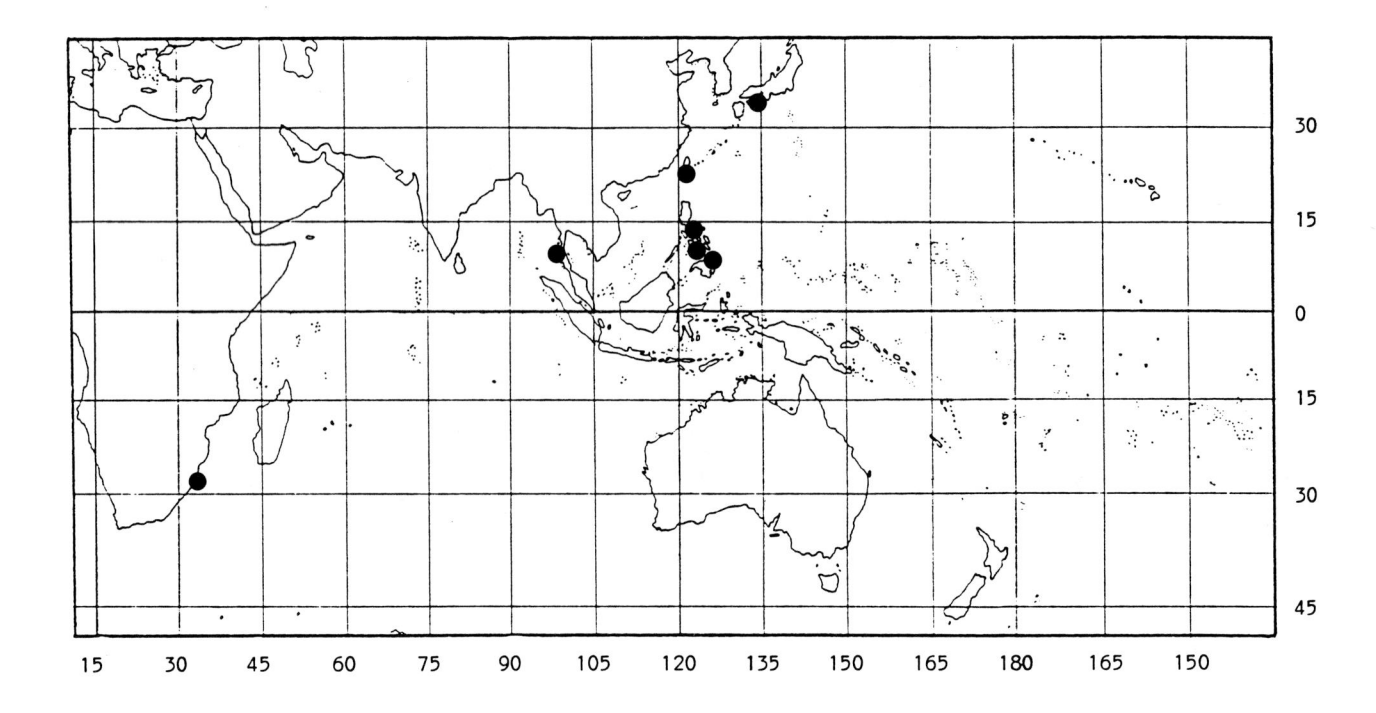

Fig. 204. — Distribution of *Chicoreus aculeatus* (Lamarck).

C. aculeatus has a rather complicated nomenclatural history. It has been regarded as a secondary homonym of *Aranea aculeata* Perry, 1811 and of *Muricites aculeatus* Schlotheim, 1820 by Vokes (1970, 1971), Fair (1976), Radwin and D'Attilio (1976), and Houart (1981, 1983), and it was renamed *Chicoreus artemis* by Radwin & D'Attilio (1976 : 32).

Cernohorsky (1985) however stated that the taxon *Murex aculeatus* Lamarck, 1822 is not a homonym and that *Chicoreus artemis* Radwin & D'Attilio, 1976 is thus an unnecessary replacement name. *Muricites aculeatus* Schlotheim is a species of *Tympanotonos* (Potamididae), while *Aranea aculeata* Perry, 1811 belongs in *Murex (s.s.)* (Ponder & Vokes, 1988 : 22-23).

Since *C. aculeatus* was originally introduced in *Murex*, some doubts persisted about a possible secondary homonymy between Lamarck's and Perry's taxon, so I requested the opinion of the International Commission on Zoological Nomenclature. ICZN Secretary, Mr. P. K. Tubbs (letters dated 6 May and 7 October 1986), confirmed *Murex aculeatus* Lamarck cannot be considered as a secondary homonym of *Aranea aculeata* Perry, since both names have never been combined with the same generic name. Therefore, the replacement name *artemis* is unnecessary and the correct name for the species is *Chicoreus (Triplex) aculeatus* (Lamarck, 1822).

Since the type specimen of *Murex aculeatus* is lost, a neotype (close to Kiener's illustration) is here designated (MNHN), to stabilise the concept of the species.

Chicoreus (Triplex) cloveri Houart, 1985
Figs 52, 131, 136, 205, 398

Chicoreus cloveri Houart, 1975b : 160, fig. 5.

ADDITIONAL REFERENCES

Chicoreus cloveri. — Houart, 1986a : 13, text fig. (paratype) ; Drivas & Jay, 1988 : 72, pl. 21, fig. 1.

TYPE LOCALITY. Dredged in 180 m, muddy sand, at Tamarin Bay, 20 miles south of Port Louis, Mauritius.

TYPE MATERIAL. Holotype MNHN and 2 paratypes.

MATERIAL EXAMINED. Type material; Mauritius, coll. J. DRIVAS (1 lv.).

DISTRIBUTION. (Fig. 205). Mauritius. Depth range: 50-180 m.

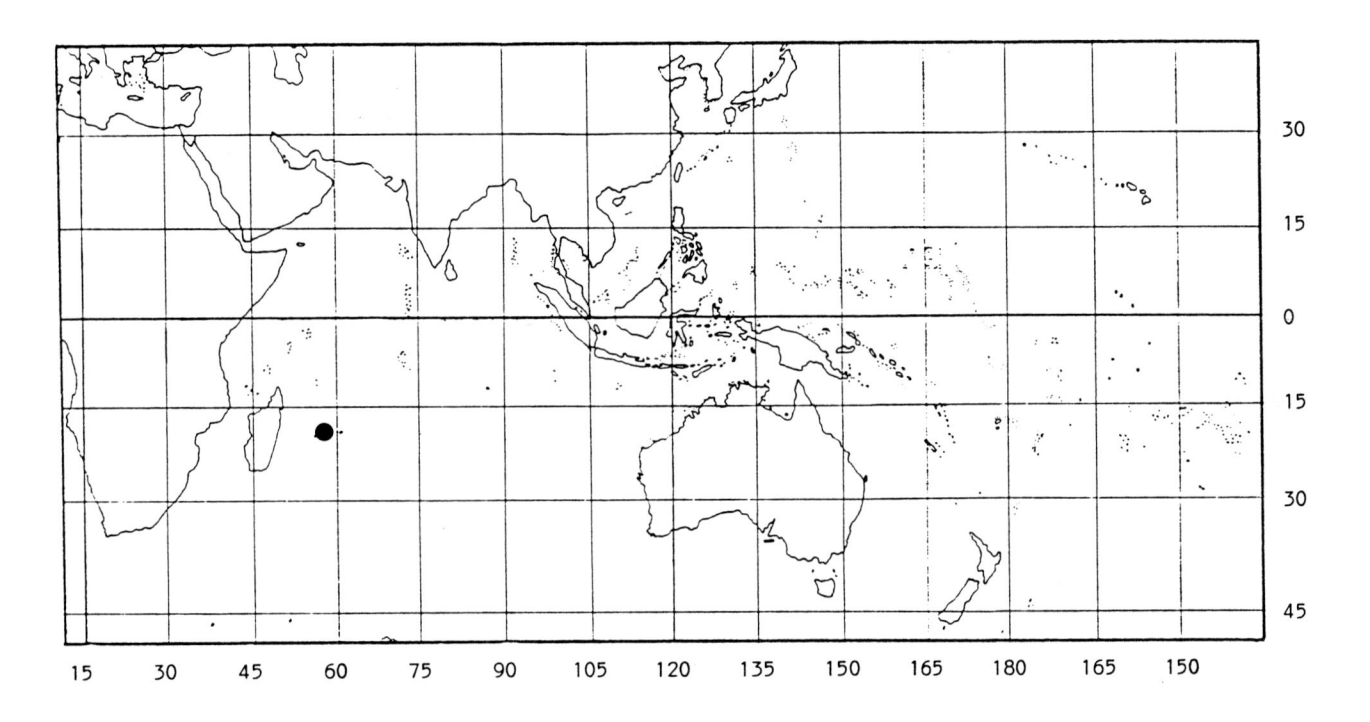

FIG. 205. — Distribution of *Chicoreus cloveri* Houart.

DESCRIPTION. Shell up to 30 mm in length, stout. Spire high, with 2 1/4 protoconch whorls and up to 5 teleoconch whorls. Suture impressed. Protoconch whorls rounded and smooth.
Last whorl with 3 rounded, frondose varices, each with 4 short frondose spines; shoulder spine broad, foliaceous, followed by 2 smaller medium-sized spines. Anterior spine shortest. A small spinelet often present between shoulder and second spine. Intervaricial axial sculpture consisting of 1 prominent node and 1 weak ridge. Spiral sculpture of numerous scabrous cords and threads.
Aperture roundly-ovate, weakly angulate adapically. Columellar lip smooth, rim adherent. Anal notch deep, rather narrow. Outer lip crenulate, lirate for short distance within. Siphonal canal long, with 3 short foliaceous, adapically bent spines.
Uniformly light brown. Aperture white.

REMARKS. Closely related to *C. nobilis* from the West Pacific, although smaller with a brownish shell and a different protoconch (Fig. 52). Little is known about its habitat.

Chicoreus (Triplex) crosnieri Houart, 1985
Figs 53-54, 141, 206, 257, 396

Chicoreus crosnieri Houart, 1985b : 161, fig. 4.

ADDITIONAL REFERENCE

Chicoreus crosnieri. — HOUART, 1986a : 13, text fig. (holotype).

TYPE LOCALITY. South of Madagascar, 26°05′ S, 44°50′ E, 100 m.

TYPE MATERIAL. Holotype MNHN and 2 paratypes.

OTHER MATERIAL EXAMINED. Off Grand Récif, Tuléar, Madagascar (RH) (1 dd), coll. BLÖCHER (3 dd), coll. MÜHLHÄUSSER (1 dd).

DISTRIBUTION. (Fig. 206). South-western Madagascar. Depth range : 85-165 m.

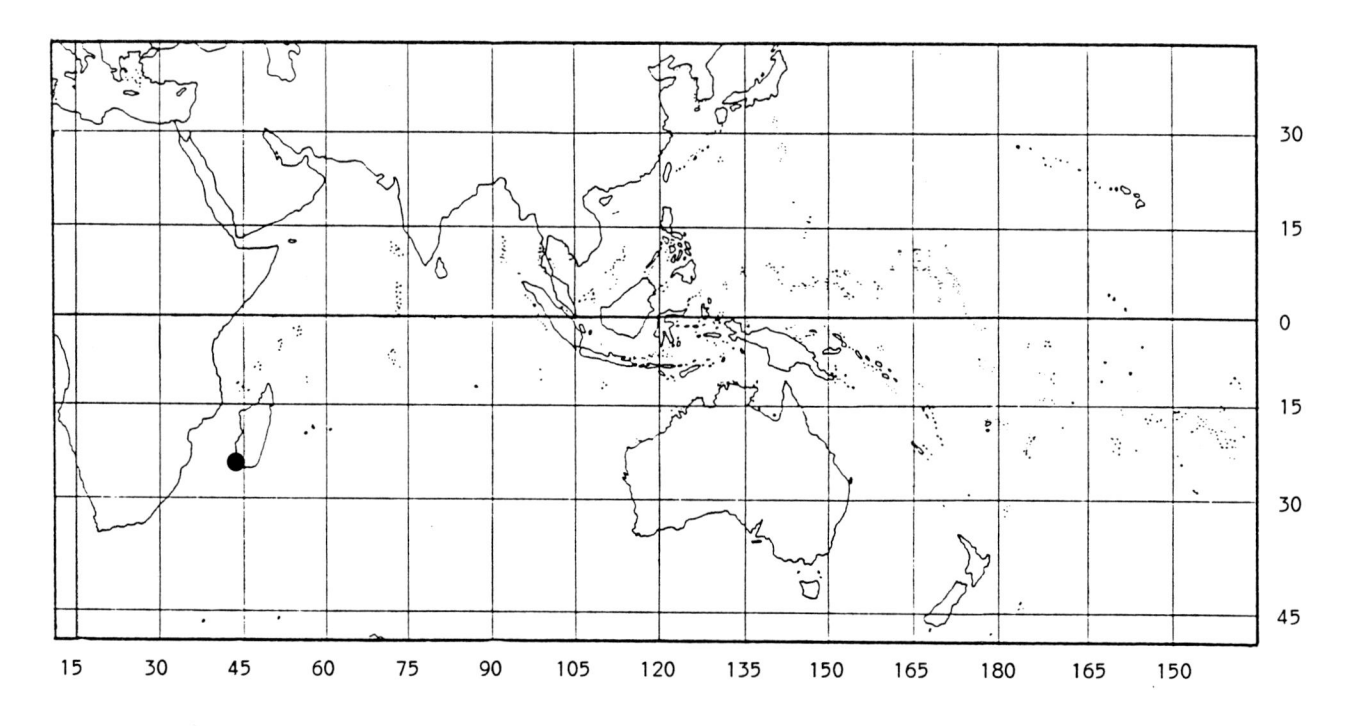

FIG. 206. — Distribution of *Chicoreus crosnieri* Houart.

DESCRIPTION. Shell up to 37 mm in length. Spire high, with 2 protoconch whorls and up to 6 weakly angulate teleoconch whorls. Suture appressed. Protoconch glossy, tip flattened.
Last whorl with 3 more-or-less frondose varices, usually with 4 medium-sized spines. Shoulder spine sometimes short or medium-sized, followed by 2 short spines, and long abapical spine. Second and third spine or second only, sometimes very short or obsolete. Intervaricial axial sculpture consisting of 1 large and 1 weak axial node. Spiral sculpture of 9 or 10 squamous cords, usually with 1 or 2 intermediate squamous threads in each interspace.
Aperture roundly-ovate. Columellar lip smooth, rim erect on one quarter of abapical part. Anal notch deep, small. Outer lip erect, crenulate, lirate for short distance within. Siphonal canal long, straight, narrowly open, weakly bent abaperturally at tip ; with 3 abapically bent spines.
Light brown, occasionally darker cords. Aperture white.

REMARKS. *C. crosnieri* is closely related to *C. rossiteri* and *C. aculeatus* (Table 3).

TABLE 3. — Comparisons of *Chicoreus* species of group 7.

Character	*C. nobilis*	*C. ryukyuensis*	*C. cloveri*	*C. aculeatus*	*C. rossiteri*	*C. crosnieri*	*C. fosterorum*	*C. zululandensis*
Protoconch	Conical, 1.5-2.25 rounded whorls. Rounded, erect terminal varix	Globose, 2.25 whorls. Flat, almost straight terminal varix	Globose, 2.25 whorls	Conical, 3-3.25 whorls, weakly carinate. Slightly rounded, erect terminal varix	Conical, 3-3.25 whorls. Weakly carinate. Slightly rounded, erect terminal varix	Flattened, 2.25 whorls. Broad, straight, terminal varix	Globose and large. 2 rounded whorls. Erect, rounded terminal varix	Rounded, 2 whorls. Erect, somewhat angulate terminal varix
Number of teleoconch whorls	8-9	6	5	8-9	8-9	5-6	5	5-6
Apertural varix and siphonal canal	See Figs 135-142							
Ornamentation of siphonal canal	3 foliaceous adaperturally recurved spines, mostly situated on abapical end of canal	3 foliaceous adaperturally recurved spines, covering entire canal	3 short, foliaceous adaperturally recurved spines, situated on abapertural end of canal	2 or 3 straight abaperturally bent spines, distributed over whole canal	3 straight, abaperturally bent spines distributed over whole canal	3 straight, abaperturally bent spines. 2 crowded spines at adapertural end, 1 between them and aperture	3 or 4 straight frondose, open, short spines, distributed over whole canal	2 similar, open, foliaceous spines. Last abapertural spine somewhat shorter
Adult shell length	37-57 mm	33.5-37.9 mm	20-29.5 mm	38-67 mm	36-58 mm	25-37 mm	41-46 mm	31.5-34.5 mm

Chicoreus (Triplex) fosterorum Houart, 1989
Figs 49, 142, 207, 401-405

Chicoreus fosterorum Houart, 1989 : 60, figs 1-5.

TYPE LOCALITY. Mzamba, Pondoland, Transkei, South Africa, c. 30°51' S, 29°46' E.

TYPE MATERIAL. Holotype NM 5343 and 3 paratypes.

MATERIAL EXAMINED. Type material.

DISTRIBUTION. (Fig. 207). Pondoland, Transkei and Aliwal Shoal, south Coast of Natal, South Africa. Depth range : 50 m.

DESCRIPTION. Shell up to 46 mm in length. Spire high, with 2 protoconch whorls and up to 5 teleoconch whorls. Suture appressed. Protoconch whorls smooth, large, globose. Last whorl with 3 frondose varices, each with 5 short, foliaceous spines. Shoulder spine broad, strongly frondose. Other spines short, broadly open. Anterior spine largest ; short intermediate spinelets. Intervaricial axial sculpture consisting of a prominent single node. Spiral sculpture consisting of numerous squamous cords and threads over entire surface.

Aperture roundly-ovate. Columellar lip smooth, rim adherent, weakly erect abapically. Anal notch narrow, rather deep. Outer lip crenulate, lirate for short distance within. Siphonal canal long, narrowly open, weakly bent abaperturally at tip, 3 or 4 short, straight, frondose, broadly open spines.

First 3 or 4 spire whorls pinkish orange, subsequent whorls white.

REMARKS. *C. fosterorum* is closely related to *C. cloveri, C. nobilis* and *C. ryukyuensis*. *C. cloveri* has a smaller shell, with a smaller protoconch, more weakly frondose spines, while the spines on the siphonal canal are smaller and situated abapically on the canal. The anal notch of *C. cloveri* is larger and the abapical end of its columellar lip is recurved and thickened, but straight and smooth in *C. fosterorum*.

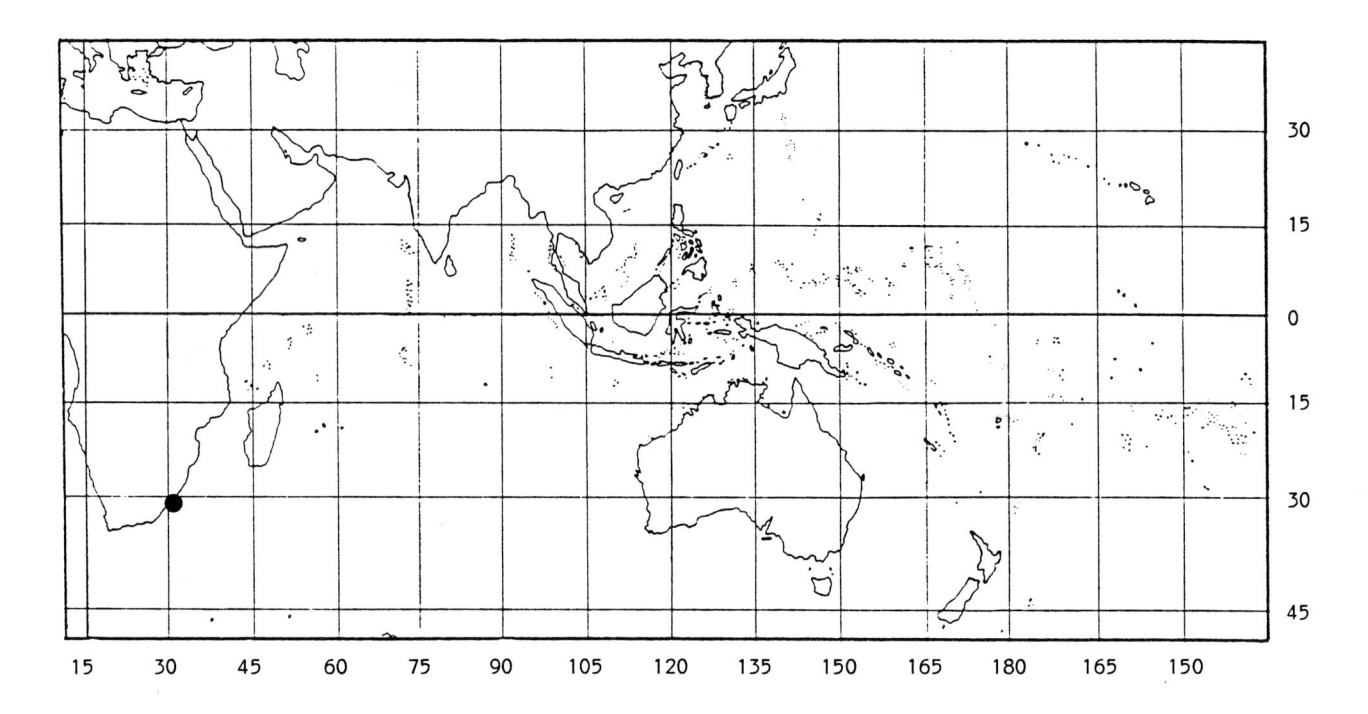

FIG. 207. — Distribution of *Chicoreus fosterorum* Houart.

From *C. nobilis*, *C. fosterorum* differs in its paucispiral, globose rather than conical protoconch. *C. nobilis* also has a broader anal notch, 2 or 3 intervaricial ridges while there are fewer and longer spines on the varices and on the siphonal canal.

C. ryukyuensis is also related but has more weakly frondose spines, a smaller protoconch with a different terminal varix (Figs 49 & 57), a broader, larger anal notch, and a more ovate aperture.

Chicoreus (Triplex) nobilis Shikama, 1977
Figs 55, 110-111, 137, 202, 208-210, 211, 251

Chicoreus nobilis Shikama, 1977 : 14, pl. 2, fig. 9, pl. 5, fig. 1.

ADDITIONAL REFERENCES

Murex aculeatus. — REEVE, 1845 : pl. 15, fig. 60 (not *Murex aculeatus* Lamarck, 1822).
Murex (Chicoreus) axicornis aculeatus. — SMITH, 1953 : 7, pl. 22, fig. 3 (not *Murex aculeatus* Lamarck, 1822).
Chicoreus aculeatus. — CERNOHORSKY, 1967a : 117, fig. 1, pl. 14, fig. 5; CERNOHORSKY, 1967b : 118, pl. 25, fig. 147; ABBOTT & DANCE, 1982 : 137, text fig.; DE COUET & MÜHLHÄUSSER, 1983 : 3, upper row of text fig.; CERNOHORSKY, 1985 : 47 (in part), fig. 3 (not *Murex aculeatus* Lamarck, 1822).
Chicoreus (Chicoreus) aculeatus. — FAIR, 1976 : 18, pl. 7, fig. 86 (not *Murex aculeatus* Lamarck, 1822).
Chicoreus (Chicoreus) rossiteri. — FAIR, 1976 : 72, pl. 6, fig. 79 (not *Murex rossiteri* Crosse, 1872).
Chicoreus sp. — LEEHMAN, 1976a : 9, text fig.
Murex artemis. — LEEHMAN, 1978b : 1, text fig., upper row, bottom row, shell on the right (not *Chicoreus artemis* Radwin & D'Attilio, 1976).
Chicoreus (Chicoreus) artemis. — SPRINGSTEEN & LEOBRERA, 1986 : 131, pl. 35, fig. 16 (not *Chicoreus artemis* Radwin & D'Attilio, 1976).
Chicoreus nobilis. — HOUART, 1981a : 8, text fig.; HOUART, 1981b : 7, fig. 3; HOUART, 1986c : 762, fig. 3; RIPPINGALE, 1987 : 3, fig. 2.
Chicoreus (Chicoreus) nobilis. — SPRINGSTEEN & LEOBRERA, 1986 : 131, pl. 35, fig. 15.

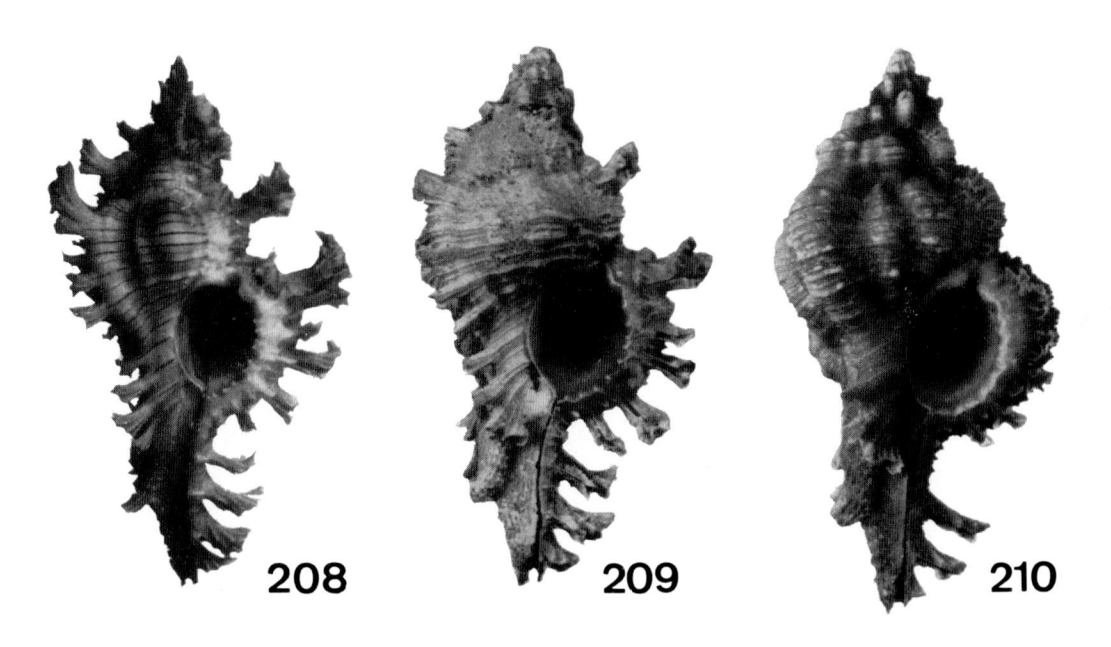

FIGS 208-210. — *Chicoreus (Triplex) nobilis* Shikama.
208, Philippines, 43 mm (coll. A. LESAGE).
209, Landsdowne-Fairway Reefs, 30.5 mm (MNHN).
210, Fiji, 32.2 mm (RH).

TYPE LOCALITY. Off Cebu I., Philippines.

TYPE MATERIAL. Holotype KPM 3280.

OTHER MATERIAL EXAMINED. C. 100 specimens from throughout the geographical range.

DISTRIBUTION. (Fig. 211). The Western Pacific : Taiwan ; the Philippine Is ; Papua New Guinea (Madang Province) ; the Coral Sea ; New Caledonia and Fiji Islands ; under and on coral rubble. Depth range : 9-100 m. A single specimen was collected in Mauritius (RH) but this record need confirmation.

DESCRIPTION. Shell up to 57 mm in length. Spire high, with 1 1/2-2 1/4 protoconch whorls and up to 9 convex teleoconch whorls. Suture appressed. Protoconch whorls rounded, glossy.
Last whorl with 3 frondose varices, each with 5 short frondose spines. Shoulder spine usually longest, shortest spine situated immediately below. Most abapical spine generally rather shorter than other spines ; presence of small intermediate spinelets. Fiji Islands specimens with very short to obsolete shoulder spine and longer abapical spines. Intervaricial axial sculpture consisting of 2 or occasionally 3 prominent ridges. Spiral sculpture of 8 or 9 major cords and 2 or 3 intermediate threads in each interspace.

Aperture roundly-ovate. Columellar lip smooth, adherent, rim slightly erect abapically. Anal notch deep, narrow, delineated by strong callus. Outer lip crenulate, lirate for short distance within. Siphonal canal long, narrowly open, bent abaperturally at tip, with 2 or 3 frondose, adapically recurved spines.
Yellowish to bright pink or orange, spiral cords usually more deeply pigmented, but uniformly coloured specimens not rare. Aperture whitish or pale violet, encircled with pale to bright pink band.
Radula (Figs 110-111).

REMARKS. Both shells illustrated by FAIR (1976 : pl. 6, fig. 79) as *C. rossiteri* and by CERNOHORSKY (1985 : 49, fig. 3) as *C. aculeatus* are *C. nobilis*.
C. nobilis was first illustrated by REEVE (1845 : fig. 60) as *Murex aculeatus* (see in remarks following *C. aculeatus*).

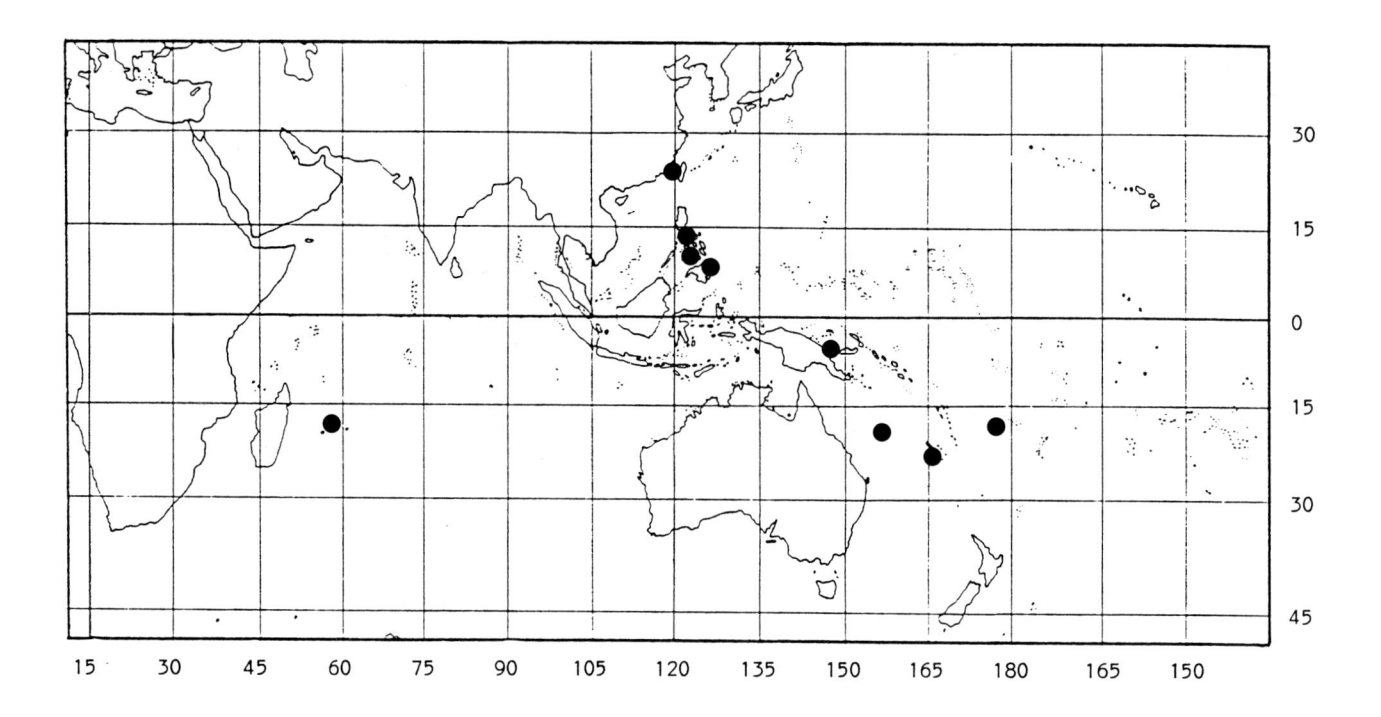

FIG. 211. — Distribution of *Chicoreus nobilis* Shikama.

Chicoreus (Triplex) rossiteri (Crosse, 1872)
Figs 56, 112-113, 138, 203, 212-213, 214, 258, 395

Murex rossiteri Crosse, 1872a : 74, 218, pl. 13, fig. 2.

Chicoreus saltatrix Kuroda, 1964 : 129, figs 1-3.

ADDITIONAL REFERENCES

Chicoreus (Chicoreus) saltatrix. — FAIR, 1976 : 74, pl. 8, fig. 105 (holotype); SPRINGSTEEN & LEOBRERA, 1986 : 130, pl. 35, fig. 10; p. 154, text fig. b (holotype).
Chicoreus saltatrix. — HOUART, 1981b : 7, fig. 1.
Chicoreus rossiteri. — RADWIN & D'ATTILIO, 1976 : 41, pl. 4, fig. 6; HOUART, 1981a : 8, text fig.; ABBOTT & DANCE, 1982 : 137, text fig.; DE COUET & MUHLHAUSSER, 1983 : 3, second and third row of text fig.; SPRINGSTEEN & LEOBRERA, 1986 : 154, text fig. a; RIPPINGALE, 1987 : 9, fig. 24.
Murex (Chicoreus) rossiteri. — EISENBERG, 1981 : 93, pl. 75, fig. 6.
Murex aculeatus. — LEEHMAN, 1978b : 1, text fig., bottom row, shell on the left (not *Murex aculeatus* Lamarck, 1822).
Murex (Chicoreus) artemis. — EISENBERG, 1981 : 87, pl. 69, fig. 10 (not *Chicoreus artemis* Radwin & D'Attilio, 1976).
NOT *Chicoreus (Chicoreus) rossiteri*. — FAIR, 1976 : 72, pl. 6, fig. 79 [= *Chicoreus (Triplex) nobilis* Shikama, 1977)].
NOT *Chicoreus rossiteri*. — DE COUET & MUHLHAUSSER, 1983 : 3, bottom row of text fig. [= *Chicoreus (Triplex) aculeatus* (Lamarck, 1822)].

TYPE LOCALITIES. *M. rossiteri* : Lifou (Loyalty Is, New Caledonia); *C. saltatrix* : Okinoshima, southwest of Tosa province, Shikoku, Japan.

TYPE MATERIAL. Not located. *C. rossiteri* : neotype MNHN, here designated (New Caledonia).

OTHER MATERIAL EXAMINED. C. 150 specimens from throughout the geographical range.

FIGS 212-213. — *Chicoreus (Triplex) rossiteri* (Crosse).
212, Philippines, 43.2 mm (MNHN).
213, Philippines, 41 mm (RH).

DISTRIBUTION. (Fig. 214). Okinoshima I. (Shikoku) and off Kyushu, Japan ; the Sulu Sea, Philippine Is ; Papua New Guinea ; New Caledonia. Depth range : 40-150 m.

DESCRIPTION. Shell up to 60 mm in length. Spire high, with 3-3 1/4 protoconch whorls and up to 9 teleoconch whorls. Suture appressed. Protoconch conical, whorls glossy, weakly subcarinate.
Last whorl with 3 rounded, frondose varices, each with 4 or 5 adapically recurved, weakly foliaceous spines. Shoulder and most abapical spine longest ; second spine shortest. Other whorls sometimes spineless. Shoulder spine occasionally very reduced to obsolete. Intervaricial axial sculpture consisting of 1 prominent node, occasionally with an additional small ridge. Spiral sculpture of numerous scabrous cords and threads.

Aperture ovate to roundly-ovate. Columellar lip with small nodule abapically and one adapically, or with a series of weak nodules along the edge ; adapically adherent and detached abapically. Anal notch broad, moderately deep, delineated by small callus. Outer lip crenulate, lirate for short distance within. Siphonal canal long, slender, narrowly open and bent abaperturally at tip, with 2 or 3 straight abapically bent spines.
Uniformly white to deep pink.
Radula (Figs 112-113).

REMARKS. Recently collected specimens in New Caledonia (MNHN) confirmed the type-locality for the species (Fig. 258). Records of *C. rossiteri* from Fiji (FAIR, 1976 : pl. 6, fig. 79) are based on *C. nobilis*. The species is probably also present in other West Pacific localities since the protoconch morphology indicates planktotrophic larval development.
It is clear that shells from New Caledonia, the Philippine Islands and Japan are conspecific since they share the same sculpture, shape, occasional absence of the shoulder spine, and exactly the same protoconch (protoconch of Japanese specimens unknown). Some forms are slightly more slender with a correspondingly higher spire, but this degree of intraspecific variation is exhibited by many species with a wide geographical range. Japanese specimens are closest to the New Caledonian form but are larger. HOUART (1981) discussed the differences between *C. aculeatus*, *C. nobilis* and

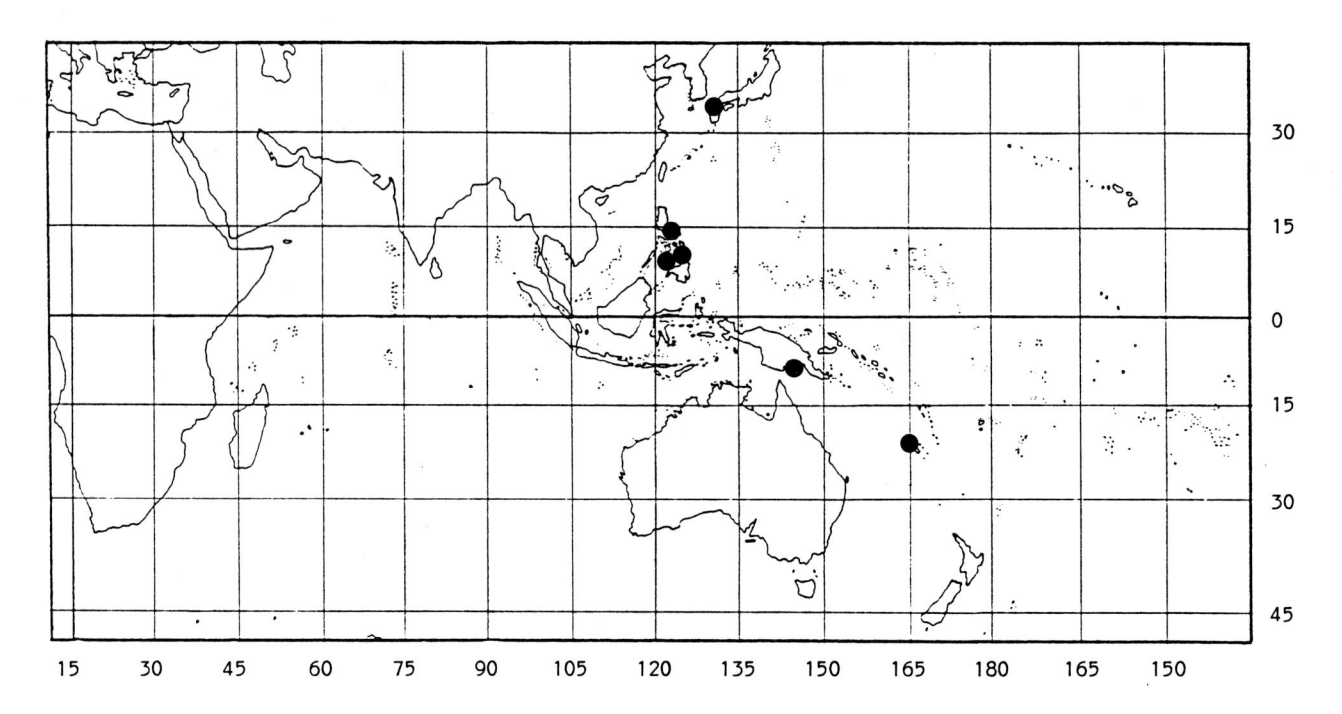

FIG. 214. — Distribution of *Chicoreus rossiteri* (Crosse).

C. rossiteri (as *C. saltatrix*). Although related to *C. aculeatus*, *C. rossiteri* differs in its varicial ornamentation and colour and by occasionally lacking the shoulder spine on the last whorl as well on the spire whorls, which, to my knowledge, is never observed in *C. aculeatus* (Table 3).

Chicoreus (Triplex) ryukyuensis Shikama, 1978
Figs 57, 130, 135, 215, 253, 359

Chicoreus (Triplex) ryukyuensis Shikama, 1978 : 35, pl. 7, figs 1-2.

ADDITIONAL REFERENCE

Chicoreus (Triplex) ryosukei. — SHIKAMA, 1978 : figs 3-4 (not *Chicoreus ryosukei* Shikama, 1978).

TYPE LOCALITY. Okinawa Islands.

TYPE MATERIAL. Holotype NSMT 60929.

MATERIAL EXAMINED. A photograph of the type material; Okinawa Is, KPM 3281 (1 lv).

DISTRIBUTION. (Fig. 215). Only known from Okinawa, the type locality.

DESCRIPTION. Shell up to 37.9 mm in length. Spire high, with 2 1/4 protoconch whorls and up to 6 rounded teleoconch whorls. Suture slightly appressed. Protoconch glossy, whorls rounded, globose.

Last whorl with 3 frondose varices, each with 6 short frondose spines. Shoulder spine largest, abapical spine shortest, short intermediate spinelets present. Other axial sculpture consisting of 1 prominent intervaricial node and 1 weak ridge. Spiral sculpture of numerous cords and threads.

Aperture roundly-ovate, weakly angulate adapically. Columellar lip smooth, rim partially erect abapically. Anal notch deep, rather large. Outer lip crenulate, lirate for short distance within. Siphonal canal moderately long, with 3 frondose, adapically bent spines.

Uniformly pinkish-white to light orange.

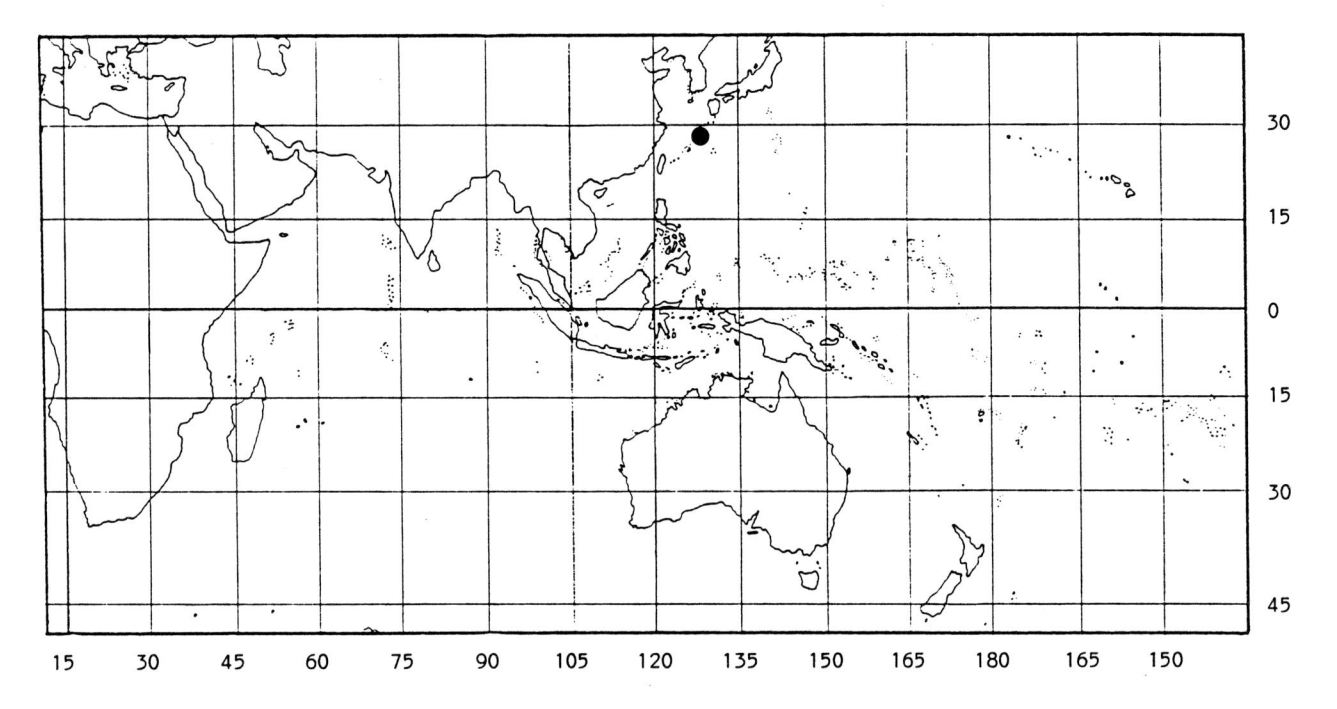

FIG. 215. — Distribution of *Chicoreus ryukyuensis* Shikama.

REMARKS. A specimen illustrated by SHIKAMA (1978 : figs 3, 4) as *C. ryosukei* represent *C. ryukyuensis* Shikama, 1978. For comparisons with other species of this group, see table 3.

Chicoreus (Triplex) zululandensis Houart, 1989
Figs 50, 140, 216, 254

Chicoreus zululandensis Houart, 1989 : 62, figs 6-8.

TYPE LOCALITY. North Zululand, South Africa : SE of Kosi River Mouth, 26°55'0" S, 32°55'8" E, 65 m, bottom of sponge, gorgonians, medium sand.

TYPE MATERIAL. Holotype NM D8049 and 2 paratypes.

MATERIAL EXAMINED. Type material.

DISTRIBUTION. (Fig. 216). Only known from the type locality and from north Zululand, off Jesser Point, bottom of sponge and coral rubble. Depth range : 65 m.

DESCRIPTION. Shell up to 34.5 mm in length. Spire acutely conical, with 2 protoconch whorls and up to 7 rounded teleoconch whorls. Suture appressed. Protoconch whorls rounded, smooth.
Last whorl with 3 frondose varices, each with 4 moderate-sized, weakly foliaceous, adapically curved, open spines. Shoulder spine shortest, sometimes obsolete. Abapical spine longest. Intervaricial axial sculpture consisting of 1 or 2 strong nodes. Spiral sculpture of numerous crowded squamous cords and threads.

Aperture rounded. Columellar lip smooth, rim erect abapically. Anal notch deep, narrow, relatively small. Outer lip weakly erect, crenulate, weakly lirate for short distance within. Siphonal canal long, narrowly open, straight, very slightly bent abaperturally at tip, with 2 equal-sized, open, foliaceous spines.
Pinkish-orange with paler maculations, especially on the varices.

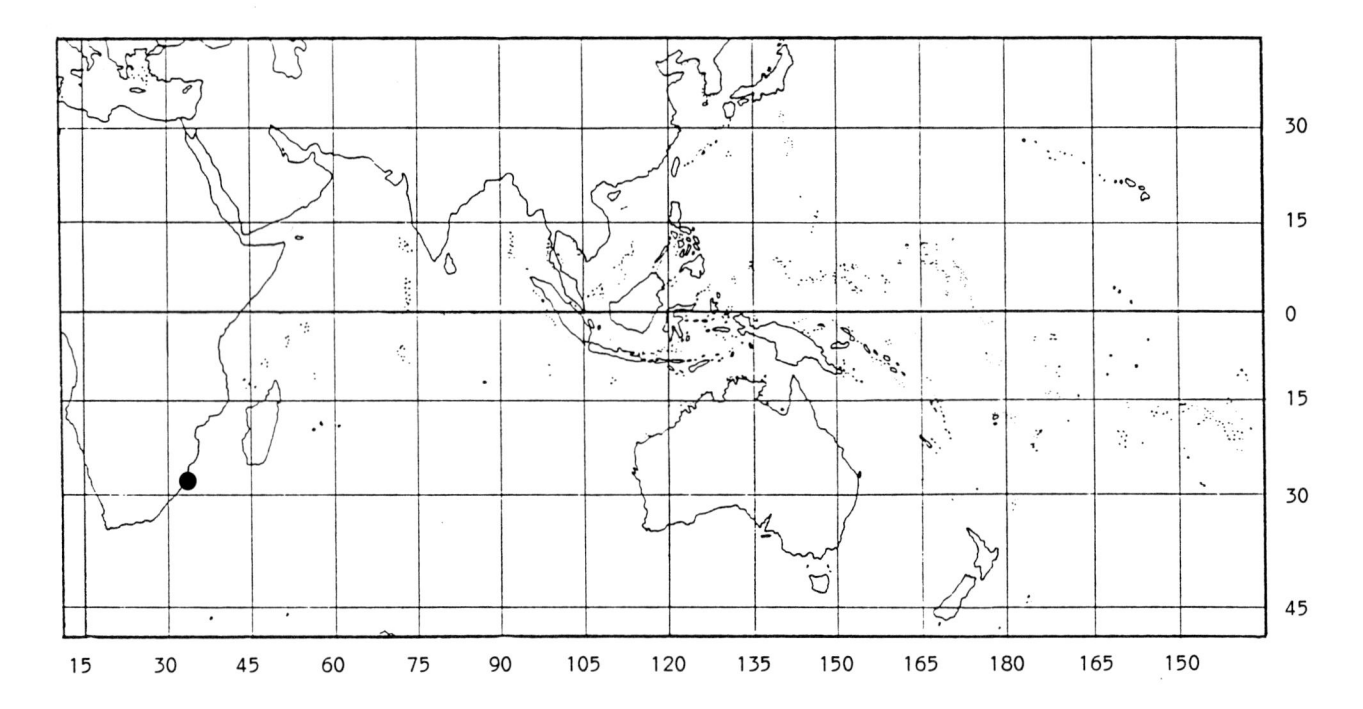

FIG. 216. — Distribution of *Chicoreus zululandensis* Houart.

DISCUSSION. This species is closely related to *C. rossiteri* and *C. crosnieri*. It differs from *C. rossiteri* in having a paucispiral rather than multispiral protoconch (Figs 50 & 56), and in having finer, more numerous spiral cords and threads on the teleoconch.

From *C. crosnieri* it differs in having a more acute protoconch with a different terminal varix (Figs 53-54 & 56), in having narrower, more numerous spiral cords, and a siphonal canal with straight spines instead of strongly abapically bent as in *C. crosnieri*.

Genus **CHICOREUS** Montfort, 1810

Subgenus **RHIZOPHORIMUREX** Oyama, 1950

Chicoreus (Rhizophorimurex) capucinus (Lamarck, 1822)
Figs 36, 105, 217, 218, 369-375

Murex capucinus Lamarck, 1822 : 164.

Murex quadrifrons Lamarck, 1822 : 170.
Murex castaneus Sowerby, 1834 : pl. 64, fig. 44.
Murex lignarius A. Adams, 1853 : 268.
Murex bituberculatus Baker, 1891 : 133, pl. 11, fig. 4.
Murex permaestus Hedley, 1915 : 745, pl. 85, fig. 91.

ADDITIONAL REFERENCES

Murex (Phyllonotus) quadrifrons. — SMITH, 1963 : 9, pl. 7, fig. 5.
Chicoreus (Chicoreus) quadrifrons. — FAIR, 1976 : 70, pl. 8, fig. 102 (syntype).
Chicoreus capucinus (Röding, 1798). — CERNOHORSKY, 1967a : 118, pl. 14, fig. 7 ; SHORT & POTTER, 1987 : 56, pl. 27, fig. 13.
Murex capucinus. — CERNOHORSKY, 1971 : 188, fig. 1 (lectotype).
Naquetia capucinus. — KAICHER, 1973 : card 168 ; HINTON, 1979 : 27, fig. 5.
Chicoreus (Naquetia) capucinus. — FAIR, 1976 : 29, pl. 14, fig. 176 ; SPRINGSTEEN & LEOBRERA, 1986 : 135, pl. 36, fig. 15.

Fig. 217. — *Chicoreus (Rhizophorimurex) capucinus* (Lamarck). Locality unknown, 124.3 mm (lectotype, MHNG 1099/23. Photo J. Dajoz).

Naquetia capucina. — Radwin & D'Attilio, 1976 : 80, pl. 15, fig. 13 ; Wells & Bryce, 1985 : 88, pl. 25, fig. 285.
Murex (Naquetia) capucinus. — Eisenberg, 1981 : 88, pl. 70, fig. 12.
Chicoreus capucinus. — Abbott & Dance, 1982 : 136, text fig.
Chicoreus (Chicoreus) capucinus. — Houart & Pain, 1983a : 18, text fig. p. 17 (syntype of *Murex quadrifrons* Lamarck).
Chicoreus permaestus. — Cernohorsky, 1967b : 120, pl. 25, fig. 149 ; Hinton, 1972 : 36, pl. 18, fig. 10.
Pterynotus (Naquetia) permaestus. — Wilson & Gillett, 1971 : 84, pl. 57, fig. 8.
Murex lignarius. — Cernohorsky, 1971 : 189, fig. 2 (lectotype).
Chicoreus banksii. — Abbott & Dance, 1982 : 136, text fig. (not *Murex banksii* Sowerby, 1841).

Type localities. *M. capucinus* : none ; *M. quadrifrons* : none ; *M. castaneus* : none ; *M. lignarius* : West Africa (error) ; *M. bituberculatus* : Australia ; *M. permaestus* : 10 fms, off Mapoon, Queensland Australia.

Type material. *M. capucinus* : lectotype MHNG 1099/23 (designated by Cernohorsky, 1971) ; *M. quadrifrons* : holotype MHNG 1099/45 ; *M. castaneus* : not located ; *M. lignarius* : lectotype BMNH 1985229 (designated by Cernohorsky, 1971) ; *M. bituberculatus* : holotype Chicago Academy of Sciences 20702 ; *M. permaestus* : holotype AMS C14130.

OTHER MATERIAL EXAMINED : c. 100 specimens from throughout the geographical range.

DISTRIBUTION. (Fig. 218). Singapore ; north western Australia to Queensland ; Papua New Guinea ; the Philippine Is ; Solomons and Fiji Is ; on muddy sand.

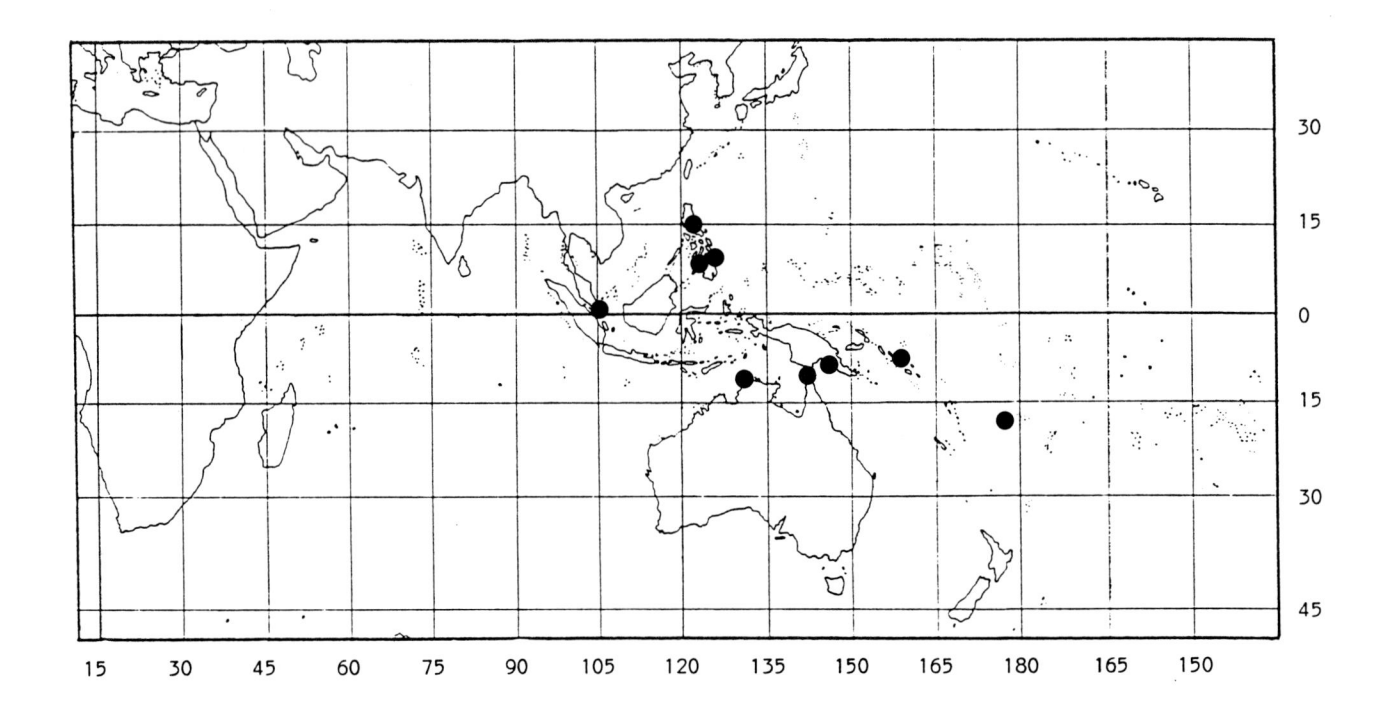

FIG. 218. — Distribution of *Chicoreus capucinus* (Lamarck).

DISTRIBUTION. Shell up to 124.3 mm in length (lectotype of *Murex capucinus*). Spire high, with 2 protoconch whorls and up to 9 convex teleoconch whorls. Suture appressed. Protoconch acutely conical.

Last whorl with 3 (very rarely 4 or 5) rounded, usually spineless varices, each with short webbed expansion on adapical section. In some specimens spiral cords extend to the varices; giving rise to short open spines. Intervaricial axial sculpture consisting of 2 or 3 low, shallow ridges. Spiral sculpture of 7-9 strong cords and 1-3 intermediate threads between each.

Aperture ovate. Columellar lip fully adherent. Anal notch shallow. Outer lip crenulate, strongly lirate for short distance within. Siphonal canal spineless, occasionally with short webbed expansion abapically ; short and broad ; weakly bent abaperturally.

Uniformly brown to dark brown, occasionally with darker spiral cords, aperture and columellar lip bluish-white, outer lip stained with brown.

Radula (Fig. 105).

REMARKS. A highly distinctive species but with an intricate nomenclature.

LAMARCK had 3 specimens when he introduced the species, the largest shell was " 4 pouces et 4 lignes " in length (approximately 117 mm), which is unusually large for the species. This is probably the only specimen remaining in the MHNG type collection, although somewhat larger than originally cited. CERNOHORSKY (1967 : 188), following HEDLEY (1915 : 746), considered that the type of *M. capucinus* was a worn *C. torrefactus* but despite its excessively large size it undoubtedly represents *C. capucinus* of authors. The overall sculpture, the rounded, spineless varices, the aperture with characteristically small callus and straight, adherent columellar lip, and the broad siphonal canal, clearly distinguish the specimen from *C. torrefactus*.

Murex quadrifrons and *Murex lignarius* do not occur off West Africa as originally stated for *M. lignarius* (A. ADAMS, 1853 : 268) and in subsequent references to *M. quadrifrons*, and the west

African locality is thus erroneous. The lectotype of *M. lignarius* is a typical specimen of *C. capucinus*. *M. quadrifrons* and *M. castaneus* are merely rare variants with 4 or 5 varices. *M. quadrifrons* was already considered by REEVE (1845) to be a variant of *M. capucinus*.

Murex bituberculatus Baker was based on a shell localized " Australia ", with no other data. The holotype is 34 mm long, has 6 teleoconch whorls (plus 2 protoconch whorls), and is undoubtedly a juvenile example of *C. capucinus*. The short siphonal canal, light chocolate coloured, as well as the shape, the 2 intervaricial axial nodes, and the spiral sculpture are characteristic of that species. *M. bituberculatus* was synonymised with *M. capucinus* by FAIR (1976 : 26).

When describing *M. permaestus*, HEDLEY (1915) considered the large type specimen of *C. capucinus* (see above) to be a probable specimen of *C. torrefactus*.

D'ATTILIO (1988) considered that *C. capucinus* (as *Murex permaestus*) was a muricopsine species, due to the projecting central denticle on the central radular teeth. Study of the radular characters of several species of *Chicoreus* s.l., including *C. capucinus*, led to the observation that this so-called " abnormal " muricine radula is present in other *Chicoreus* species, including *C. brunneus*, and *C. torrefactus* (Figs 94-95, 102-103). Classification of *C. capucinus* in the Muricopsinae only based on radular characters is thus inadequate.

Genus *CHICOREUS* Montfort, 1810

Subgenus *SIRATUS* Jousseaume, 1880

Chicoreus (Siratus) alabaster (Reeve, 1845)
Figs 74, 219, 414-415

Murex alabaster Reeve, 1845 : pl. 10, fig. 39.

ADDITIONAL REFERENCES

Murex (Siratus) alabaster. — SHIKAMA, 1963 : 71, pl. 55, fig. 3 ; EISENBERG, 1981 : 87, pl. 69, fig. 5.
Siratus alabaster. — KAICHER, 1973 : card 162 ; RADWIN & D'ATTILIO, 1976 : 103, pl. 17, fig. 10 ; ABBOTT & DANCE, 1982 : 134, text fig. ; OKUTANI, 1983 : 8, pl. 23, fig. 4 ; KOSUGE, 1985a : 25, pl. 11, fig. 5 ; SPRINGSTEEN & LEOBRERA, 1986 : 130, pl. 35, fig. 7 ; RIPPINGALE, 1987 : 31, fig. 86.
Chicoreus (Siratus) alabaster. — FAIR, 1976 : 19, pl. 5, fig. 60 ; LAI, 1987 : 57, pl. 27, fig. 3.

TYPE LOCALITY. Island of Cagayan, Province of Misamis, Mindanao, Philippine Islands ; found on the beach.

TYPE MATERIAL. Holotype BMNH 1974079.

MATERIAL EXAMINED. Ca. 20 specimens from the Philippine Is and one specimen from Taiwan.

DISTRIBUTION. (Fig. 219). Only known from the Philippine Is, Taiwan and southeastern Japan.

DESCRIPTION. Shell up to 184 mm in length. Spire high, with 3 protoconch whorls and up to 9 teleoconch whorls. Suture impressed. Protoconch conical, broad, rounded, smooth.
Last whorl with 3 winged varices, each with webbing extending to the adapical part of siphonal canal. Intervaricial axial sculpture consisting of 2 or 3 low elongate nodes. Spiral sculpture of numerous major and minor threads that extend on to varical wings.

Aperture rounded, large. Columellar lip smooth, rim detached abapically, adherent on a small portion adapically. Anal notch obsolete. Outer lip weakly crenulate, weakly lirate for short distance within. Siphonal canal long, slender, narrowly open, bent abaxially, with 1 or 2 broadly open spines.
White to ivory-white with sometimes some pale brown traces on the spiral threads and on the intervaricial node.

REMARKS. A highly distinctive species, and the largest of the subgenus. RADWIN & D'ATTILIO (1976 : 103) incorrectly stated that the protoconch comprises 1 1/2 nuclear whorls ; in fact there are 3 whorls. The limited geographical range is quite unusual for a species that probably has

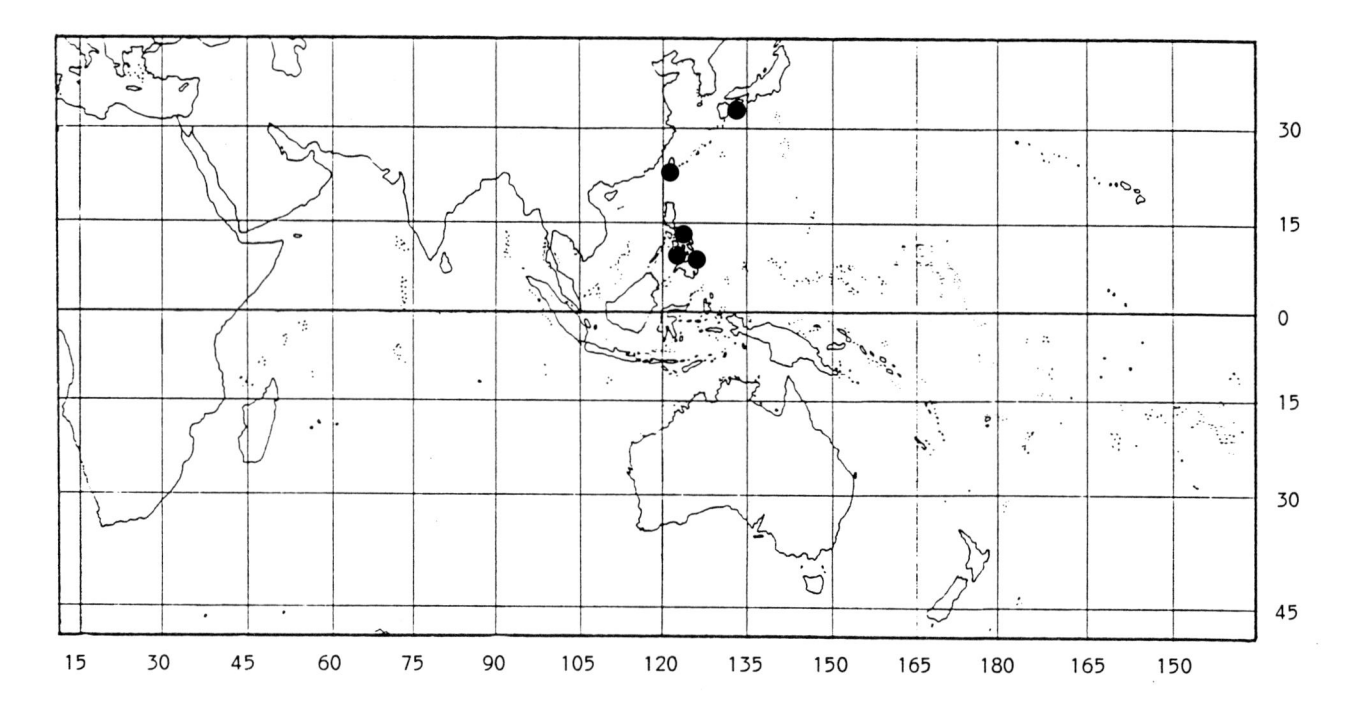

FIG. 219. — Distribution of *Chicoreus alabaster* (Reeve).

planktotrophic larval development, although the larval shell is large and certainly heavier than other multispiral protoconchs of species treated in this revision. Nevertheless it would be not surprising if the geographical distribution of this species proves to be wider than is currently known.

Chicoreus (Siratus) pliciferoides Kuroda, 1942
Figs 75, 114-115, 220, 416-420

Murex pliciferus Sowerby, 1841 : pl. 195, fig. 101 ; 1841b : 38.

Chicoreus pliciferoides Kuroda, 1942 : 81, new name for *Murex pliciferus* Sowerby, 1841 non *Murex pliciferus* Bivona-Bernardi, 1832.

Murex propinquus Kuroda & Azuma, in AZUMA, 1961 : 300, text fig. 7, 10.
Siratus hirasei Shikama, 1977 : 5, pl. 2, figs 9-12.
Siratus vicdani Kosuge, 1980 : 55, pl. 14, figs 1-2, 4.

ADDITIONAL REFERENCES

Murex (Siratus) pliciferoides propinquus. — SHIKAMA, 1964 : 119, pl. 64, fig. 9.
Siratus pliciferoides. — KIRA, 1965 : 61, pl. 23, fig. 16 ; KURODA, HABE & OYAMA, 1971 : 140, pl. 41, fig. 7 ; KAICHER, 1974 : card 539 ; RADWIN & D'ATTILIO, 1976 : 107, pl. 17, fig. 17 ; ABBOTT & DANCE, 1982 : 134, text fig. ; KOSUGE, 1985b : 59, pl. 23, fig. 3 ; SPRINGSTEEN & LEOBRERA, 1986 : 132, pl. 36, fig. 4.
Chicoreus (Siratus) pliciferoides. — FAIR, 1976 : 68, pl. 5, fig. 59 ; LAI, 1987 : 57, pl. 27, fig. 3.
Murex pliciferoides. — LEEHMAN, 1978c : 3, fig. 1.
Murex (Siratus) pliciferoides. — EISENBERG, 1981 : 92, pl. 74, fig. 4.
Siratus vicdani. — SPRINGSTEEN & LEOBRERA : 150, pl. 41, fig. 3.

TYPE LOCALITIES. *M. pliciferus* : None ; *M. propinquus* : Fishing area " Renkoba " off Tosa, Japan, 80 fms (146 m) ; *S. hirasei* : None ; *S. vicdani* : Punta Engano, Mactan Island, Philippines.

TYPE MATERIAL. *S. hirasei* : Holotype KPM 3326 ; *S. vicdani* : Holotype IMT 80-58 ; other material not located.

MATERIAL EXAMINED. C. 100 specimens from throughout the geographical range.

DISTRIBUTION. (Fig. 220). Northern New Caledonia (250-290 m) ; the Solomon Is (120 m) ; off Port Hedland, western Australia (500 m) ; the Philippine Is ; Taiwan ; southeastern and middle Japan with Boso Peninsula as north limit. Depth range : 50-500 m.

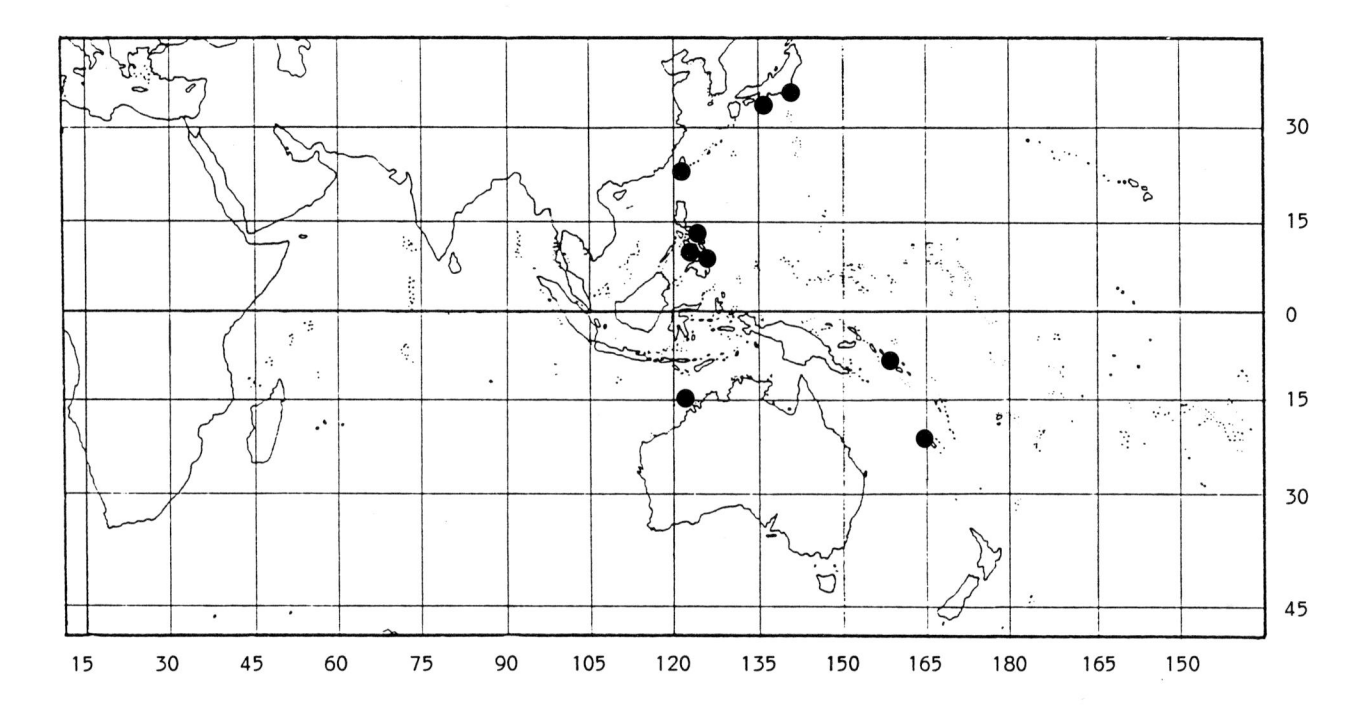

FIG. 220. — Distribution of *Chicoreus pliciferoides* Kuroda.

DESCRIPTION. Shell up to 146 mm in length, stout. Spire high, with 2 3/4-3 protoconch whorls and 9 up to teleoconch whorls. Suture deeply impressed. Protoconch conical, whorls rounded, smooth.

Last whorl with 3 rounded, spinose varices, each with sharp open spines, number of spines varying from one single small shoulder spine to one long shoulder spine followed by 5 or 6 other spines that progressively lengthen abapically ; some specimens completely spineless, others with webbed flange on each varix on last whorl. Intervaricial axial sculpture consisting of 2 or 3 shallow nodes, or sometimes 2 or 3 prominent ridges. Spiral sculpture of numerous primary and secondary cords.

Aperture rounded. Columellar lip smooth, rim erect abapically and adherent at adapically extremity. Anal notch moderately deep, sometimes delineated by small callus. Outer lip weakly crenulate, lirate for short distance within. Siphonal canal short to long, narrowly open, bent abaxially, with 1-3 broadly open spines, or smooth.

Ivory-white, sometimes with 2 or 3 bands of darker spiral cords, especially when juvenile, but they may persist in adult forms.

Radula (Figs 114-115).

REMARKS. *M. propinquus* is based on a variant with large body whorl, short siphonal canal, and short shoulder spines, while *S. hirasei* is a form with brown spiral cords and thick varices. Such forms, however intergrade with typical examples of the species. The holotype of *S. vicdani* is a juvenile specimen of *C. pliciferoides* with only 6 teleoconch whorls. D'ATTILIO & MYERS (1988) discussed and illustrated specimens from Japan, the Philippine Is, the Solomon Is and northwest Australia.

Genus *CHICOREUS* Montfort, 1810

Subgenus *CHICOPINNATUS* n. subgen.

Chicoreus (Chicopinnatus) guillei (Houart, 1985)
Figs 76, 221, 263, 411

Pterynotus (Pterynotus) guillei Houart, 1985 : 162, figs 1-1A.

ADDITIONAL REFERENCE

Pterynotus guillei. — HOUART, 1986a : 13, text figs (holotype and paratype).

TYPE LOCALITY. Off Réunion, 20°52′ S, 55°38′ E, 110 m.

TYPE MATERIAL. Holotype and 1 paratype MNHN.

MATERIAL EXAMINED. Type material.

DISTRIBUTION. (Fig. 221). Known only from the type locality.

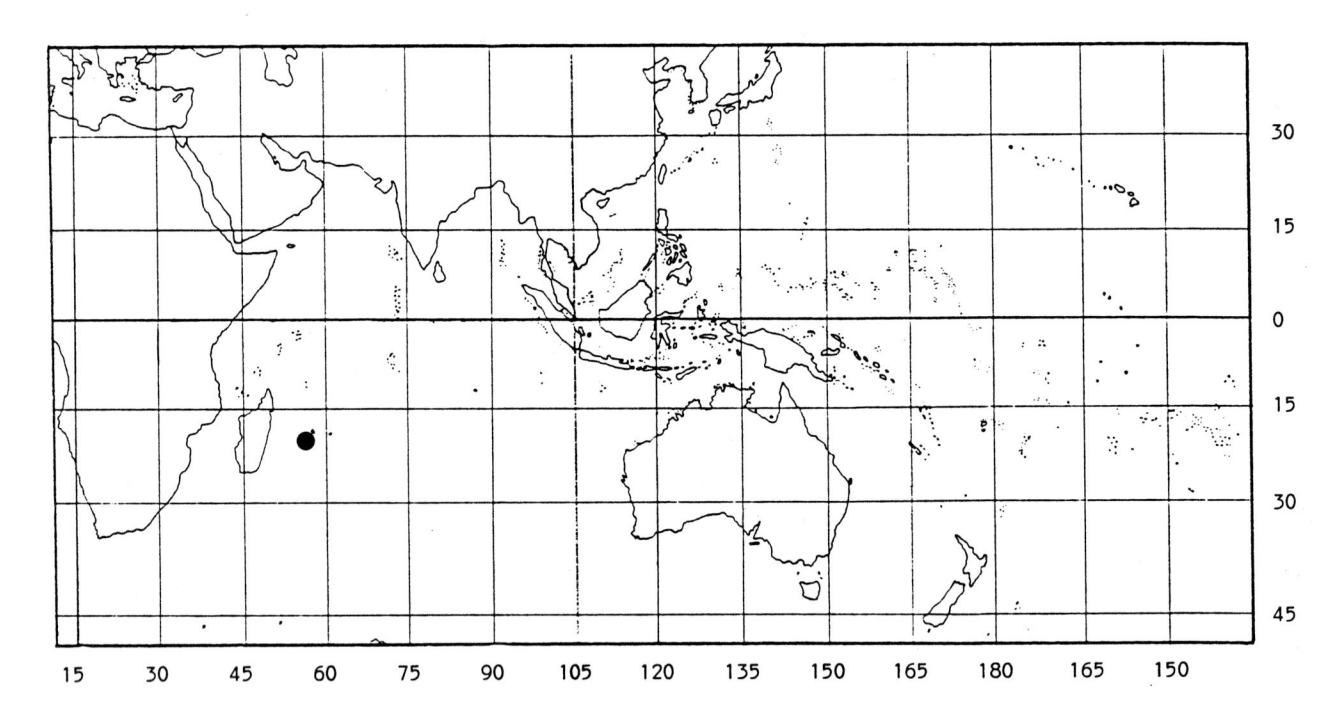

FIG. 221. — Distribution of *Chicoreus guillei* (Houart).

DESCRIPTION. Shell up to 33.2 mm in length. Spire high, with 2 protoconch whorls and up to 6 weakly shouldered teleoconch whorls. Suture appressed. Protoconch whorls rounded, smooth.

Last whorl with 3 broad varices, apertural varix with short spines at intersections with the spiral cords, connected by thin webbing that extends onto siphonal canal. Intervaricial axial sculpture consisting of a single strong node. Spiral sculpture of 12-14 cords, flanked by 2 minor threads.

Aperture ovate. Columellar lip smooth, rim adherent adapically, weakly erect abapically. Anal notch shallow, broad. Outer lip slightly erect, crenulate, lirate for short distance within. Siphonal canal long, straight, narrowly open, tip slightly bent abaperturally, with 2 or 3 short, broadly open spines, connected by thin webbing.

Light brown with darker coloured intervaricial nodes and medium part of the varices. Aperture white.

REMARKS. *C. guillei* differs from *C. orchidiflorus* in having a straight siphonal canal with thin webbing instead of long spines, in having a single, strong intervaricial node, in having more numerous spiral cords, and in the shape of the terminal varix of nuclear whorls (Figs 76 & 78).

Chicoreus (Chicopinnatus) laqueatus (Sowerby, 1841)
Figs 222, 412-413

Murex laqueatus Sowerby, 1841 : pl. 190, fig. 78 ; 1841b : 142.

ADDITIONAL REFERENCES

Pterynotus laqueatus. — FAIR, 1976 : 53, pl. 13, fig. 168 ; D'ATTILIO, 1981 : 78, figs 1-5 (syntype) ; ABBOTT & DANCE, 1982 : 141, text fig.
Marchia laqueata. — RADWIN & D'ATTILIO, 1976 : 58, pl. 9, fig. 5 ; MAC DONALD, 1979 : 8, fig. 3 ; KOSUGE, 1985a : 24, pl. 11, fig. 4.
Murex laqueatus. — LEEHMAN, 1977 : 4, text fig. ; HIGA, 1978 : 5 ; LEEHMAN, 1978d : 5, text fig. ; LEEHMAN, 1979 : 6, text fig.
Murex (Marchia) laqueata. — EARLE, 1980 : 1, text fig.
Naquetia laqueata. — KAICHER, 1980 : card 2587.

TYPE LOCALITY. Unknown.

TYPE MATERIAL. Holotype UMZ.

Material examined. Holotype (photographs only) ; Guam, RH (1 lv).

DISTRIBUTION. (Fig. 222). Off Okinawa ; Taiwan ; Kwajalein ; Guam ; Tahiti ; Hawaii ; under large boulders. Depth range : 15-45 m.

DESCRIPTION. Shell up to 39 mm in length. Spire high, protoconch unknown, up to 7 weakly angulate teleoconch whorls. Suture slightly appressed.

Last whorl with 3 winglike varices, adaperturally squamous, strongly folded. Intervaricial axial sculpture consisting of a single, strong node. Spiral sculpture of 7 or 8 cords, flanked on each side by a squamous thread. Cords and threads extending onto varices.

Aperture rounded. Columellar lip with 1 or 2 small nodules abapically, rim adherent. Anal notch narrow, shallow, delineated by small callus. Outer lip strongly denticulate, 10 strong similar lirae within. Siphonal canal narrowly open, long, straight, slightly bent abaperturally at tip, with 2 or 3 abapically bent, short, blunt spines.

Pink with shades of orange, yellow and violet on the spiral sculpture, spines and siphonal canal. Aperture pale violet. Outer and columellar lips pink, except the crenulations and the columellar knobs which are pale violet.

REMARKS. The (worn) protoconch and early teleoconch morphology suggest a placement in *Chicopinnatus* rather than in *Pterynotus*, where it is usually placed. D'ATTILIO (1981) illustrated the holotype and other shells from Guam, and gave a detailed description. He also illustrated the radula, which has a strongly projecting central cusp as in Muricopsinae. Radulae with a projecting central cusp, however, occur in several other species of *Chicoreus* (*s.l.*).

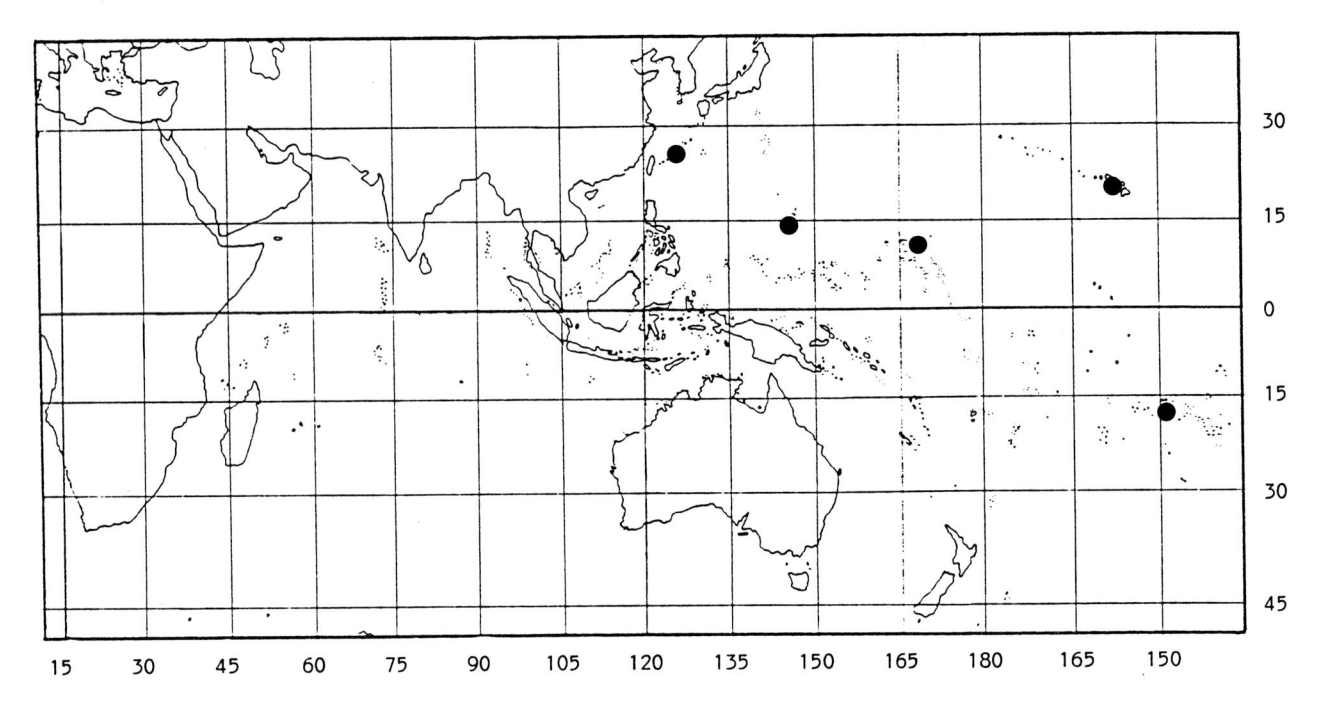

FIG. 222. — Distribution of *Chicoreus laqueatus* (Sowerby).

Chicoreus (Chicopinnatus) orchidiflorus (Shikama, 1973)
Figs 77-78, 129, 223, 262, 406-410

Pterynotus orchidiflorus Shikama, 1973 : 5, pl. 2, figs 7-8.

Chicoreus subtilis Houart, 1977 : 13, figs 1-5.
Pterynotus cerinamarumai (incorrect original spelling) Kosuge, 1980 : 53, pl. 14, figs 3, 5-9, pl. 15, figs 1-2.
Pterynotus celinamarumai (justified emendation) Kosuge, 1985 : 27.

ADDITIONAL REFERENCES
Pterynotus orchidiflorus. — LAN, 1980 : 71, pl. 29, figs 63-65; OKUTANI, 1983 : 8, pl. 24, fig. 4 (holotype); RIPPINGALE, 1987 : 29, fig. 82.
Murex orchidifloris (sic). — LEEHMAN, 1980b : 11, text fig.
Chicoreus orchidifloris (sic). — HOUART, 1981a : 9, text fig.; HOUART, 1981c : 16, text fig. p. 17.
Chicoreus orchidiflorus. — HOUART, 1986c : 763, figs 9-9b.
Chicoreus (Chicoreus) orchidiflorus. — SPRINGSTEEN & LEOBRERA, 1986 : 130, pl. 30, fig. 11.
Chicoreus cerinamarumai. — HOUART, 1981c : 16, text fig. p. 17.
Chicoreus (Chicoreus) celinamarumai. — SPRINGSTEEN & LEOBRERA, 1986 : 130, pl. 30, fig. 12.
Pterynotus celinamarumai. — DRIVAS & JAY, 1988 : 70, pl. 20, fig. 7.
Pterynotus orchidiformis (sic). — ABBOTT & DANCE, 1982 : 138, text fig.

TYPE LOCALITIES. *P. orchidiflorus* : none, Taiwan here designated; *C. subtilis* : northern-east of Taiwan; *P. celinamarumai* : off Bohol Island, Philippine Islands, 100-150 m.

TYPE MATERIAL. *P. orchidiflorus* : Holotype NSMT 60927; *C. subtilis* : Holotype IRSNB IG 25708; *P. celinamarumai* : Holotype IMT 80-55.

OTHER MATERIAL EXAMINED. C. 40 specimens from throughout the geographical range.

DISTRIBUTION. (Fig. 223). Réunion and Mauritius; Taiwan; Philippine Is; New Caledonia; Tubuaï, Austral Is, French Polynesia. Depth range : c. 150 m (sublittoral-minimum depth unknown).

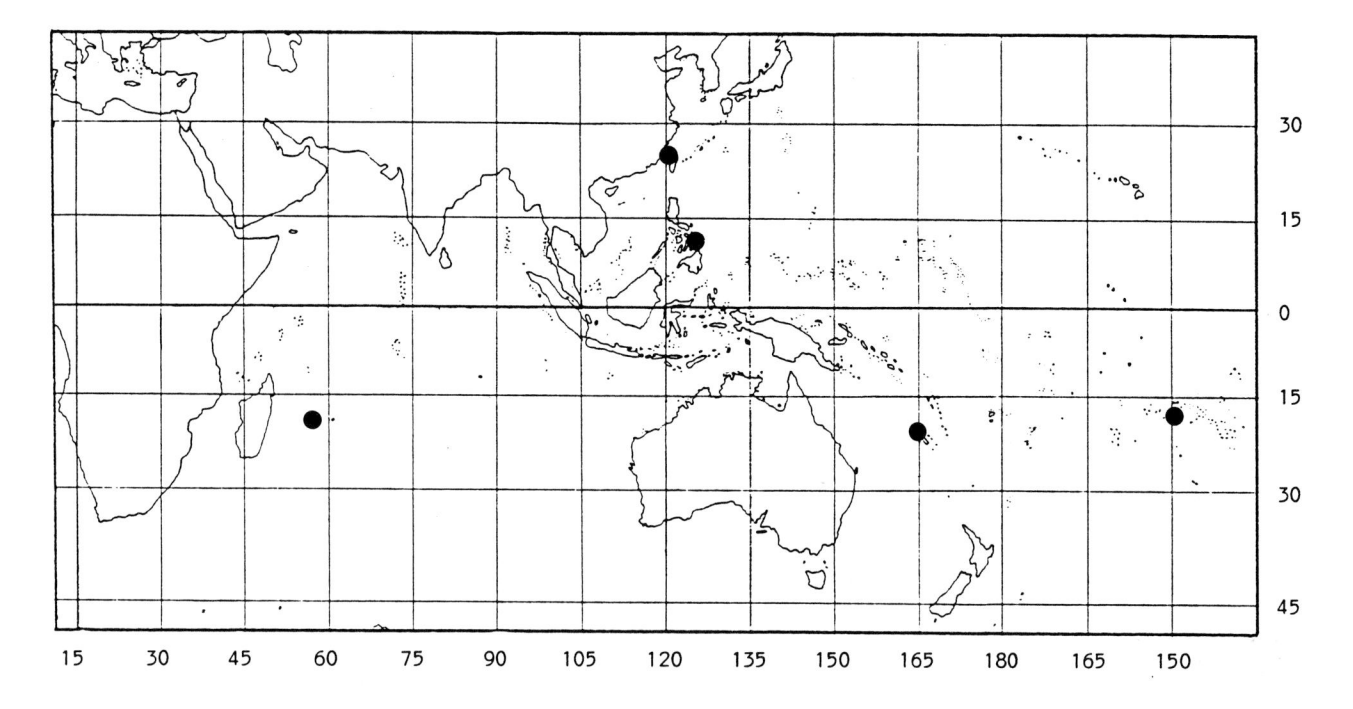

FIG. 223. — Distribution of *Chicoreus orchidiflorus* (Shikama).

DESCRIPTION. Shell up to 40 mm in length, fragile. Spire high, with 1 3/4-2 protoconch whorls and up to 7 weakly convex teleoconch whorls. Suture impressed or slightly appressed. Protoconch whorls rounded and glossy.
Last whorl with 3 winglike, finely folded varices, adapical and abapical parts divided or joined by a thin webbing. Intervaricial axial sculpture consisting of 2 or 3 low ridges. Spiral sculpture of 7-9 cords and 2 or 3 intermediate squamous threads in each interspace.

Aperture roundly-ovate. Columellar lip smooth or with 1 or 2 small nodules abapically, rim adherent, weakly detached abapically. Anal notch large, rather shallow. Outer lip erect, denticulate, strongly lirate for short distance within. Siphonal canal long, curved, tip bent abaperturally, narrowly open, with 2 or 3 abapically bent, open, squamous spines.
Varying from uniform white to yellowish or pure orange. Radula (Fig. 129).

REMARKS. It is here appropriate to designate Taiwan as type locality, which is also the type locality of the synonymous *C. subtilis* Houart.

Pterynotus celinamarumai is the form with non-divided varicial wings.

Specimens dredged off New Caledonia (HOUART : 1987) (AMS and MNHN) have heavier shells with shorter varicial flanges (Fig. 409), but the spiral and axial sculpture, as well as the protoconch, are the same as in the typical form.

Genus **CHICOMUREX** Arakawa, 1964

Chicomurex Arakawa, 1964 : 361.

Type-species (by original designation) : *Murex superbus* Sowerby, 1889.

DESCRIPTION. Shell up to 85 mm in length, with 3 more or less spinose varices, adapically webbed ; aperture rounded, with a small anal sulcus ; outer lip striate for short distance within ; siphonal canal medium-sized.

REMARKS. The decision to consider *Chicomurex* and *Naquetia* as genera separate from *Chicoreus* is based on differences in radular morphology. Both genera have crowded rows of teeth along the radula with a large and broad triangular central cusp on the rachidian, unlike those in *Chicoreus* (*s.s.*) and its subgenera.

I hesitated to treat some currently recognized species of *Chicomurex* as distinct species, because they have identical protoconchs and are otherwise extremely similar, the decision to treat *C. problematicus*, *C. superbus* and *C. venustulus* as distinct is still a matter of personal opinion.

Chicomurex elliscrossi (Fair, 1974)
Figs 224, 269

Chicoreus elliscrossi Fair, 1974 : 1, text fig. 2.

ADDITIONAL REFERENCES

Naquetia cf. *laciniatus*. — D'ATTILIO, 1966: 4, fig. 2 (not *Murex laciniatus* Sowerby, 1841).
Chicoreus laciniatus. — HABE, 1968 : 80, pl. 25, fig. 16 (not *Murex laciniatus* Sowerby, 1841).
Chicoreus elliscrossi. — KAICHER, 1974 : card 511.
Chicoreus (Chicomurex) elliscrossi. — FAIR, 1976 : 39, pl. 14, fig. 173.
Murex elliscrossi. — FAIR, 1981 : fig. 5.
Siratus elliscrossi. — ABBOTT & DANCE, 1982 : 133, text fig.
Chicomurex elliscrossei (sic). — RIPPINGALE, 1987 : 11, fig. 28.
Phyllonotus superbus. — RADWIN & D'ATTILIO, 1976 : 92 (in part), pl. 6, fig. 1 (not *Murex superbus* Sowerby, 1879).

TYPE LOCALITY. Off Kii Peninsula, Japan.

TYPE MATERIAL. Holotype USNM 709574 (not seen).

MATERIAL EXAMINED. Off Saeki City, Japan, 100 m, RH (1 lv); off Nada, Gobo City, Wakayama Pref., Japan, in lobster net, RH (4 lv).

DISTRIBUTION. (Fig. 224). Off south-eastern Japan. Depth range : 50-100 m.

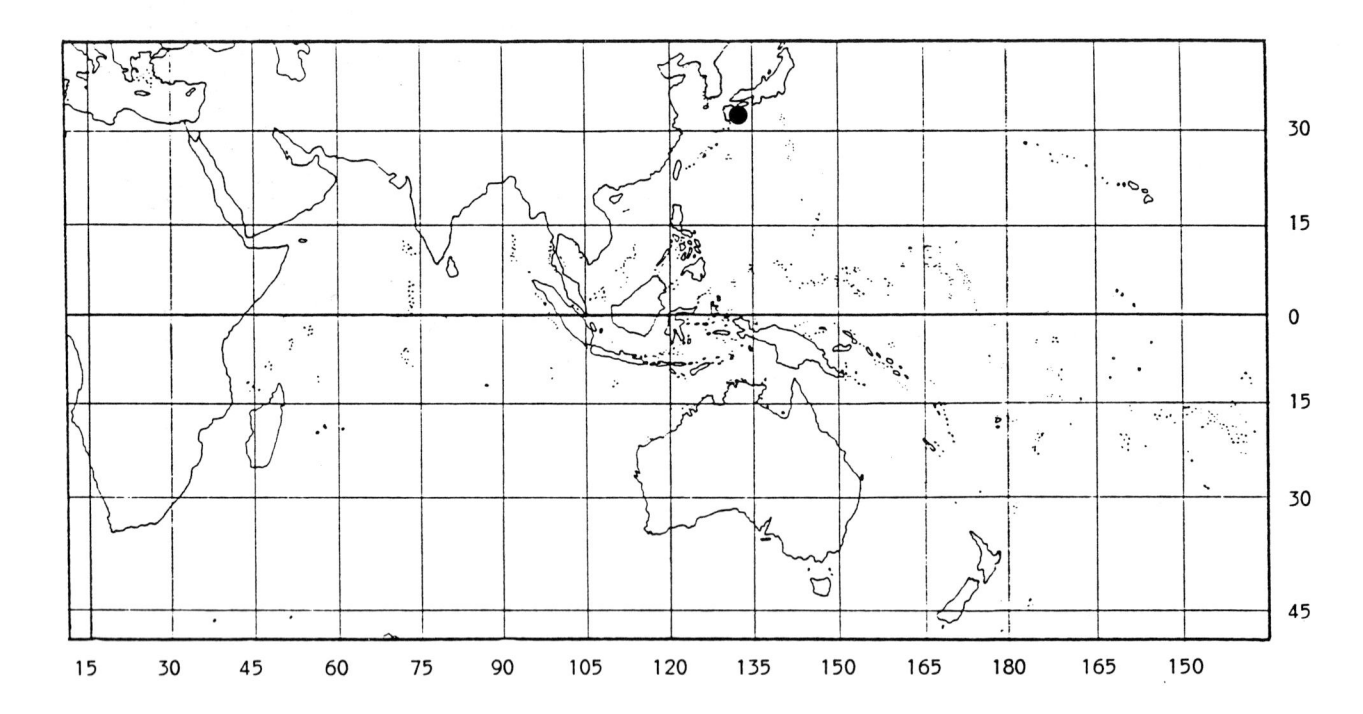

FIG. 224. — Distribution of *Chicomurex elliscrossi* (Fair).

DESCRIPTION. Shell up to 78 mm in length, stout. Spire moderately high, up to 8 teleoconch whorls, protoconch unknown. Suture slightly appressed.

Last whorl with 3 rounded varices, adapically spinose and abapically webbed between spines; spines usually short to obsolete. Intervaricial axial sculpture consisting of 2 or 3 low, broad ridges. Spiral sculpture of 7 or 8 obsolete cords and numerous obsolete threads, strongest and squamous on varices.

Aperture rounded, broad. Columellar lip smooth, rim erect abapically, adherent adapically. Anal notch narrow, shallow, delineated by small, elongate callus. Outer lip finely denticulate; strongly lirate for short distance within. Siphonal canal moderately long, broad, narrowly open, with 3 abapically bent open spines, most adapical spine bent abaperturally.

Whitish or cream with pale orange or brown spots. Aperture porcellaneous white.

REMARKS. This species originally was misidentified as *Chicomurex laciniatus* (Sowerby) by japanese authors. RADWIN & D'ATTILIO (1976 : pl. 6, fig. 1) identified a specimen of *C. elliscrossi* as *C. superbus*, but *C. elliscrossi* has a broader shell, with stronger intervaricial nodes, a broader aperture, more shouldered and thicker shell, narrower and straighter columellar lip, smaller anal notch, delineated with a small callus, a lower spire and broader siphonal canal.

FIGS 225-227. — Genus *Chicomurex*.
225, *C. venustulus* (Rehder & Wilson). Off south-western coasts of Tahuata, Marquesas Is, 40.5 mm (holotype, USNM 707241. Courtesy of H. Rehder).
226, *C. laciniatus* (Sowerby). Philippines, 54.5 mm (RH).
227, *C. superbus* (Sowerby). Taiwan, 80 mm (RH).

Chicomurex laciniatus (Sowerby, 1841)

Figs 71, 120-121, 226, 228, 267, 421-423

Murex laciniatus Sowerby, 1841 : pl. 187, fig. 59.
Murex scabrosus Sowerby, 1841 : pl. 189, fig. 73 ; 1841b : 140.
Chicoreus filialis Shikama, 1971 (*Chicoreus filiaris* on plate) : 29, pl. 3, figs 3, 4.

ADDITIONAL REFERENCES

Chicoreus (Chicomurex) laciniatus. — ARAKAWA, 1964 : 362, text fig. 2 ; HOUART, 1986c : 762,fig. 4 ; SPRINGSTEEN & LEOBRERA, 1986 : 132, pl. 36, fig. 2.
Chicoreus laciniatus. — CERNOHORSKY, 1967a : 119, fig. 3, pl. 4, fig. 9 ; FAIR, 1974a : 1, fig. 1 (lectotype) ; CERNOHORSKY, 1978b : 65, pl. 18, fig. 3.
Naquetia laciniatus. — KAICHER, 1973 : card 169.
Chicoreus (Chicoreus) laciniatus. — FAIR, 1976 : 53, pl. 7, fig. 934 (lectotype).
Phyllonotus laciniatus. — RADWIN & D'ATTILIO, 1976 : 89, pl. 6, fig. 3 ; HINTON, 1979 : 27, fig. 1.
Murex (Phyllonotus) laciniatus. — EISENBERG, 1981 : 90, pl. 72, fig. 17.
Siratus laciniatus. — ABBOTT & DANCE, 1982 : 133, text fig.
Phyllonotus superbus. — HINTON, 1979 : 27, fig. 2 (not *Murex superbus* Sowerby, 1889).
Chicoreus filiaris. — OKUTANI, 1983 : 8, pl. 24, figs 2, 3.
NOT *Naquetia* cf. *laciniatus.* — D'ATTILIO, 1966 : 4, fig. 2 [= *Chicomurex elliscrossi* (Fair, 1974)].
NOT *Chicoreus laciniatus.* — HABE, 1968 : 80, pl. 25, fig. 16 [= *Chicomurex elliscrossi* (Fair, 1974)].

TYPE LOCALITIES. *Murex laciniatus* and *Murex scabrosus* : none ; *C. filialis* : off Taiwan.

TYPE MATERIAL. *M. laciniatus* : lectotype BMNH 1974072/1, here selected from 2 syntypes ; *M. scabrosus* : not located ; *C. filialis* : holotype KPM 3334 (not seen).

OTHER MATERIAL EXAMINED. C. 100 specimens from throughout the geographical range.

DISTRIBUTION. (Fig. 228). Southern Africa, north Zululand (NM D6819) ; Seychelles ; Sri Lanka ; Indonesia ; the Philippine Is ; Taiwan and southern Japan ; Marshall Is ; Papua New Guinea ; Queensland, Australia ; New Caledonia ; Fiji Is ; under coral and on Sand. Depth range : 40-200 m.

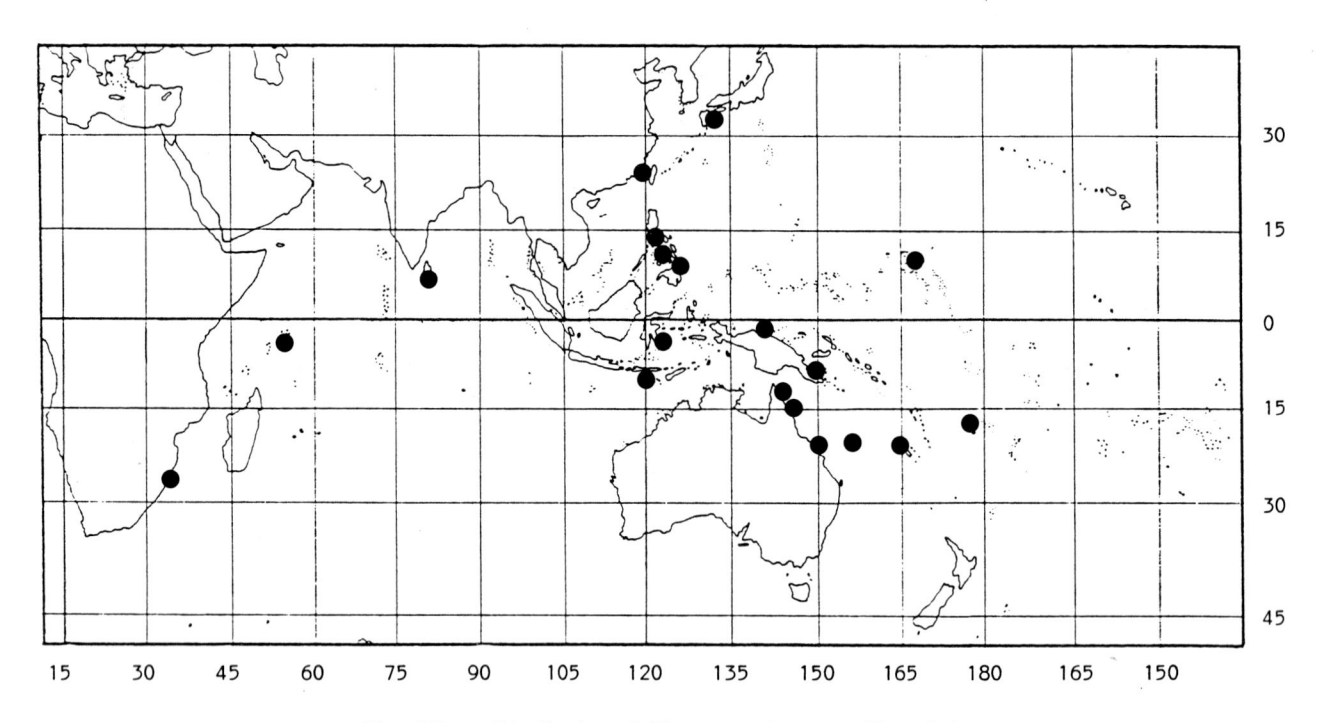

FIG. 228. — Distribution of *Chicomurex laciniatus* (Sowerby).

DESCRIPTION. Shell up to 77 mm in length. Spire high, with 2 3/4 protoconch whorls and up to 8 teleoconch whorls. Suture impressed. Protoconch conical, whorls rounded, glossy.

Last whorl with 3 rounded, squamous varices. Intervaricial axial sculpture consisting of 2 moderately high to low elongate nodes. Spiral sculpture of 7 squamous primary cords and intermediate secondary cords and threads in each interspace, small, open, squamous spinelets, strongest on abapical part of each varix.

Aperture rounded. Columellar lip smooth, rim weakly erect abapically, adherent adapically. Anal notch narrow, shallow, delineated by elongate callus. Outer lip denticulate, lirate for short distance within. Siphonal canal large, moderately short and narrowly open, tip bent abaperturally, with 3 short spines, adapical or two adapical spines weakly bent abaperturally.

Generally light brown with darker varices, occasionally orange, white, or pale brown, 3 paler bands frequently present, most conspicuous on varices. Aperture white, columellar lip violet or pink.

Radula (Figs 120-121).

REMARKS. A highly distinctive species. Some shells, particularly from off Queensland and the Coral Sea may be paler coloured and more narrowly elongate, but exhibit no other differences, and seemingly, all present *C. laciniatus*. Specimens from off Queensland closely resembles *C. venustulus* (Fig. 423) from which they differ, however, in having a larger aperture, lower axial nodes, a narrower columellar lip and a broader siphonal canal.

C. filialis is a synonym and was introduced as the result of confusion between *C. laciniatus* and *C. elliscrossi* by Japanese authors.

Chicomurex problematicus (Lan, 1981)
Figs 229, 266

Phyllonotus superbus problematicum Lan, 1981 : 11, figs 1-4.

ADDITIONAL REFERENCES

Siratus superbus. — ABBOTT & DANCE, 1982 : 133, text fig. (not *Murex superbus* Sowerby, 1889).
Murex superbus. — COUCOM, 1983 : 1, text fig. (not *Murex superbus* Sowerby, 1889).
Phyllonotus superbus. — OKUTANI, 1983 : 8, pl. 24, fig. 10.
Chicoreus (Chicomurex) superbus problematicum. — HOUART, 1984a : 12, text fig.
Chicoreus (Chicomurex) superbus problematicus. — SPRINGSTEEN & LEOBRERA, 1986 : 151, text fig. (holotype).
Chicoreus superbus problematicus. — LAI, 1987 : 63, pl. 30, fig. 2.
Chicomurex problematica. — RIPPINGALE, 1987 : 11, fig. 29.

TYPE LOCALITY. Off Bohol, Cebu, Philippine Island, 300 m.

TYPE MATERIAL. Holotype Taiwan Museum, Tapei (not seen).

MATERIAL EXAMINED. C. 50 specimens from throughout the geographical range.

DISTRIBUTION. (Fig. 229). Philippine Is and Taiwan, generally associated with corals; Australia : North Queensland and Capricorn Channel, Queensland. Depth range : 80-100 m.

DESCRIPTION. Shell up to 75 mm in length. Spire moderately high, with 3-3 1/2 protoconch whorls and up to 8 rounded teleoconch whorls. Suture impressed. Protoconch conical, whorls smooth, glossy.

Last whorl with 3 thick, rounded varices; adapically spinose, abapically webbed between spines. Intervaricial axial sculpture consisting of 2, occasionally 3, elongate ridges. Spiral sculpture of 15 or 16 cords and intermediate threads in each interspace.

Aperture roundly-ovate. Columellar lip weakly folded, rim erect abapically, adherent at adapical extremity. Anal notch small shallow, delineated by small elongate callus. Outer lip denticulate, lirate for short distance within. Siphonal canal long, narrowly open, bent abaperturally, with 3 or 4 broadly open spines that are webbed between.

Pale brown or ochre with darker coloured spiral cords and abaperturally side of siphonal canal.

REMARKS. Although originally proposed as a subspecies of *C. superbus*, they are clearly distinct species since they are sympatric and to my knowledge, do not intergrade. *C. problematicus* differs from *C. superbus* in having a broader, more angulate shell, a smaller aperture and straighter columellar lip, while the spiral cords are more numerous and finer. The siphonal canal is ornamented with thin convoluted webbing instead of 3 open spines as in *C. superbus*.

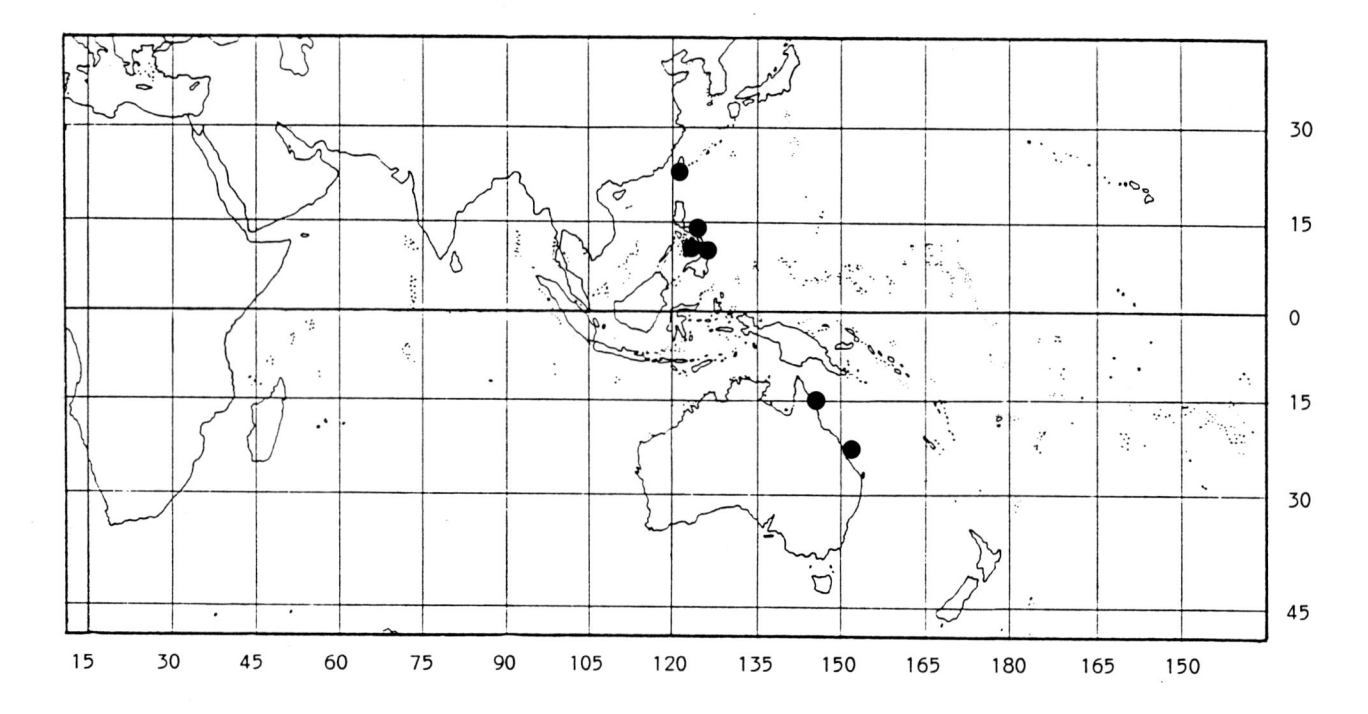

FIG. 229. — Distribution of *Chicomurex problematicus* (Lan).

C. problematicus differs from *C. elliscrossi* in its smaller aperture, folded columellar lip, more numerous spiral cords, and longer, narrower siphonal canal, which is ornamented with folded webbing instead of spines.

Chicomurex protoglobosus n. sp.
Figs 72-73, 126-127, 230, 427

TYPE MATERIAL : Holotype MNHN.

TYPE LOCALITY : Off SW New Caledonia, 22°46' S, 167°20' E, 300 m, MUSORSTOM 4, stn DW 227, collected by Bouchet and Richer de Forges, 30 September 1985 aboard R. V. *Vauban*.

MATERIAL EXAMINED : only known from the holotype.

DESCRIPTION. Shell 30.1 mm in length, broad. Spire high with 1 1/2 protoconch whorls and 4 1/4 teleoconch whorls. Suture impressed. Protoconch large, whorls rounded and glossy ; terminal varix weakly curved and erect. Teleoconch whorls convex. First teleoconch whorl with 15 rounded axial ridges ; second whorl with 3 varices and 4 or 5 intervaricial axial ridges ; last whorl with 3 varices and 2 or 3 intervaricial axial ridges. Varices with small fronds or spines ; shoulder spine longest. Spiral sculpture of first teleoconch whorl consisting of 4 cords ; second with 7 unevenly spaced cords ; penultimate whorl with 15 irregular cords and threads, and last whorl with many nodulose cords and threads. Aperture rather large, roundly-ovate, outer lip thin (immature). Siphonal canal moderately long, open and slightly bent abaperturally.

Ochre with some brown maculations on shoulder, varices, and spiral sculpture.

Radula (Figs 126-127). With along crowded rows of teeth, typical for *Chicomurex*, lateral tooth sickle-shaped, central tooth with a large central triangular cusp, two smaller marginal cusps, and two minor lateral cusps.

REMARKS. *Chicomurex protoglobosus* differs from all other species of *Chicomurex* in its large and globose protoconch, most probably indicating intracapsular metamorphosis. *C. turschi* has a

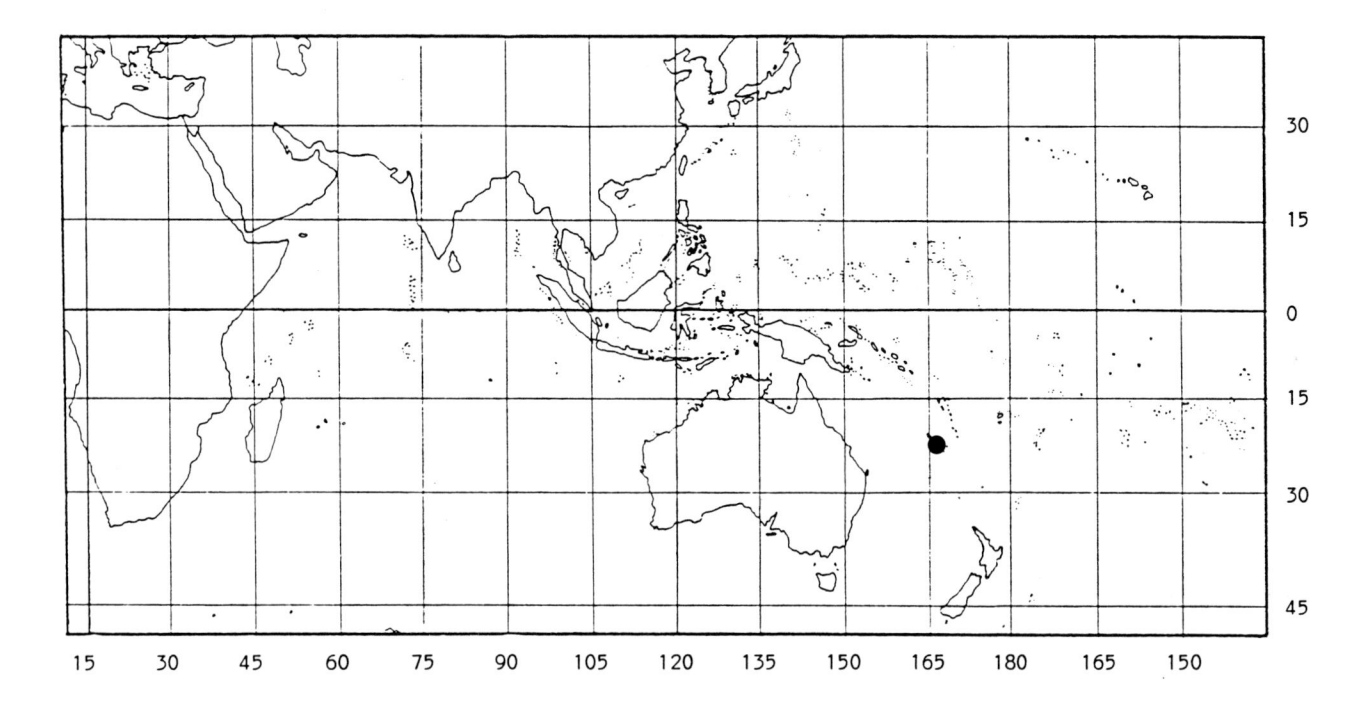

FIG. 230. — Distribution of *Chicomurex protoglobosus* n.sp.

small, rounded, paucispiral protoconch, and shell with 4 1/4 teleoconch whorls like the holotype of *C. protoglobosus* are only 13 or 14 mm in length. All other species of *Chicomurex* have conical, multispiral protoconchs. Judging from its paucispiral protoconch, *C. protoglobosus* is probably endemic to the New Caledonian area.

Chicomurex superbus (Sowerby, 1889)
Figs 67, 124-125, 227, 231, 424

Murex superbus Sowerby, 1889 : 565, pl. 28, figs 10-11.

ADDITIONAL REFERENCES

Murex (Chicoreus) supersus (sic). — SHIKAMA, 1963 : 70, pl. 54, fig. 8.
Chicoreus (Chicomurex) superbus. — ARAKAWA, 1964 : 361, pl. 21, fig. 5-6 (radula) : 362, text fig. 1 (radula); FAIR, 1976 : 79, pl. 14, fig. 174; SPRINGSTEEN & LEOBRERA, 1986 : 132, pl. 36, fig. 1.
Chicoreus superbus. — D'ATTILIO, 1966 : 4, fig. 1; FAIR, 1974a : 1, fig. 3.
Phyllonotus superbus. — RADWIN & D'ATTILIO, 1976 : 92 (in part), pl. 6, fig. 2.
Murex (Phyllonotus) superbus. — EISENBERG, 1981 : 94, pl. 76, fig. 4.
Murex superbus. — LAN, 1981 : fig. 6.
Chicomurex superbus. — RIPPINGALE, 1987 : 11, fig. 30.
NOT *Chicoreus superbus.* — KAICHER, 1973 : card 139; LAI, 1987 : 63, pl. 30, fig. 1; DRIVAS & JAY, 1988 : 70, pl. 20, fig. 3 [= *Chicomurex venustulus* (Rehder & Wilson, 1975)].
NOT *Phyllonotus superbus.* — RADWIN & D'ATTILIO, 1976 : 92 (in part), pl. 6, fig. 1 [= *Chicomurex elliscrossi* (Fair, 1974)].
NOT *Phyllonotus superbus.* — HINTON, 1979 : 27, fig. 2 [= *Chicomurex laciniatus* (Sowerby, 1841)].
NOT *Siratus superbus.* — ABBOTT & DANCE, 1982 : 133, text fig. [= *Chicomurex problematicus* (Lan, 1981)].
NOT *Murex superbus.* — COUCOM, 1983 : 1, text fig. [= *Chicomurex problematicus* (Lan, 1981)].
NOT *Phyllonotus superbus.* — OKUTANI, 1981 : 8, pl. 24, fig. 10 [= *Chicomurex problematicus* (Lan, 1981)].

TYPE LOCALITY. Hong Kong.

TYPE MATERIAL. Holotype NMW 1955.158.16 (not seen).

MATERIAL EXAMINED. C. 30 specimens from throughout the geographical range.

DISTRIBUTION. (Fig. 231). Southeastern Japan ; northeastern Taiwan ; the Philippine Is ; New Caledonia ; Australia : off Lord Howe I. ; Capricorn Channel, Queensland. Depth range : 80-325 m.

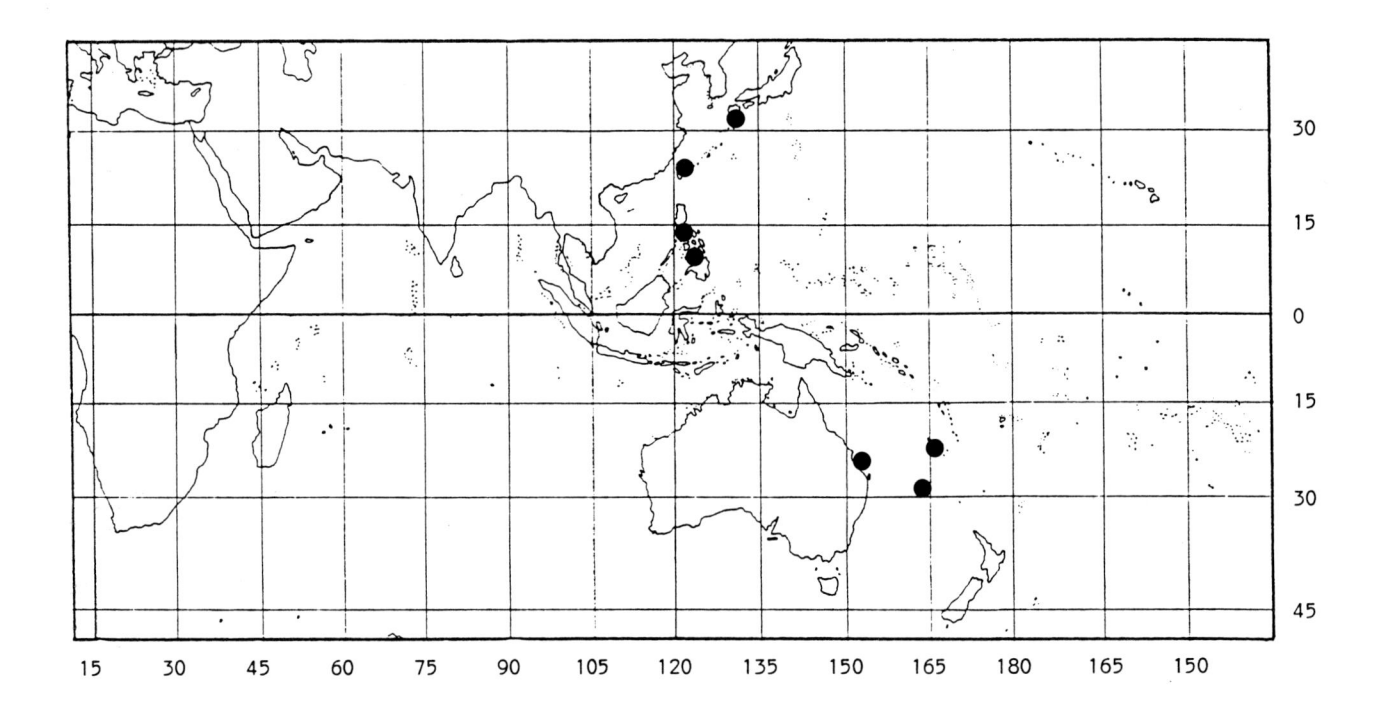

FIG. 231. — Distribution of *Chicomurex superbus* (Sowerby).

DESCRIPTION. Shell up to 85 mm in length. Spire high, narrowly conical, with 3-3 1/2 protoconch whorls and up to 8 rounded teleoconch whorls. Suture impressed. Protoconch conical, whorls smooth, glossy.
Last whorl with 3 rounded varices ; each adapically spinose and abapically webbed. Intervaricial axial sculpture consisting of 2, occasionally 3 nodes or axial costae. Spiral sculpture of 7 or 8 major cords (including shoulder sculpture) with minor squamous cords and threads in each interspace.
Aperture roundly-ovate. Columellar lip broad, smooth, rim erect, adherent adapically. Anal notch broad, shallow. Outer lip erect, weakly undulate, lirate for short distance within. Siphonal canal moderately long, slightly bent abaperturally, narrowly open, with 3 slightly abapical bent spines, most abapical spine sometimes very weakly bent abaperturally.
White or cream with orange or brown spots. Aperture white.
Radula (Figs 124-125).

REMARKS. This is a highly distinctive species. For comparisons with *C. elliscrossi*, *C. problematicus* and *C. venustulus*, see remarks on these species.

Chicomurex turschi (Houart, 1981)
Figs 69-70, 122-123, 232, 264, 431-433

Chicoreus (Chicomurex) turschi Houart, 1981 : 186, figs 1-6.

ADDITIONAL REFERENCES

Phyllonotus sp. — HINTON, 1979 : 27, fig. 3.
Chicoreus (Chicomurex) turschi. — HOUART, 1984a : 12, text fig.
Chicomurex turschi. — RIPPINGALE, 1987 : 13, fig. 31.

TYPE LOCALITY. Off Durangit, Hansa Bay, Papua New Guinea, 45-60 m, bottom with sponges.

TYPE MATERIAL. Holotype IRSNB 374.

OTHER MATERIAL EXAMINED. C. 40 specimens from throughout the geographical range.

DISTRIBUTION. (Fig. 232). Zululand, South Africa (NM E4300); Madagascar; the Philippine Is (no other data); Papua New Guinea; south of New Caledonia. Depth range : 45-75 m.

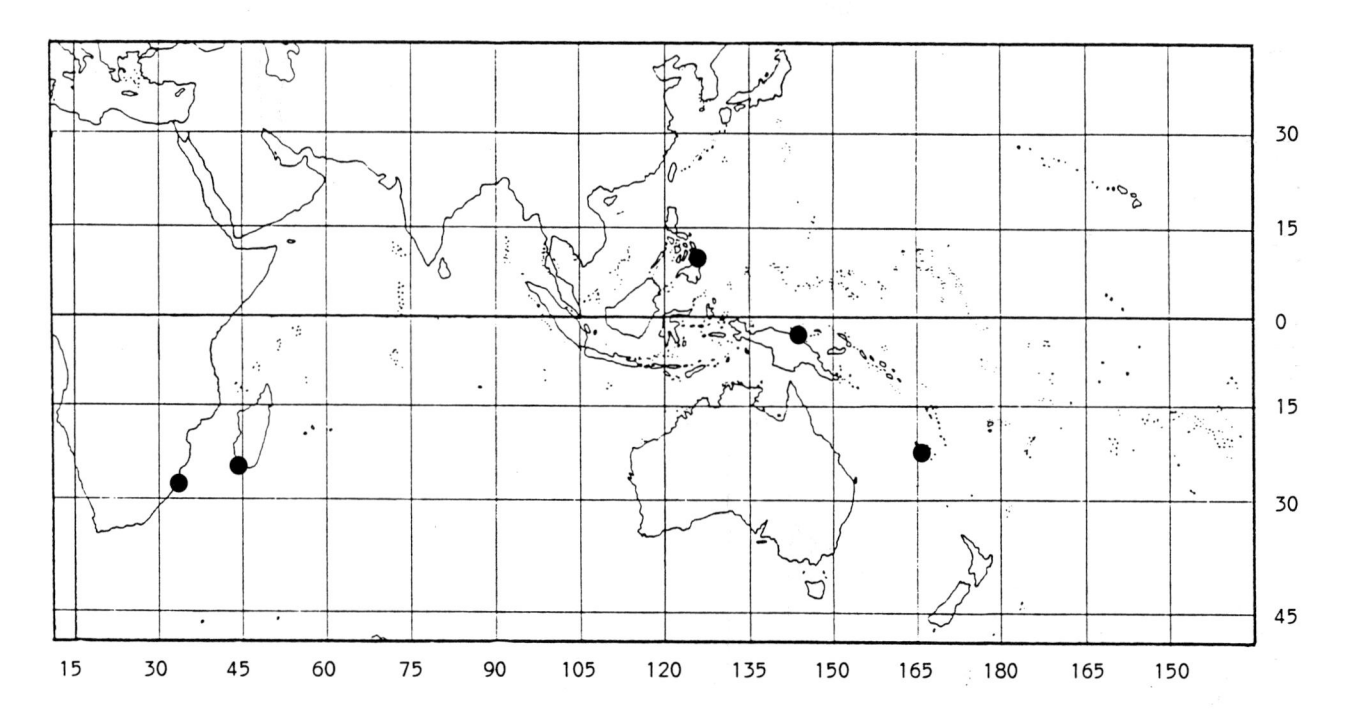

FIG. 232. — Distribution of *Chicomurex turschi* (Houart).

DESCRIPTION. Shell up to 40 mm in length, stout. Spire high, with 1 1/2 protoconch whorls and up to 7 teleoconch whorls. Suture impressed. Protoconch whorls smooth, glossy.

Last whorl with 3 rounded, squamose varices, slightly webbed between spines abapically. Intervaricial axial sculpture consisting of 2 or 3 elongate nodes, strongest on shoulder. Spiral sculpture of 7 or 8 primary cords, secondary cords, and numerous threads that extend onto varices.

Aperture ovate, weakly angulate. Columellar lip abapically with 3 or 4 small denticles, folded adapically, rim adherent adapically, detached abapically. Anal notch shallow, delineated by elongate callus. Outer lip denticulate, lirate within. Siphonal canal narrowly open, of moderate length, weakly bent abaperturally at tip, with 3 short, open spines, adapical spine slightly bent abaperturally.

Cream or light brown with 3 darker bands, most apparent on varices; brown spots on suture and sometimes on axial node. Occasionally entirely white or orange. Aperture bluish-white.

Radula (Figs 122-123).

REMARKS. *Chicomurex turschi* was originally compared with *C. venustulus* and is clearly distinguishable from this and other species of the genus by its paucispiral protoconch and small adult shell size.

Chicomurex venustulus (Rehder & Wilson, 1975)
Figs 68, 116-119, 225, 233, 265, 268, 425-426, 428-430

Chicoreus (Chicomurex) venustulus Rehder & Wilson, 1975 : 7, figs 4, 5, frontispiece figs 2, 3.

Chicoreus gloriosus Shikama, 1977 : 14, pl. 2, fig. 8.

ADDITIONAL REFERENCES

Chicoreus superbus. — KAICHER, 1973 : card 139 ; LAI, 1987 : 63, pl. 30, fig. 1 ; DRIVAS & JAY, 1988 : 70, pl. 20, fig. 3 [not *Chicoreus (Chicomurex) venustulus* REHDER & WILSON, 1975].
Chicoreus venustulus. — CERNOHORSKY, 1978b : 65, pl. 18, fig. 5 (holotype) ; KAICHER, 1979 : card 1596.
Chicomurex venustulus. — HOUART, 1981a : 10, text fig. p. 7 ; RIPPINGALE, 1987 : 13, fig. 32.
Chicoreus (Chicomurex) venustulus. — HOUART, 1981d : figs 7-9 (holotype) ; SPRINGSTEEN & LEOBRERA, 1986 : 132, pl. 36, fig. 3.
Siratus venustulus. — ABBOTT & DANCE, 1982 : 133, text fig.
Phyllonotus sp. — HINTON, 1979 : 27, fig. 4.

TYPE LOCALITIES. *C. venustulus* : off south-western coasts of Tahuata, Marquesas Islands, 36-39 fms (66-71 m) ; *C. gloriosus* : off Cebu island, Philippine Islands.

TYPE MATERIAL. *C. venustulus* : holotype USNM 707241 ; *C. gloriosus* : holotype KPM 3277.

OTHER MATERIAL EXAMINED. C. 100 specimens from throughout the geographical range.

DISTRIBUTION. (Fig. 233). Réunion and Mauritius ; the Philippines is ; Taiwan ; Papua New Guinea ; North Queensland, Australia ; New Caledonia (MNHN) ; the Marquesas. Depth range : 45-100 m.

DESCRIPTION. Shell up to 55 mm in length. Spire high, with 3-3 1/2 protoconch whorls and up to 8 teleoconch whorls. Suture impressed. Protoconch conical, whorls smooth, glossy.

Last whorl with 3 rounded varices, each adapically spinose and abapically webbed. Intervaricial axial sculpture generally consisting of 1 strong and 1 weak elongate axial node ; some specimens from Papua New Guinea or New Caledonia with lower axial sculpture, and generally 2 or 3 nodulose axial costae. Spiral sculpture of 6 or 7 primary cords, 5-8 secondary cords, and numerous threads in each interspace. Primary and secondary cords ending as short, open and adapically bent spinelets on varices.

Aperture roundly-ovate. Columellar lip usually smooth, rarely with weak narrow folds adapically. Rim erect abapically, adherent adapically. Anal notch narrow, shallow, delineated by small elongate callus. Outer lip denticulate, lirate for short distance within. Siphonal canal moderately strong, narrowly open, bent abaxially, with 3 or 4 spiral cords ending as open spines ; 2 adapical spines strongly bent abaperturally.

Pink or light orange, occasionally with darker spiral bands, generally one large or two narrow bands on last whorl, extending onto varices. Aperture white.

Radula (Figs 116-119).

REMARKS. Specimens from off New Guinea and New Caledonia attain smaller size (length 35-40 mm) than those from elsewhere, but apart from size and the usually lower axial ridges, all other features are identical.

C. venustulus differs from *C. superbus* in its smaller size, the strongly abaperturally bent spines on the siphonal canal, the relatively smaller aperture, and the lower spire.

C. venustulus may be also compared with some forms of *C. laciniatus* but *C. laciniatus* has lower intervaricial axial cords, a narrower, coloured, more completely adherent columellar lip, a larger aperture, more strongly fimbriate, squamous varices, and a broader and shorter siphonal canal.

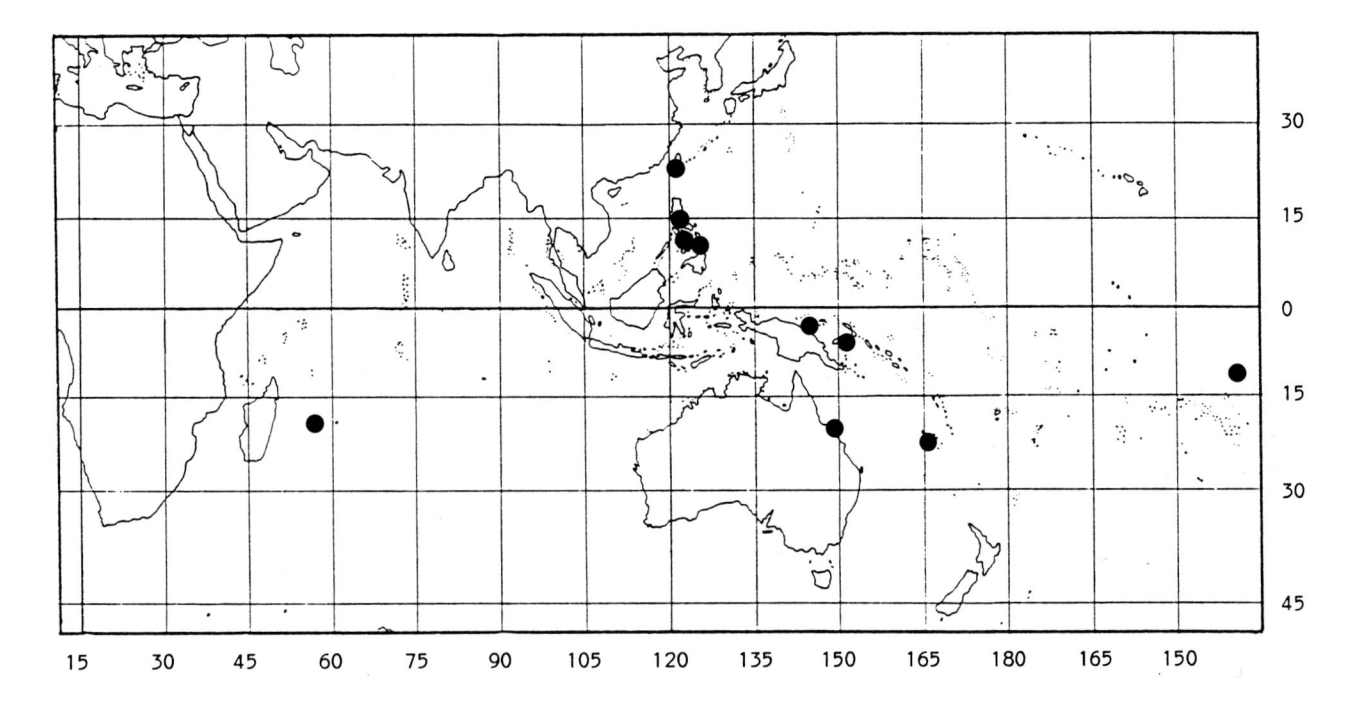

FIG. 233. — Distribution of *Chicomurex venustulus* (Rehder & Wilson).

Genus *NAQUETIA* Jousseaume, 1880

Naquetia Jousseaume, 1880 : 335.

Type-species (by original designation) : *Murex triqueter* Born, 1778.

Triplex Humphrey in HARRIS, 1897 : 172. Type-species (by original designation) : *Triplex flexuosa* Perry, 1811 (= *Murex triqueter* Born, 1778) (not *Triplex* Perry, 1810).

DESCRIPTION. Shell up to 104 mm in length, spire high, with 3 winglike or rounded varices on last whorl ; aperture ovate ; outer lip undulate, weakly or strongly lirate within. Anal notch rather deep, often delineated by a callus. Siphonal canal medium sized to long.

GEOGRAPHICAL DISTRIBUTION OF THE GENUS. Indo West-Pacific.

REMARKS. Recent authors have used *Naquetia* at the generic or subgeneric level (RADWIN & D'ATTILIO, 1976 ; FAIR, 1976 ; HOUART, 1985a ; D'ATTILIO, 1987a & b). Like D'ATTILIO (1987a & b) I prefer to treat *Naquetia* as a genus for species that lack the foliaceous varicial spines, characteristic of *Chicoreus* species. Members of both groups have shells with 3 rounded varices on the last whorl, and a medium-sized siphonal canal, most species having similar ornamentation on the early teleoconch whorls. Furthermore, some young specimens of *Naquetia* have small shoulder spines as in *Chicoreus*. Nevertheless, radulae of *Naquetia* and *Chicomurex* species are closely similar (Figs 116-127 and 128) and are quite different from those of species of *Chicoreus* and its subgenera (Figs 92-115 and 129).

Naquetia barclayi (Reeve, 1858)
Figs 58, 234-236, 237, 434

Murex barclayi Reeve, 1858 : 209, pl. 3, fig. 2.

Pteronotus (sic) *annandalei*Preston, 1910 : 119, fig. 3.

FIGS 234-236. — *Naquetia barclayi* (Reeve).
234, St. Brandon Shoal, near Mauritius, 83.2 mm (lectotype BMNH 196277. Courtesy of E.H. VOKES).
235, Reunion, 89.5 mm (coll. P. BERT).
236, Taiwan, 88 mm (RH).

ADDITIONAL REFERENCES

Murex (Naquetia) barcleyi (sic). — SHIKAMA, 1964 : 119, pl. 64, fig. 10.
Naquetia barclayi. — D'ATTILIO, 1966 : 4-5, figs 1,2; RADWIN & D'ATTILIO, 1976 : 80, pl. 15, fig. 8; KAICHER, 1980 : card 2515 (syntype); ABBOTT & DANCE, 1982 : 133, text fig. (paralectotype); D'ATTILIO & HERTZ, 1987a : 58, figs 7-8 (lectotype), 9 (sculpture), 10 (anal notch); DRIVAS & JAY, 1988 : 68, pl. 19, fig. 16.
Pterynotus barclayi. — WILSON & GILLETT, 1971 : 83, pl. 56.
Murex barclayi. — HARASEWYCH, 1973 : 4 (in part), figs 1, 2; LEEHMAN, 1974 : 6, text fig.; LEEHMAN, 1980a : 14, text fig.
Murex (Latirus) barclayi. — LEEHMAN, 1973a : 8, text fig.
Chicoreus (Naquetia) barclayi. — FAIR, 1976 : 24, pl. 14, fig. 172 (syntype); HOUART, 1985 : 8, figs 1-4, 6.
Murex barclayi. — LEEHMAN, 1978a : 9, text fig.
Murex (Naquetia) barclayi. — EISENBERG, 1981 : 87, pl. 69, fig. 14.
Naquetia annandalei. — KAICHER, 1973 : card 165; D'ATTILIO & HERTZ, 1987a : 56, figs 1-2 (holotype), 3-6, 11; D'ATTILIO & HERTZ, 1987b : fig. 7 (holotype), 8-9; RIPPINGALE, 1987 : 25, fig. 67.
Chicoreus (Naquetia) annandalei. — VOKES, 1973 : 5, fig. 3; VOKES, 1978 : 396; SPRINGSTEEN & LEOBRERA, 1986 : 130, pl. 35, fig. 6.
Naquetia annadalei (sic). — OKUTANI, 1983 : 8, pl. 23, fig. 3.

not *Murex (Pteronotus) trigonulus*. — SMITH, 1953 : 4, pl. 8, figs 8, 12 (= *Naquetia fosteri* D'Attilio & Hertz, 1987).
not *Murex annandalei*. — LEEHMAN, 1973b : 3, fig. 3 (= *Naquetia fosteri* D'Attilio & Hertz, 1987).
not *Naquetia annandalei*. — RADWIN & D'ATTILIO, 1976 : 80 (in part), pl. 15, figs 9-10 (= *Naquetia fosteri* D'Attilio & Hertz, 1987).
not *Naquetia annandalei*. — ABBOTT & DANCE, 1982 : 133, text fig. [= *Chicoreus (Siratus) consuela* (Verrill, 1950)].
not *Naquetia annandalei*. — SHARABATI, 1984 : pl. 18, fig. 3 [= *Naquetia cumingii* (A. Adams, 1853)].
not *Chicoreus (Naquetia) barclayi*. — HOUART, 1985 : 10, fig. 5 (= *Naquetia fosteri* D'Attilio & Hertz, 1987).

TYPE LOCALITIES. *M. barclayi* : St. Brandon Shoal, near Mauritius; *P. annandalei* : off Gopalpore [Gopalpur], Bay of Bengal, India, 30-38 fms (55-69 m).

TYPE MATERIAL. *M. barclayi* : lectotype BMNH 196277 (designated by D'ATTILIO & HERTZ, 1987a), paralectotype NMW 155.158.15; *P. annandalei* : holotype ZSI 4708/1 (not seen).

OTHER MATERIAL EXAMINED. C. 50 specimens from throughout the geographical range.

DISTRIBUTION. (Fig. 237). Off Durban, South Africa; Barazulo I., Mozambique; Mauritius and Réunion; the Bay of Bengal; southern Queensland, Australia; the Philippine Is; Taiwan, and southern Japan (Kii). Depth range : 70-100 m.

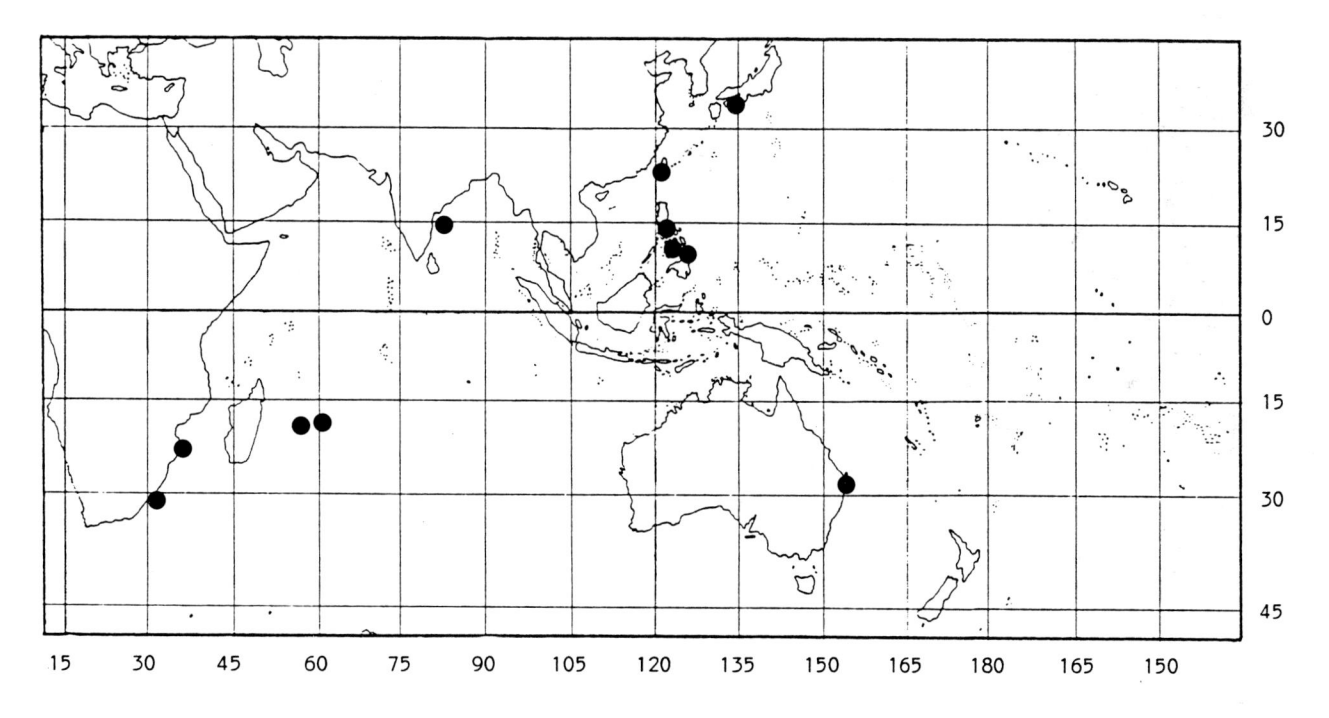

FIG. 237. — Distribution of *Naquetia barclayi* (Reeve).

DESCRIPTION. Shell up to 104 mm in length, elongate or broadly elongate. Spire high, acutely conical, with 3 1/2 protoconch whorls and up to 8 teleoconch whorls. Suture impressed. Protoconch conical, glossy.
Last whorl with 3 winglike varices, each finely plicate, joining the siphonal canal on its midline. Intervaricial axial sculpture consisting generally of 3 elongate ridges. Spiral sculpture of 8 or 9 cords and of numerous intermediate threads in each interspace.

Aperture ovate to roundly-ovate. Columellar lip smooth, rim detached abapically, adherent adapically. Anal notch deep, delineated by small callus and 2 strong denticles on outer lip. Outer lip smooth, lirate within. Siphonal canal relatively long, narrowly open, strongly bent abaperturally.
Light brown with darker spiral bands and spots. Aperture white.

REMARKS. Due to its confusion with *N. fosteri* D'Attilio & Hertz, 1987, *N. barclayi* has often been confused with *Murex consuela* Verrill, 1950 (see FINET & HOUART, 1989). Indeed, *C. consuela*, a Caribbean *Siratus* species, resembles a small *N. fosteri* but, with a same number of teleoconch whorls, *C. consuela* has a smaller shell with rounded aperture, and small denticles on the columellar lip.

Murex trigonulus Lamarck, 1822 has been included as a synonym of *N. barclayi* (mainly due to the confusion existing between *N. barclayi* and *N. fosteri*) by many authors (*e.g.* SOWERBY, 1841; TAPPARONE-CANEFRI, 1875; POIRIER, 1883; SMITH, 1953; VOKES, 1971; FAIR, 1976; HOUART, 1985a) but they are clearly distinct species (see below).

Pteronotus annandalei Preston, 1910, described from a specimen from off Gopalpur (Bay of Bengal) differs from typical *N. barclayi* in having a narrower shell and weaker sculpture. Typical *N. barclayi* also has more pronounced denticles on the inner side of the outer lip and a broader aperture. Nevertheless there are intermediates, and these differences are not considered sufficient to separate the two forms at species or subspecies level. Unfortunately the protoconch of the Indian form is unknown, *N. annandalei* of authors from the Red Sea, confined in the Gulf of Aqaba, is *N. fosteri* D'Attilio & Hertz, 1987.

Naquetia cumingii (A. Adams, 1853)
Figs 61-64, 238, 437-445

Murex cumingii A. Adams, 1853 : 270.

Murex trigonulus Lamarck, 1822 : 167 (not *M. trigonulus* Lamarck, 1816).
Murex (Chicoreus) triqueter var. *amanuensis* Couturier, 1907 : 142.
Murex jickelii Tapparone-Canefri, 1875 : 582, pl. 19, fig. 6.

ADDITIONAL REFERENCES

Murex (Pteronotus) (sic) *triqueter*. — SMITH, 1953 : 4, pl. 4, fig. 4 (not *Murex triqueter* Born, 1778).
Pterynotus triqueter. — CERNOHORSKY, 1967b : 126, pl. 26, fig. 160 (not *Murex triqueter* Born, 1778).
Pterynotus (Naquetia) triqueter. — WILSON & GILLETT, 1971 : 88, pl. 59, fig. 5 (not *Murex triqueter* Born, 1778).
Pterynotus (Naquetia) trigonulus. — CERNOHORSKY, 1967a : 124, fig. 6, pl. 15, fig. 15.
Murex trigonulus. — CERNOHORSKY, 1971 : 189, fig. 3 (lectotype); PURTYMUN, 1981 : 6, text fig. (bottom fig., shell on the left).
Naquetia trigonulus. — KAICHER, 1973 : card 166; RADWIN & D'ATTILIO, 1976 : 81, fig. 46, pl. 15, fig. 12; ABBOTT & DANCE, 1982 : 133, text fig.
Chicoreus (Naquetia) trigonulus. — VOKES, 1974 : figs 1, 3; FAIR, 1976 : 83, pl. 14, fig. 179; VOKES, 1978 : pl. 5, fig. 5; HOUART, 1985 : 11, figs 11-13; SPRINGSTEEN & LEOBRERA, 1986 : 132, pl. 36, fig. 6.
Naquetia trigonula. — MAC DONALD, 1979 : 8, fig. 1; D'ATTILIO & HERTZ, 1987b : figs 10-13.
Murex cumingii. — CERNOHORSKY, 1971 : 189, fig. 4 (lectotype).
Chicoreus (Naquetia) cumingii. — FINET & HOUART, 1989 : 7, figs 1-3.
Pterynotus (Naquetia) jickelii. — FAIR, 1973 : 8, text fig.
Naquetia jickelii. — KAICHER, 1973 : card 167; SHARABATI, 1984 : pl. 18, fig. 1.
Chicoreus (Naquetia) jickelii. — FAIR, 1976 : 51, pl. 14, fig. 178; VOKES, 1978 : 396, pl. 5, figs 6-7; HOUART, 1985 : 10, fig. 7.
Phyllonotus laciniatus. — RADWIN & D'ATTILIO, 1976 : 89 (in part) (not *Murex laciniatus* Sowerby, 1841).
Naquetia annandalei. — SHARABATI, 1984 : pl. 18, fig. 3 (not *Pteronotus annandalei* Preston, 1910).
NOT *Murex trigonulus*. — PURTYMUN, 1981 : 6, bottom fig., shell on the right [= *Naquetia triqueter* (Born, 1778)].

TYPE LOCALITIES. *M. cumingii* : Philippine Islands; *M. trigonulus* none; *Murex triqueter* var. *amanuensis* : Amanu (Tuamotu); *M. jickelii* : Suakin, Sudan, Red Sea.

TYPE MATERIAL. *M. cumingii* : lectotype BMNH 1963.817 (designated by CERNOHORSKY, 1971); *M. trigonulus* : lectotype and paralectotype MHNG 1099/35 (designated by FINET & HOUART, 1989); *M. triqueter* var. *amanuensis* : not located (not in MNHN); *M. jickelii* : holotype ZMB 37370 (not seen).

OTHER MATERIAL EXAMINED. C. 150 specimens from throughout the geographical range.

DISTRIBUTION. (Fig. 238). South-western Madagascar; the Seychelles; the southern Red Sea (from the Dahlak Archipelago to Suakin); Sri Lanka; Celebes Sea; the Moluccas; the Philippine Is; Okinawa; Middle Japan; Guam; Marshall Is; Wallis; Samoa Is, and the Tuamotus as eastern limit.

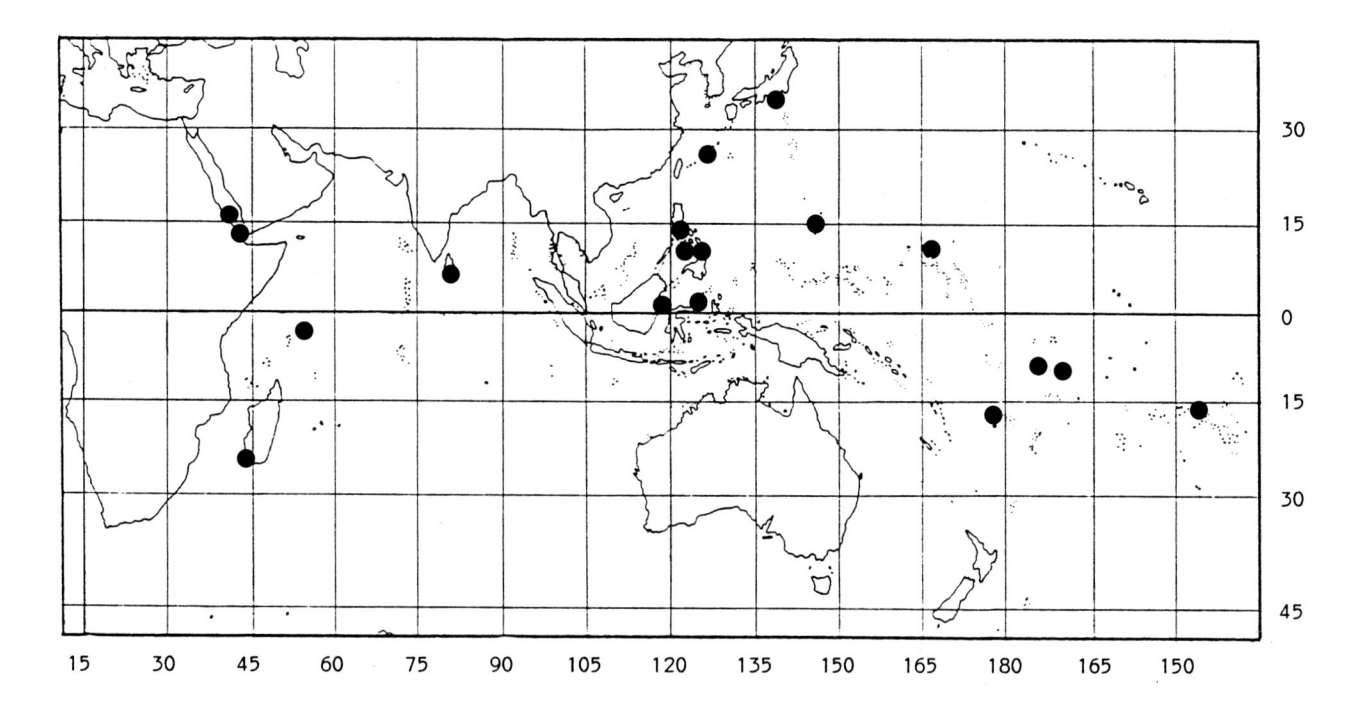

FIG. 238. — Distribution of *Naquetia cumingii* (A. Adams).

DESCRIPTION. Shell up to 63 mm in length, stout. Spire high, with 2-2 1/4 protoconch whorls and up to 7 elongate, weakly convex teleoconch whorls. Suture impressed. Protoconch whorls smooth, last whorl shouldered.

Last whorl with 3 rounded, frondose, webbed varices, webbing strongest abapically. Intervaricial axial sculpture consisting of 2 or 3 nodulose cords. Spiral sculpture of 6 nodulose cords and numerous threads.

Aperture ovate. Columellar lip smooth, sometimes with a small callus adapically; rim adherent. Anal notch rather deep, sometimes delineated by shallow callus and 2 small denticles. Outer lip undulate, lirate for short distance within. Siphonal canal short to moderate in length, broad, narrowly open, ornamented with 2-4 spiral cords.

Yellowish to pale brown with 2 or 3 brown spiral bands on last whorl and numerous brown blotches on shell surface. Aperture white.

REMARKS. FINET & HOUART (1989) presented a historical review of this species and, on basis of new evidence, they concluded that *Murex trigonulus* Lamarck, 1816 is probably a synonym of *Murex triqueter* Born, 1778, and that the 2 syntypes of *M. trigonulus* (MHNG 1099/35) are *M. trigonulus* Lamarck, 1822 (not Lamarck, 1816). *M. trigonulus* Lamarck, 1822 is a homonym of the prior *M. trigonulus* Lamarck, 1816 and the next available name for the species is *M. cumingii* A. Adams, 1853, the lectotype for which was designated by CERNOHORSKY (1971, fig. 3). See discussion of *N. triqueter* above, for a comparison of these two species.

Murex jickelii was incorrectly synonymised with *Murex laciniatus* Sowerby, 1841 by RADWIN & D'ATTILIO (1976 : 89). *M. jickelii* is rather a synonym of *M. cumingii*, since shells from Madagascar (Fig. 443), the Moluccas and elsewhere are quite indistinguishable from the Red Sea form. Moreover, the protoconch (partially broken) of a Red Sea shell (Fig. 64) is the same as that of a typical specimen from the Moluccas (Fig. 63).

Naquetia fosteri D'Attilio & Hertz, 1987
Figs 60, 239, 435-436

Naquetia fosteri D'Attilio & Hertz, 1987 : 190, figs 1-6.

ADDITIONAL REFERENCES

Murex (Pteronotus) (sic) *trigonulus*. — SMITH, 1953 : 4, pl. 8, figs 8, 12 (not *Murex trigonulus* Lamarck, 1822).
Murex annandalei. — LEEHMAN, 1973b : 3, fig. 3 (not *Pteronotus annandalei* Preston, 1910).
Chicoreus (Naquetia) annandalei. — FAIR, 1976 : 21, pl. 14, fig. 171 (not *Pteronotus annandalei* Preston, 1910).
Naquetia annandalei. — RADWIN & D'ATTILIO, 1976 : 80 (in part), pl. 15, figs 9-10 ; MIENIS, 1983 : 538, text fig. (not *Pteronotus annandalei* Preston, 1910).
Chicoreus (Naquetia) barclayi. — HOUART, 1985 : 10, fig. 5 (not *Murex barclayi* Reeve, 1858).

TYPE LOCALITY. Off Eilat, Gulf of Aqaba, Red Sea.

TYPE MATERIAL. Holotype SDNHM 91996.

MATERIAL EXAMINED. Holotype (photographs) ; Eilat, Red Sea, 20-25 m, coll. D. Peled (2 dd) ; Eilat, 30 m, RH (1 dd).

DISTRIBUTION. (Fig. 239). Northern end of the Gulf of Aqaba, off Eilat, Red Sea.

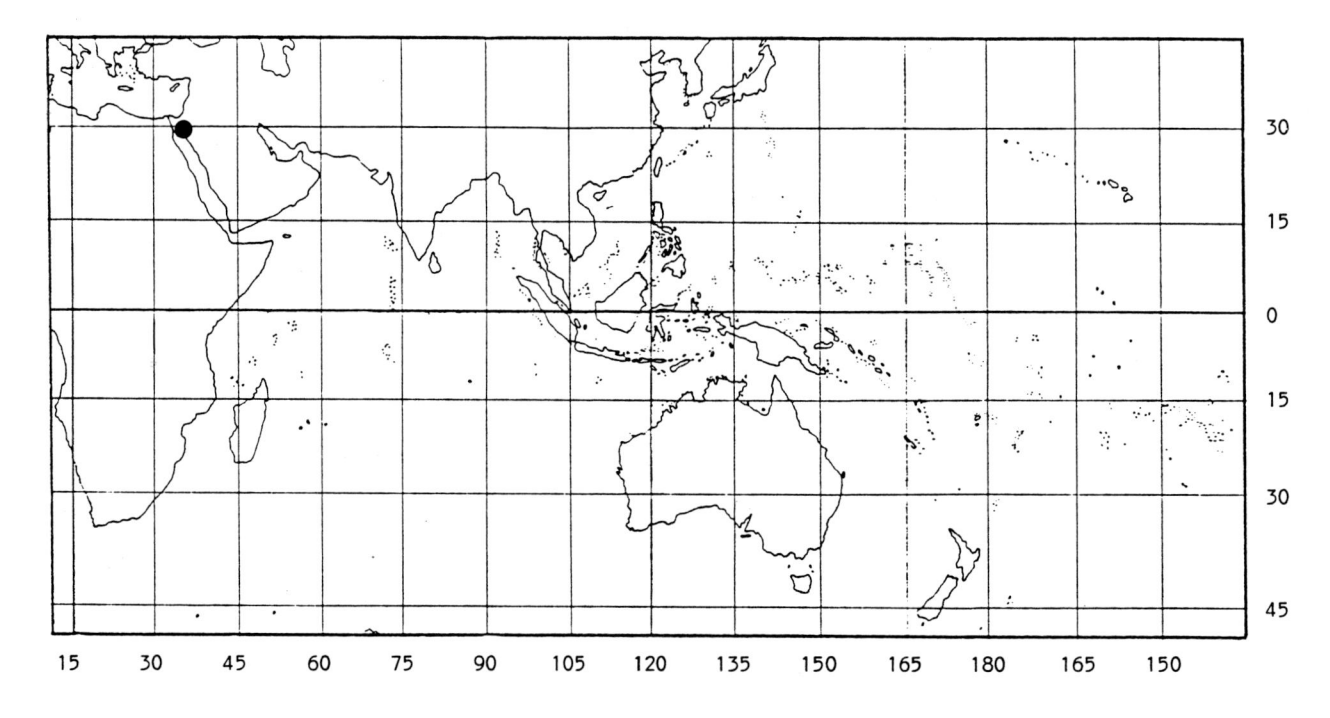

FIG. 239. — Distribution of *Naquetia fosteri* D'Attilio & Hertz.

DESCRIPTION. Shell up to 94.5 mm in length, narrowly fusiform, stout. Spire high, with 1 1/2 protoconch whorls and up to 8 weakly convex teleoconch whorls. Suture impressed. Protoconch whorls convex, smooth.

Last whorl with 3 rounded, spineless varices, each abapically ornamented with varicial flange that extend onto siphonal canal. Intervaricial axial sculpture consisting of 3-5 elongate, irregular ridges crossed by 8 spiral cords and 1 intermediate thread in each interspace, giving a nodulose appearance.

Aperture ovate. Columellar lip smooth, rim adherent adapically, weakly erect abapically. Anal notch narrow, deep. Outer lip erect, strongly lirate within. Siphonal canal medium-sized, narrowly open, weakly bent abaperturally at tip, with 2 short, broadly open triangular spines.

First 4 or 5 teleoconch whorls bright pink or pale orange; teleoconch light brown with 2 darker brownish bands on last whorl and 1 on early whorls.

REMARKS. *N. fosteri* has long been interpreted as a slender form of *N. barclayi*. It has also been erroneously synonymised with *Murex trigonulus* Lamarck, 1822 (not 1816) by several authors (see discussion on *N. barclayi*) and confused also with *Naquetia annandalei* (ABBOTT & DANCE, 1982 : 133, text fig.) and with *Murex consuela* Verrill, 1950 (REEVE, 1845 : pl. 22, fig. 87 ; KOBELT & KUSTER *in* MARTINI & CHEMNITZ, 1870 : 121, pl. 36, fig. 9 ; TRYON, 1880 : 84, pl. 11, fig. 120).

N. fosteri differs from *N. barclayi* in its more narrowly fusiform shape, fewer intermediate spiral threads, heavier, more strongly rounded varices, paired lirations within the outer apertural lip and especially in its protoconch (1 1/2 smooth and convex whorls, vs 3 1/2 whorls).

N. fosteri seems to be restricted to the Gulf of Aqaba in the Red Sea, while *N. barclayi* is widely distributed in the Indo West-Pacific, from southern Africa to south of Japan, probably reflecting respectively non planktotrophic and planktotrophic larval development as suggested by protoconch morphology.

Naquetia triqueter (Born, 1778)

Figs 59, 128, 240, 446-447

Murex triqueter Born, 1778 : 288 (reference to MARTINI, 1777 : fig. 1038).

Purpura cancellata Röding, 1798 : 143.
Purpura variegata Röding, 1798 : 143.
Triplex flexuosus Perry, 1811 : pl. 7, fig. 1.
Murex trigonulus Lamarck, 1816 (not 1822) : pl. 417, fig. 4.
Murex roseotinctus Sowerby, 1860 : 429, pl. 49, fig. 6.

ADDITIONAL REFERENCES

Murex (Naquetia) triqueter. — SHIKAMA, 1963 : 71, pl. 54, fig. 7.
Pterynotus triqueter. — HINTON, 1972 : 36, pl. 18, fig. 12 ; SALVAT & RIVES, 1975 : 312, fig. 196.
Naquetia triqueter. — KAICHER, 1973 : card 164 (in part); HINTON, 1979 : 27, fig. 6 ; DRIVAS & JAY, 1988 : 68, pl. 19, fig. 9.
Chicoreus (Naquetia) triqueter. — VOKES, 1974 : fig. 2 ; FAIR, 1976 : 83, pl. 14, fig. 177 ; VOKES, 1978 : 394 (in part) ; HOUART, 1985 : 12, figs 8-10 ; HOUART, 1986b : figs 3-3b.
Naquetia triquetra (sic). — RADWIN & D'ATTILIO, 1976 : 82 (in part) ; MAC DONALD, 1979 : 8, fig. 2 ; ABBOTT & DANCE, 1982 : 133, text fig.
Murex triqueter. — JAZWINSKI, 1979 :11, text fig. ; PURTYMUN, 1981 : 6, upper text fig.
Chicoreus (Naquetia) triquetra (sic). — SPRINGSTEEN & LEOBRERA, 1986 : 132, pl. 36, fig. 7.
Murex trigonulus. — PURTYMUN, 1981 : 6, bottom fig., shell on the right (not *Murex trigonulus* Lamarck, 1822).
NOT *Murex triqueter*. — BORN, 1780 : 291, pl. 11, figs 1, 2 [= *Naquetia cumingii* (A. Adams, 1853)].
NOT *Murex (Pteronotus) triqueter*. — SMITH, 1953 : 4, pl. 4, fig. 4 [= *Naquetia cumingii* (A. Adams, 1853)].
NOT *Pterynotus triquetor* (sic). — SPRY, 1961 : 19, pl. 4, fig. 139 [= *Naquetia vokesae* (Houart, 1985)].
NOT *Pterynotus triqueter*. — CERNOHORSKY, 1967b : 126, pl. 26, fig. 160 [= *Naquetia cumingii* (A. Adams, 1853)].
NOT *Pterynotus (Naquetia) triqueter*. — WILSON & GILLETT, 1971 : 88, pl. 59, fig. 5 [= *Naquetia cumingii* (A. Adams, 1853)].
NOT *Naquetia triqueter*. — KAICHER, 1973 : card 164 (in part) [= *Naquetia vokesae* (Houart, 1986)].
NOT *Naquetia triquetra* (sic). — RADWIN & D'ATTILIO, 1976 : 82 (in part), pl. 15, fig. 11 [= *Naquetia vokesae* (Houart, 1986)].
NOT *Chicoreus (Naquetia) triqueter*. — VOKES, 1978 : 394 (in part), pl. 5, fig. 4 ; HOUART, 1985a : 12 (in part) [= *Naquetia vokesae* (Houart, 1986)].

TYPE LOCALITIES. *M. triqueter* : East Indies and Tranquebar (MARTINI, 1777), restricted to Tranquebar, India (VOKES, 1974 : 259) ; *M. roseotinctus* : Philippines.

TYPE MATERIAL. *M. triqueter* : VOKES (1974) designated lectotype the shell on fig. 1038 of MARTINI (1777) ; this shell cannot presently be located (see discussion below) ; *M. roseotinctus* : holotype BMNH 1974100. No material for the other names.

OTHER MATERIAL EXAMINED. C. 100 specimens from throughout the geographical range.

DISTRIBUTION. (Fig. 240). Christmas I., Indian Ocean; Straits of Makassar; the Moluccas; Okinawa; Papua New Guinea; Marshall Is; Samoa Is, and the Tuamotus as eastern limit.

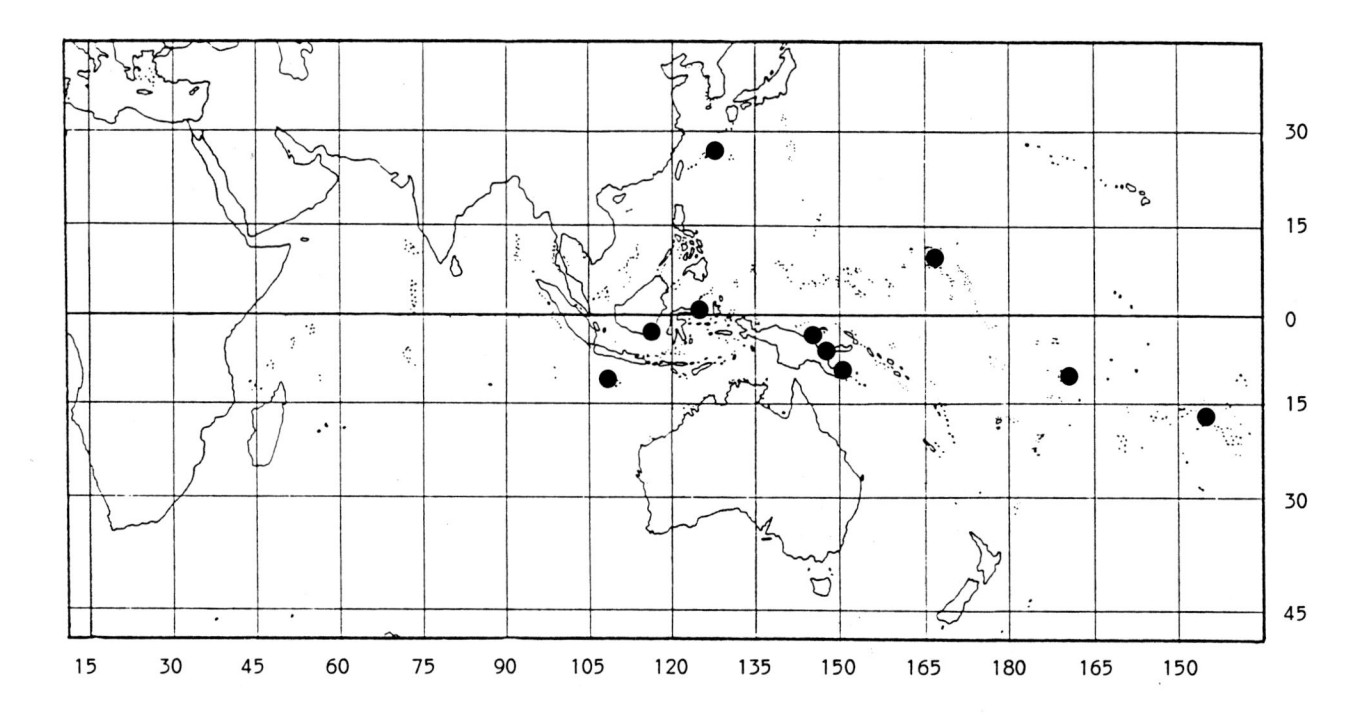

FIG. 240. — Distribution of *Naquetia triqueter* (Born).

DESCRIPTION. Shell up to 60 mm in length, stout. Spire high, acutely conical, with 3 1/2 protoconch whorls and up to 9 elongate teleoconch whorls. Suture impressed. Protoconch conical, glossy.

Last whorl with 3 rounded squamous varices, each more developed abapically. Intervaricial axial sculpture consisting of 2 or 3, occasionally 4, rounded, nodulose cords, crossed by 14 or 15 squamous, similar sized spiral cords.

Aperture ovate. Columellar lip smooth, small elongate callus adapically, rim adherent. Anal notch deep. Outer lip denticulate, strongly lirate for short distance within. Siphonal canal broad, short, narrowly open, slightly bent abaxially, ornamented with 4 or 5 squamous spiral cords, tip bent abaperturally.

White to pale brown with darker bands, especially visible on varices, darker maculations on axial ridges. Aperture white.

Radula (Fig. 128).

REMARKS. In a historical review of *Murex trigonulus* Lamarck, 1816 and *Murex trigonulus* Lamarck, 1822, FINET & HOUART (1989) concluded that *M. trigonulus* Lamarck, 1816 was synonymous with *M. triqueter* Born, 1778.

BORN (1780) illustrated *M. cumingii* erroneously identified by him as *M. triqueter*. In her paper on the identity of *Murex triqueter* Born, VOKES (1974) designated MARTINI's figure 1038 as lectotype for that species in accordance with ICZN [article 73c (i) and 74b]. She also illustrated the specimen figured in BORN (1780), now in NHMW.

Houart (1985a) erroneously included the later named *N. triqueter vokesae* in the geographical distribution map of *N. triqueter* and confused both species.

Since *N. cumingii* and *N. triqueter* are frequently confused, the two species are contrasted (Table 4).

TABLE 4. — Comparisons of *Naquetia triqueter* and *N. cumingii*.

Character	*N. triqueter*	*N. cumingii*
Protoconch	Conical, multispiral, 3 1/2 whorls	Weakly angulate, last whorl keeled. 2-2 1/4 whorls
Number of teleoconch whorls	9	6-7
Outer lip and inner side	Rim denticulate, internally striate	Rim slightly crenulate. Internally with short, elongate denticles
Spiral sculpture	12-14 squamous cords of approximately similar size	6 strong spiral cords and 3-4 intermediate threads in each interspace
Colour	Pale brown to white with darker spiral bands and brown maculations on the axial costae	Pale brown to white or yellowish with 3 darker bands on last whorl and on varices

Naquetia vokesae (Houart, 1986)
Figs 65-66, 241, 448-449

Chicoreus (Naquetia) triquiter (sic) *vokesae* Houart, 1986 : 95, figs 1-2.

ADDITIONAL REFERENCES

Pterynotus triquetor (sic). — SMITH, 1961 : 19, pl. 4, fig. 139 (not *Murex triqueter* Born, 1778).
Naquetia triqueter. — KAICHER, 1973 : card 164 (in part) (not *Murex triqueter* Born, 1778).
Naquetia triquetra (sic). — RADWIN & D'ATTILIO, 1976 : 82 (in part), pl. 15, fig. 11 (not *Murex triqueter* Born, 1778).
Chicoreus (Naquetia) triqueter. — VOKES, 1978 : 394 (in part), pl. 5, fig. 4 (holotype) ; HOUART, 1985a : 12 (in part) (not *Murex triqueter* Born, 1778).

TYPE LOCALITY. South east Nacala Bay, northern Mozambique, dredged from gravelly bottom with sparse *Cymodocea*, 9 m.

TYPE MATERIAL. Holotype NM H213.

OTHER MATERIAL EXAMINED. C. 60 specimens from throughout the geographical range.

DISTRIBUTION. (Fig. 241). N. Zululand (NM D7448) and Natal (SAM A35983), South Africa ; Mozambique ; Nossi-Bé and Tulear, Madagascar ; Comoros I., and southern Tanzania.

DESCRIPTION. Shell up to 70 mm in length, stout. Spire high, with 2 protoconch whorls and up to 9 elongate, slightly convex teleoconch whorls. Suture impressed. Protoconch whorls rounded, smooth.
Last whorl with 3 low, rounded varices. Terminal varix adaperturally squamous, abaperturally ornamented with varicial flange that extends along siphonal canal. Other sculpture consisting of 3-5 low axial ridges, crossed by 12-15 spiral cords, each flanked by fine spiral threads. Axial varices occasionally with small open, shoulder spines.
Aperture roundly-ovate. Columellar lip rim adherent, briefly detached abapically, smooth. Anal notch small, shallow. Outer lip weakly denticulate, lirate for short distance within. Siphonal canal short and broad, narrowly open, slightly bent abaperturally.
Cream to light brown with darker spiral bands. Axial ribs also darker coloured. Aperture white.

REMARKS. Although introduced as a subspecies of *N. triqueter* because of extreme similarities in teleoconch morphology, the protoconch is entirely different, paucispiral (non planktotrophic) in *N. vokesae*, conical and multispiral (planktotrophic) in *N. triqueter*. However, as indicated by BOUCHET (1989), different types of larval development have never been observed in a single species, so it is preferable to treat it as a distinct species.
Further differences from *N. triqueter* include more convex spire outline, the less prominent axial sculpture, and the weaker spiral sculpture.

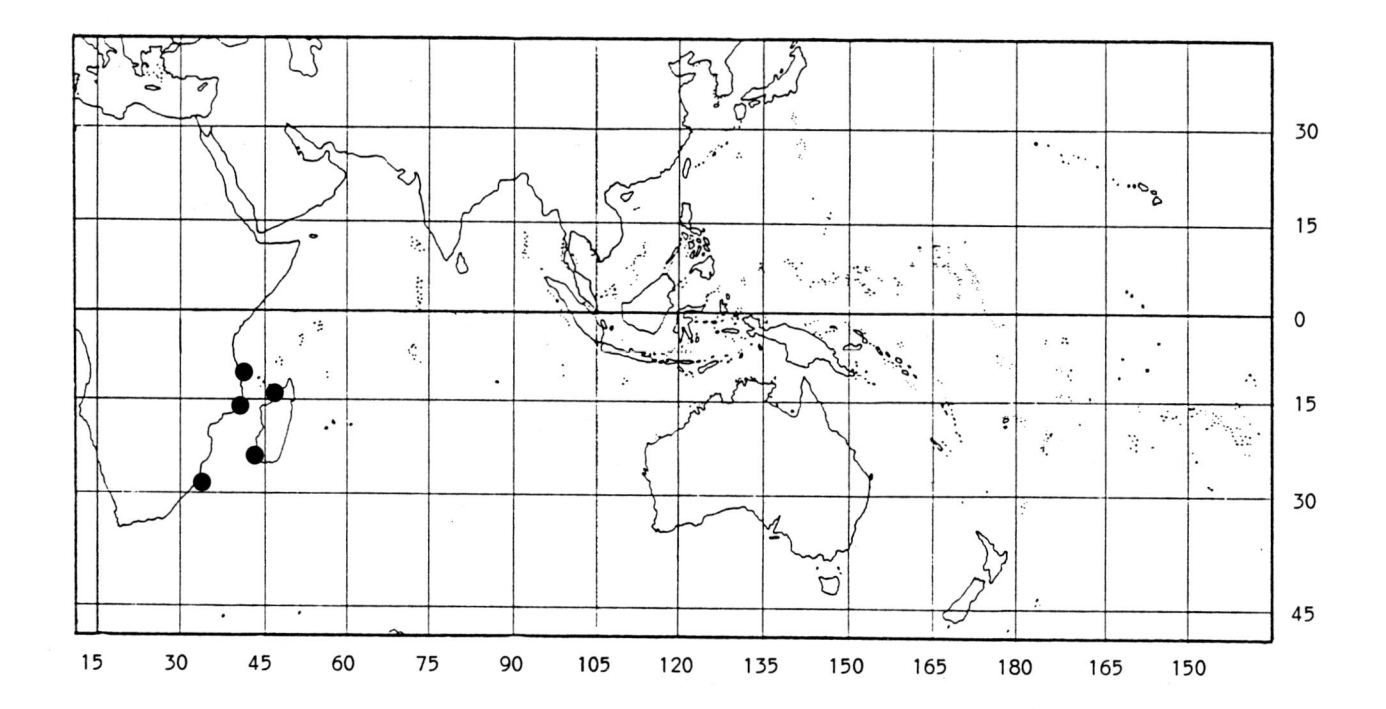

FIG. 241. — Distribution of *Naquetia vokesae* (Houart).

FOSSIL SPECIES

Genus *CHICOREUS* Montfort, 1810

Subgenus *TRIPLEX* Perry, 1810

Chicoreus (Triplex) altenai (Cox, 1948)
Figs 242-243

Murex altenai Cox, 1948 : 46, pl. 4, fig. 2.

TYPE LOCALITY. North Borneo, Dent Peninsula, Sg. Togopi, Neogene.

TYPE MATERIAL. Holotype NHMB.

REMARKS. Probably related to *Chicoreus microphyllus* (Lamarck, 1822). Although the absence of protoconch whorls makes a better comparison difficult, *C. altenai* has more globose teleoconch whorls, fewer spines on varices and more numerous axial ridges between each pair of varices.

FIGS 242-245. — Subgenus *C. (Triplex)*.
242-243, *C. altenai* (Cox). North Borneo, Neogene, 39.1 mm (holotype, NHMB).
244-245, *C. dennanti* (Tate). 244, Clifden Bank, Muddy Creek, Victoria, Australia, Miocene, 22.1 mm (MV 112052). 245, near Hamilton, Victoria, Australia, Miocene, 30.1 mm (lectotype, SAM T416B).

Chicoreus (Triplex) amblyceras (Tate, 1888)
Figs 82, 450-451

Murex amblyceras Tate, 1888 : 101, pl. 2, fig. 12.

TYPE LOCALITY. Fyansford Formation, Schnapper Point, Victoria, Australia (here designated); Miocene.

TYPE MATERIAL. Lectotype SAM T428A, here selected from 10 syntypes.

REMARKS. One specimen (specimen H in the type material) is *Murex basicinctus* Tate, 1888; 3 specimens (C, D and F) although somewhat different (see remarks under *Chicoreus* cf. *amblyceras*) are not definitely separated.
TATE noted 2 localities : " Lower Beds at Muddy Creek " and " Blue Clays at Schnapper Point, Victoria ". This latter is here restricted as type locality, being the locality of specimen A (noted as holotype on the plaquette, by a previous revisor), and here designated as lectotype. *C. amblyceras* is related to the Recent Australian species *C. territus*, *C. damicornis* and *C. denudatus*.

Chicoreus (Triplex) cf. *amblyceras* (Tate, 1888)
Figs 81, 454-456

(See remarks under *Chicoreus amblyceras* (Tate, 1888).

REMARKS. Unlike *C. amblyceras*, this form has no axial ornamentation on the first three teleoconch whorls, while *C. amblyceras* has prominent axial costae. The spiral sculpture on the last whorl is also more regular, and the spiral cords are heavier abapically whereas these are approximately equal-sized in *C. amblyceras*. Although there are no other noteworthy differences, these specimens may represent another species.

Chicoreus (Triplex) basicinctus (Tate, 1888)
Figs 80, 452-453

Murex basicinctus Tate, 1888 : 99, pl. 2, fig. 9.

TYPE LOCALITY. Codell Marl, Murray River Cliffs, near Morgan, South Australia; Miocene.

TYPE MATERIAL. Lectotype SAM T417A, here selected from 6 syntypes; specimen 8 on the plaquette is *Typhis philippensis* Watson, 1883.

REMARKS. Distinct from any Recent and fossil species and related to the Recent Australian *C. damicornis*.

Chicoreus (Triplex) batavianus (Martin, 1884)
Figs 465-467

Murex batavianus Martin, 1884 : 97, pl. 6, fig. 9.

TYPE LOCALITY. Batavia, Java, Indonesia; Miocene.

TYPE MATERIAL. Lectotype RML 9689, here selected from 7 syntypes.

REMARKS. This fossil clearly belongs to the Recent group including *C. axicornis*, it could be the ancestor of this species.

Chicoreus (Triplex) dennanti (Tate, 1888)
Figs 79, 244-245

Murex dennanti Tate, 1888 : 98, pl. 2, fig. 7.

TYPE LOCALITY. Muddy Creek Marl, near Hamilton, Victoria, Australia; Miocene.

TYPE MATERIAL. Lectotype SAM T416B, here selected from 5 syntypes.

REMARKS. Distinct from any Recent species, it has bulbous paucispiral nuclear whorls as in other Australian fossil *Chicoreus* species. It seems to belong in the *C. microphyllus* group.

Chicoreus (Triplex) juttingae (Beets, 1941)
Figs 83, 469-471

Murex (Chicoreus) juttingae Beets, 1941 : 95, pl. 5, figs 207-211.

TYPE LOCALITY. Borneo (precise locality unstated); Miocene.

TYPE MATERIAL. Holotype RML 312.451 and 5 paratypes 312.453.

REMARKS. From the shell contour, lack of spines and protoconch morphology, this distinctive species seems to be related to the Recent *C. capucinus* (Lamarck, 1822).

Chicoreus (Triplex) karangensis (Martin, 1895)
Fig. 474

Murex karangensis Martin, 1895 : 130, pl. 20, fig. 295.

TYPE LOCALITY. Djilintung, Java, Indonesia; Miocene.

TYPE MATERIAL. Holotype RML 9683.

REMARKS. This species is based on a small shell (length 16 mm), and with 4 or 5 teleoconch whorls lacking the protoconch. It is apparently distinct from any Recent and fossil species, but is probably a juvenile.

Chicoreus (Triplex) komiticus (Suter, 1917)
Fig. 468

Murex zelandicus var. *komiticus* Suter, 1917 : 37, pl. 4, fig. 21.

TYPE LOCALITY. Komiti Bluff, Kaipara Harbour, North Auckland, New Zealand; Lower Miocene.

TYPE MATERIAL. Holotype OM C16.46.

REMARKS. This species apparently belongs to the group that includes the Recent species *C. cervicornis* and *C. boucheti* (group 6). The specimen illustrated by SUGGATE et al. (1978 : fig. 11.12 : 7) and by SPEDEN & KEYES (1981 : pl. 29, fig. 7) is not this species in my opinion but is related to *Chicoreus syngenes*, from which it differs, however, in having a wider aperture, a longer and sharper siphonal canal, and smaller varices and intervaricial nodes. This specimen (Figs 460-461) may represents an undescribed species. FINLAY (1930 : pl. 1, fig. 13) illustrated another specimen of this form that was also misidentified as *C. komiticus*.

Chicoreus (Triplex) lawsi (Maxwell, 1971)
Figs 475-476

Murex lawsi Maxwell, 1971 : 758, figs 11, 12.

TYPE LOCALITY. Pakaurangi Point, Kaipara Harbour, New Zealand; Lower Miocene.

TYPE MATERIAL. Holotype NZGS, TM4912.

REMARKS. Although its outline resembles that of some *Siratus* species, *Murex lawsi* is closely related to *Chicoreus batavanius* (Martin, 1884) and *C. amblyceras* (Tate, 1888), which also have a relatively long siphonal canal. Accordingly it is here included in the subgenus *Triplex*.

Chicoreus (Triplex) lundeliusae Ludbrook, 1978
Figs 479-480

Chicoreus (Chicoreus) lundeliusae Ludbrook, 1978 : 140, pl. 16, figs 1-8.

TYPE LOCALITY. Roe Plain, Eucla Basin-Hampton Microwave Repeater Tower, South Australia, 31°57'57" S, 127°34'45" E, Roe Calcarenite; Early Pleistocene.

TYPE MATERIAL. Holotype WAM 70.1137.

REMARKS. Although compared originally with *Chicoreus (Chicoreus) ramosus* (Linné, 1758) it differs mainly from that species in its smaller size, more elongate shape, and by the absence of a labral tooth. It was also compared with *Chicoreus (Triplex) denudatus* (Perry, 1811), but that species is smaller with more palmate fronds and the aperture is denticulate instead of crenulate.

Chicoreus (Triplex) naricus (Vredenburg, 1925)
Figs 463-464

Murex (Haustellum) naricus Vredenburg, 1925 : 213, pl. 7, fig. 13.

TYPE LOCALITY. Nari, Bagothoro Hili in Sind, India; ?Oligocene.

TYPE MATERIAL. Holotype GSI 12571.

REMARKS. Although originally compared with species of *Murex* (*s.s.*), the shell is incomplete, consisting of 2 whorls and part of last whorl and aperture. I agree with VOKES (1971 : 73) that this species probably belongs in *Chicoreus*. It may be related to *C. altenai* and *C. timorensis*.

Chicoreus (Triplex) rutteni (Beets, 1950)
Figs 477-478

Murex (Phyllonotus) rutteni Beets, 1950 : 308, figs. 3, 4.

TYPE LOCALITY. Coral limestone, hill near Sekoerau, East Borneo; Pliocene.

TYPE MATERIAL. Holotype RML 113.166.

REMARKS. The remaining parts of the incomplete holotype are sufficient to separate if from any hitherto known species.

Chicoreus (Triplex) syngenes Finlay, 1930
Figs 457-459

Chicoreus ?syngenes Finlay, 1930 : 76, pl. 1, figs 3, 4.

TYPE LOCALITY. Clifden, New Zealand ; [Hutchinsonian] Miocene.

TYPE MATERIAL. Holotype AIM TM-164.

REMARKS. The nearest Recent species is *C. virgineus* (Röding, 1798) that ranges from the Red Sea to Sri Lanka, but *C. syngenes* lacks the typical labral tooth of this and other species of *Chicoreus* (*s.s.*).

Chicoreus (Triplex) tateiwai Hatai & Kotaka, 1952
Figs 472-473

Chicoreus tateiwai Hatai & Kotaka, 1952 : 78, pl. 7, figs 13-14.

TYPE LOCALITY. Paiponchon, Shinsoruton, San-u-nanmyon, Myonchon District, Hamukyon-pukuton, Korea ; Heiroku formation, Lower Miocene.

TYPE MATERIAL. Holotype IGPS 74352.

REMARKS. The holotype is a somewhat immature specimen, probably related to the *C. microphyllus* group. The protoconch is unknown. VOKES (1971 : 105) erroneously credited authorship to Kuroda & Kotaka.

Chicoreus (Triplex) timorensis (Tesch, 1915)
Fig. 462

Murex timorensis Tesch, 1915 : 64, pl. 82 (10), fig. 141.

TYPE LOCALITY. Timor ; Miocene.

TYPE MATERIAL. Holotype MGM 13886.

REMARKS. The shell has a narrow aperture, broad and spineless varices, and a short, rather sharply pointed siphonal canal. It is possibly the ancestor of the Recent *C. thomasi*, which is endemic to the Marquesas Islands.

Chicoreus (Triplex) totomiensis (Makiyama, 1927)

Murex totomiensis Makiyama, 1927 : 126, pl. 6, figs 20, 21.

TYPE LOCALITY. Dainiti, Japan ; Pliocene.

TYME MATERIAL. Not seen (not available).

REMARKS. Although originally compared with *C. penchinati* (= *C. strigatus*) and *Murex borni* (an European Miocene species), from the original illustrations, the species is nearer to the *C. microphyllus* group (group 2) as well as to some forms of *C. brunneus*. Moreover, it strongly resembles *C. tateiwai*.

Genus *CHICOMUREX* Arakawa, 1964

Chicomurex kendengensis (van Regteren Altena, 1950)
Figs 248-249

Chicoreus kendengensis van Regteren Altena, 1950 : 206, text fig. 1, 2.

TYPE LOCALITY. Poetjangan layers (volcanic facies). — Kendeng beds, East Java, Indonesia ; Miocene.

TYPE MATERIAL. Holotype and 4 paratypes RML.

DISCUSSION. This species was originally refered to *Chicoreus* (*s.s.*) but it has the outline of a *Chicomurex* species. It is distinctive in having 4 varices on the last whorl. This species is not referable to *Chicoreus* (*s.l.*), and it is here tentatively included in the genus *Chicomurex*.

Chicomurex lophoessus (Tate, 1888)
Figs 84-85, 246-247

Murex lophoessus Tate, 1888 : 98, pl. 2, fig. 5.

TYPE LOCALITY. Fyansford Formation, Schnapper point, Victoria, Australia ; Miocene.

TYPE MATERIAL. Lectotype SAM T424, here selected from 8 syntypes.

REMARKS. Classified in *Chicoreus* (*s.s.*) by VOKES (1971 : 67) it is here transferred to the genus *Chicomurex*.

FIGS 246-247. — Genus *Chicomurex*.
246-247, *C. lophoessus* (Tate). 246, Schnapper Point, Victoria, Australia, Miocene, 39 mm (lectotype, SAM T424). 247, Balcombe Bay, Mornington, Victoria, Australia, Miocene, 35.5 mm (MV 110690-1).

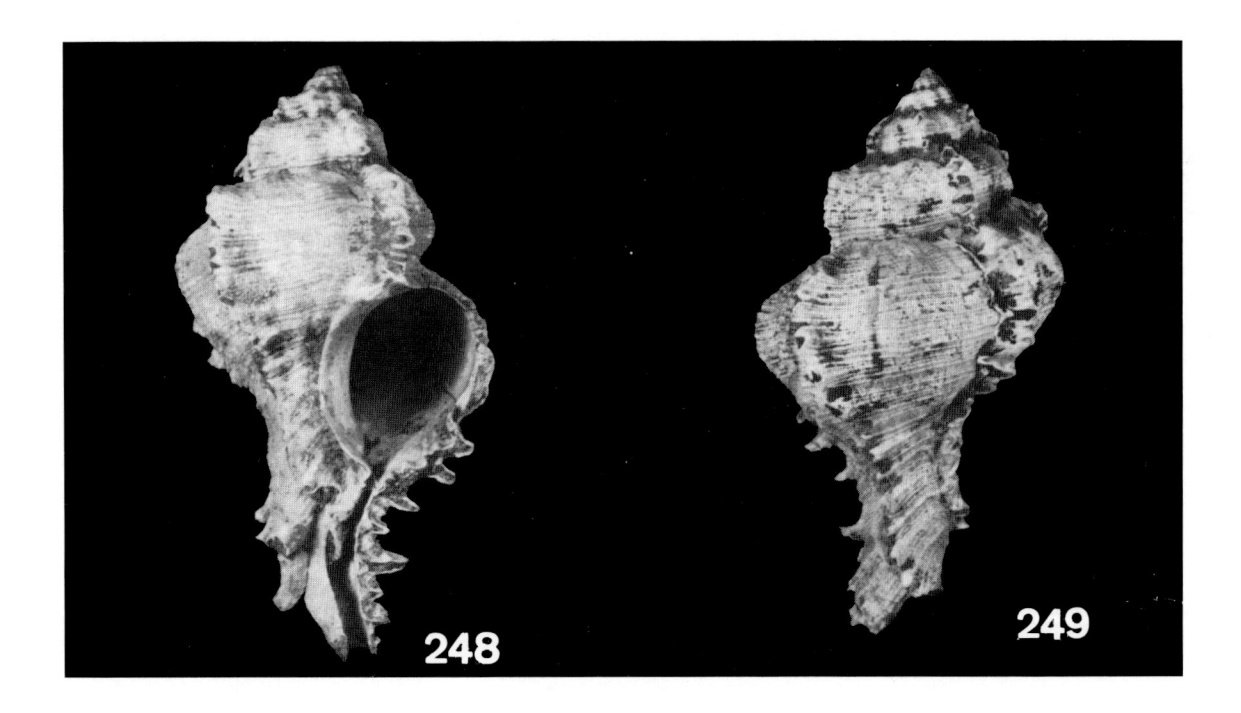

FIGS 248-249. — *Chicomurex kendengensis* (van Regteren Altena), East Java, Indonesia, Miocene, 46 mm (holotype, RML).

ACKNOWLEDGEMENTS

This work has been possible due to the help of many people and I would like to thank here all of those who permitted me to examine type material or who provided useful information, photographs or other material for examination, or who helped me in one way or another. Their help and collaboration is most appreciated : Dr G. Arbocco (MCST), Mr P. Bert (France), Dr M. Bishop (UMZ), Mr M. Blöcher (W. Germany), Dr J. Boscheinen (LM), Dr P. Bouchet (and staff) (MNHN), Dr S. Boyd (MV), Dr J. Breurois (MHNM), Prof. J. D. Campbell (UO), Dr W. O. Cernohorsky (AIM), Dr H. E. Coomans (ZMA), Mr J. Dajoz (MHNG), Mr A. D'Attilio (SDNHM), Mr K. DeWolfe (Chicago Academy of Sciences, Chicago, USA), Mr J. Drivas (Reunion Id.), Mr C. P. Fernandes (Portugal), Dr H. Feustel (HLD), Dr Y. Finet (MHNG), Mr R. Foster (Calif., U.S.A.), Mr A. Garback (ANSP), Dr E. Gittenberger (RMNH), Mr C. Glass (Calif., U.S.A.), Dr P. G. Heltne (Chicago Academy of Sciences, Chicago, U.S.A.), Mr R. Isaacs (U.K.), Dr A. W. Janssen (RML), Dr R. Janssen (SMF), Dr P. Jung (NHMB), Dr Y. Kanie (YCM), Dr R. N. Kilburn (NM), Dr R. Kilias (ZMB), Dr S. Kosuge (IMT), Prof. T. Kotaka (IGPS), Mr T. C. Lan (Taiwan), Mrs A. Lesage (Belgium), Mr G. Lévêque (New Caledonia), Mr A. Lievrouw (IRSNB), Mr I. Loch (AMS), Dr. C. Maugenest (MGM), Dr P. A. Maxwell (NZGS), Mr C. Muhasay (Philippines), Mr H. Mühläusser (Germany), Dr K. Muraoka (KPM), Prof. T. Okutani (NSMT), Dr P. G. Oliver (NMW), Mr T. Pain (U.K.), Mr D. Peled (Israël), Dr N. Pledge (SAM), Dr W. F. Ponder (AMS), Dr D. Potter (QM), Mrs A. Richards (New Guinea), Dr R. Robertson (ANSP), the late Dr J. Rosewater (USNM), Dr C. Ruggieri (IMG), Dr S. C. Shah (GSI), Mr F. J. Springsteen (Australia), Mr V. Sweetman (Australia), Mr R. J. Symonds (UMZ), Dr A. Trew (NMW), Mr P. K. Tubbs (ICZN), Dr J. van Goethem and staff (IRSNB), Mr A. van Walleghem (Belgium), Mr R. Vasile (Chicago Academy of Sciences), Dr C. Vaucher (MHNG), Prof. E. H. Vokes (TU), Ms K. M. Way (BMNH), Dr F. Wells (WAM) and Mrs T. Whitehead (Australia). I also thank Mrs J. Buyle (Belgium) for processing some of my photographs.

Dr P. Bouchet and Dr A. Warén (Stockholm) were most helpful for preparation and SEM work of the radulae. I would also like to thank Dr P. Bouchet, Prof. E. H. Vokes (Tulane University), and especially Mr B. A. Marshall (National Museum of New Zealand, Wellington) for considerable editorial work on the manuscript.

FIGS 250-254. — Subgenus *C. (Triplex)*.
250, *C. peledi* Vokes. Eilat, Red Sea, 65 mm (RH).
251, *C. nobilis* Shikama. Off Cebu I., Philippines, 55.6 mm (holotype, KPM 3280).
252, *C. rubescens* (Broderip). ?Tahiti, 48 mm (RH).
253, *C. ryukyuensis* Shikama. Okinawa Is, 33 mm (KPM 3281).
254, *C. zululandensis* Houart. North Zululand, South Africa, 31.5 mm (holotype, NM D8049).

FIGS 255-259. — Subgenus *C. (Triplex)*.
255, *C. boucheti* Houart. South of New Caledonia, 31.2 mm (holotype, MNHN).
256, *C. subpalmatus* Houart. South of New Caledonia, 30 mm (holotype, MNHN).
257, *C. crosnieri* Houart. South of Madagascar, 37 mm (holotype, MNHN).
258, *C. rossiteri* (Crosse). New Caledonia, 45.1 mm (MNHN).
259, *C. paucifrondosus* Houart. New Caledonia, off Grand Récif Sud, 28.7 mm (holotype, MNHN).

FIGS 260-264. — Subgenera *C. (Triplex)* and *C. (Chicopinnatus)*, genus *Chicomurex*.
260, *C. (Triplex) dovi* Houart. Mombasa, Kenya, 77 mm (paratype, RH).
261, *C. (T.) bourguignati* (Poirier). Nossi-Bé, Madagascar, 81 mm (lectotype, MNHN).
262, *C. (Chicopinnatus) orchidiflorus* (Shikama). Tubuaï I., 25.8 mm (MNHN).
263, *C. (C.) guillei* (Houart). Off Réunion, 33.2 mm (holotype, MNHN).
264, *Chicomurex turschi* (Houart). Off Durangit, Papua New Guinea, 35.5 mm (paratype, RH).

FIGS 265-269. — Genus *Chicomurex*.
265, *C. venustulus* (Rehder & Wilson). Taiwan, 54 mm (RH).
266, *C. problematicus* (Lan). Philippines, 70 mm (RH).
267, *C. laciniatus* (Sowerby). Chesterfields, 38.5 mm (MNHN).
268, *C. venustulus* (Rehder & Wilson). Off Cebu I., Philippines, 49.5 mm (holotype of *Chicoreus gloriosus* Shikama, KPM 3277).
269, *C. elliscrossi* (Fair). Wakayama, Japan, 73 mm (RH).

FIGS 270-273. — Subgenus *C. (Chicoreus)*.
270, *C. ramosus* (Linné). New Caledonia, 180 mm (RH).
271, *C. austramosus* Vokes. S. Africa, 65.2 mm (coll. GLASS & FOSTER).
272, *C. ramosus* (Linné). Natal, South Africa, 107 mm (RH).
273, *C. litos* Vokes. South Africa, 73 mm (coll. GLASS & FOSTER).

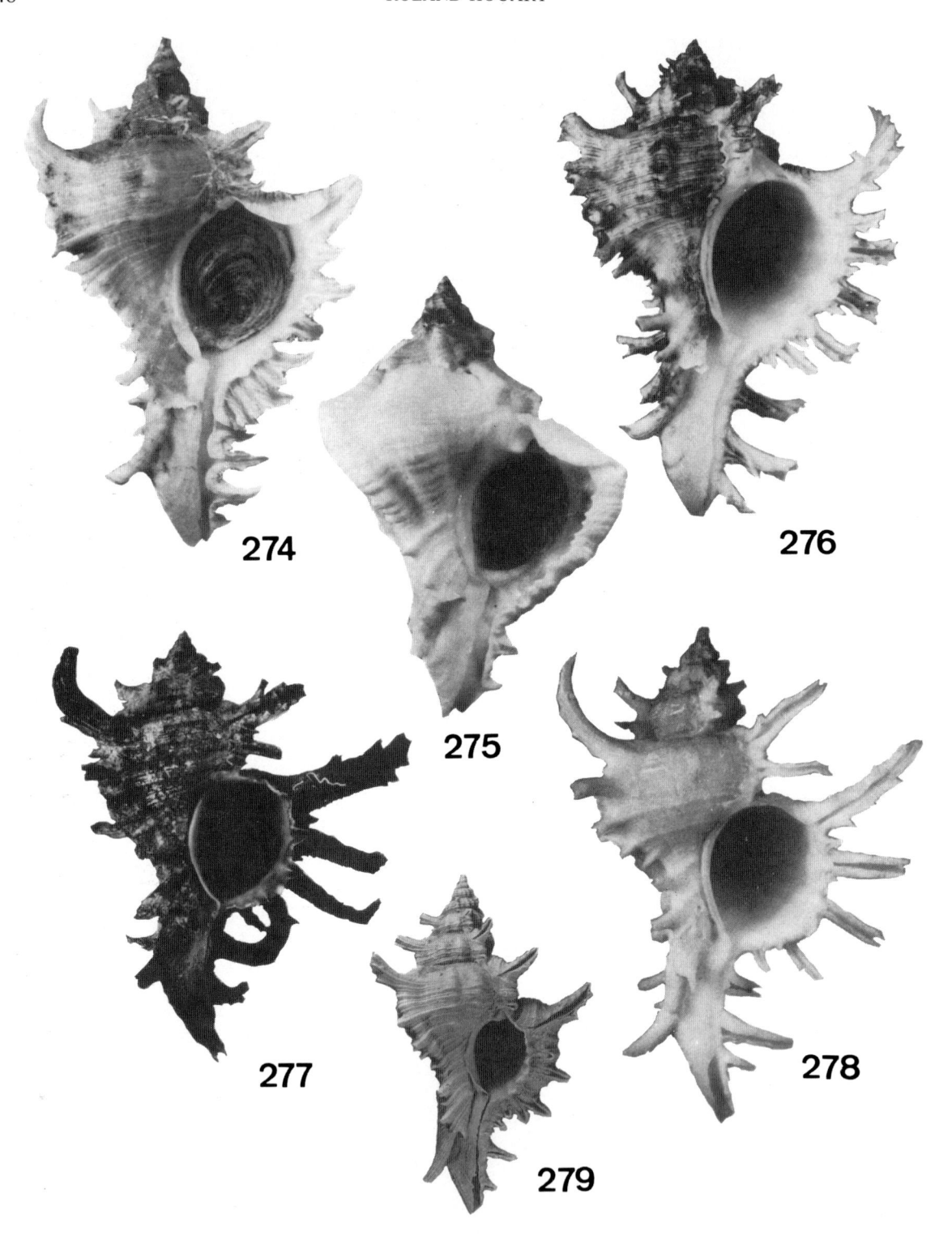

FIGS 274-279. — Subgenus *C. (Chicoreus)*.
274-275, *C. virgineus* (Röding). 274, Dahlak, Red Sea, 73 mm (MNHN). 275, Sri Lanka, 53.5 mm (MNHN).
276, *C. ramosus* (Linné). New Caledonia, juvenile, 90 mm (MNHN).
277-278, *C. cornucervi* (Röding). 277, West Australia, 78 mm (MNHN). 278, Arafura Sea, 87 mm (RH).
279, *C. litos* Vokes. North-east of Beira, Mozambique, 55.7 mm (holotype, NM G8656/T2130. Courtesy of E.H. VOKES).

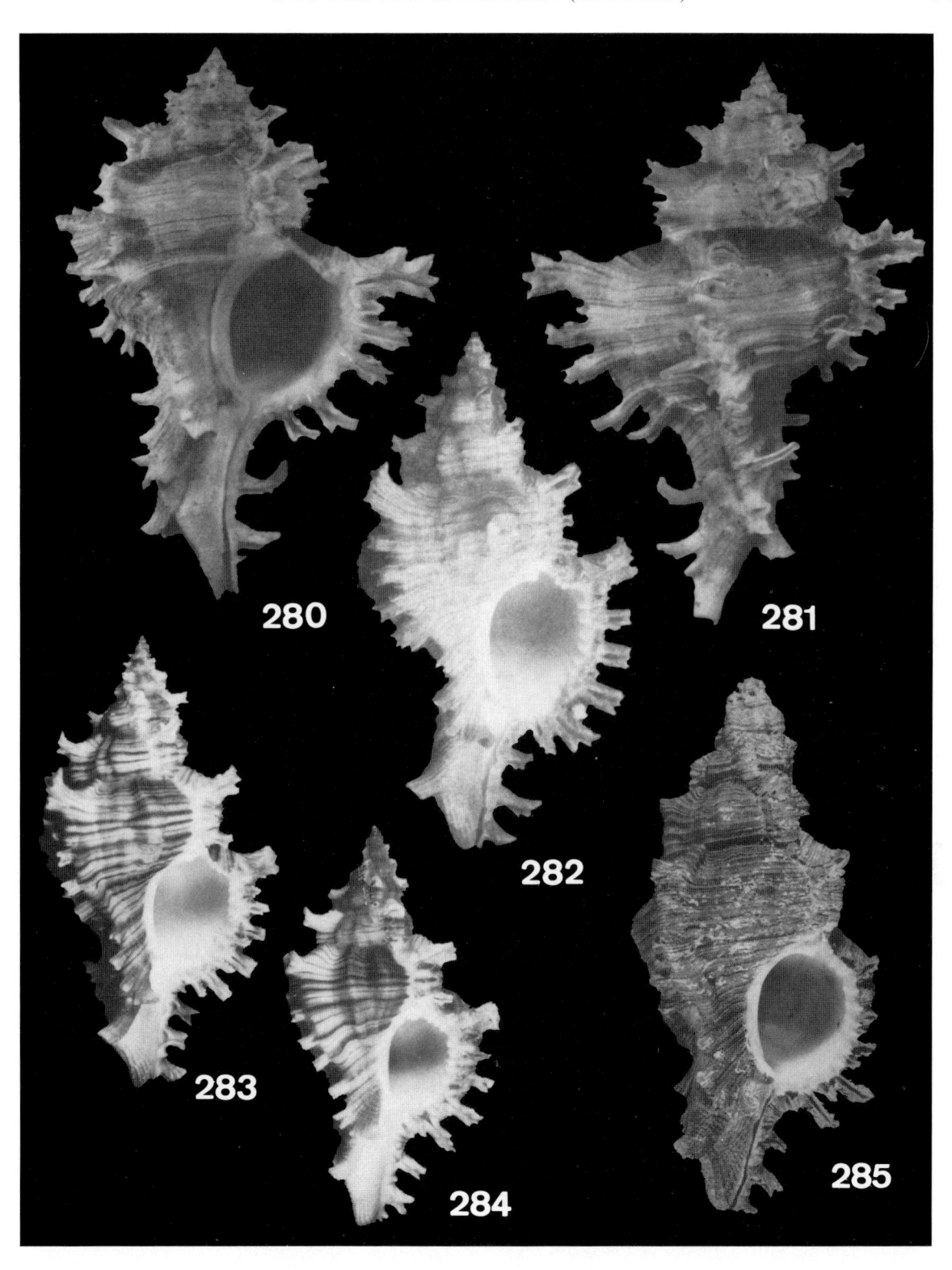

FIGS 280-285. — Subgenera *C. (Chicoreus)* and *C. (Triplex)*.
280-281, *C. (Chicoreus) bundharmai* Houart. Banjarmasin, South Borneo, 63.8 mm (holotype, MNHN).
282-285, *C. (Triplex) torrefactus* (Sowerby). 282, Philippines Is, 68.1 mm (lectotype of *Murex rubiginosus* Reeve, BMNH 197475). 283, locality unknown, 83.1 mm (lectotype of *Murex affinis* Reeve, BMNH 197499). 284, Papua New Guinea, juvenile, 30.1 mm (MNHN). 285, Diego-Suarez, 76 mm (holotype of *Murex rochebruni* Poirier, MNHN).

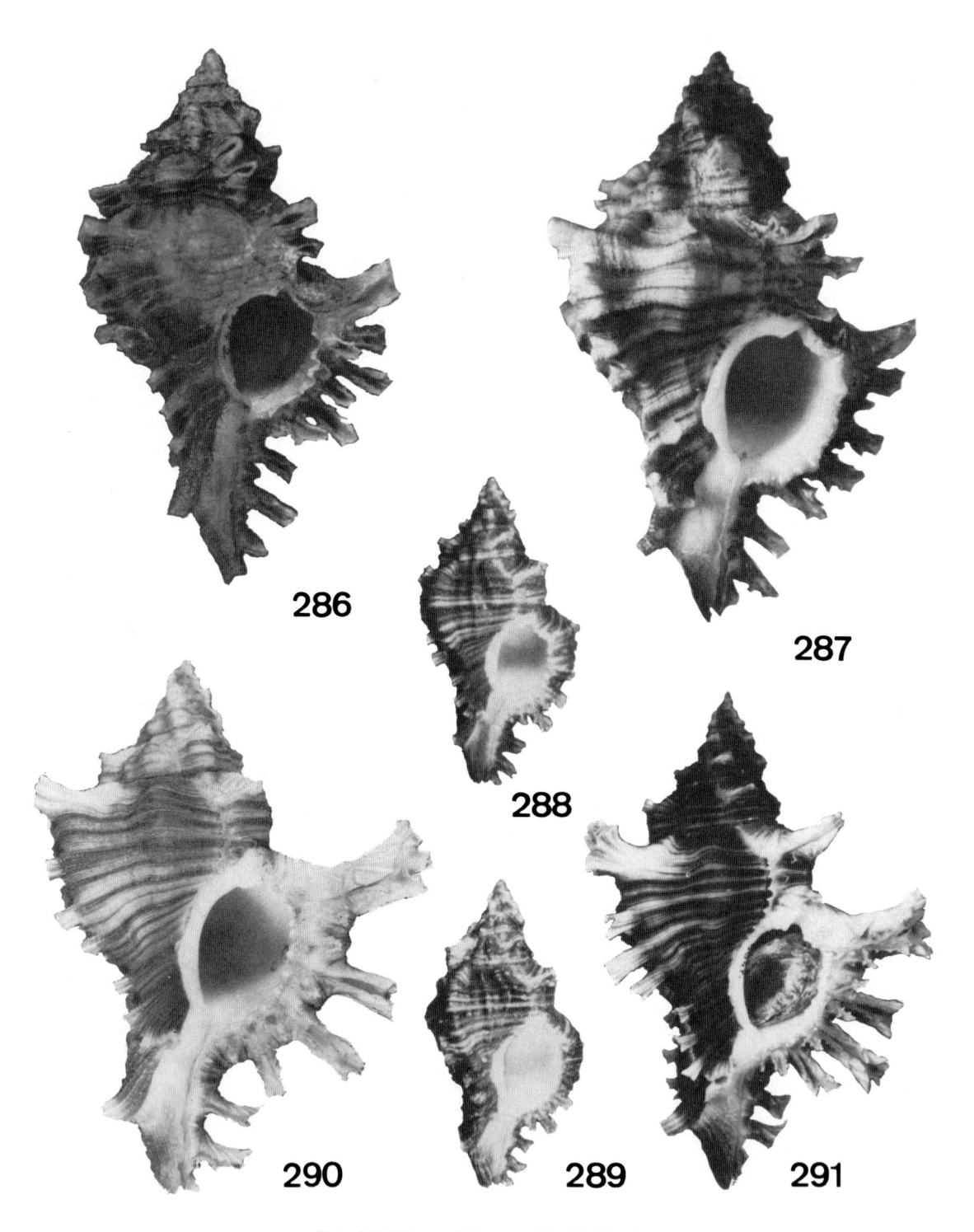

FIGS 286-291. — Subgenus *C. (Triplex)*.
286-287, *C. insularum* (Pilsbry). 286, off Waikiki, Oahu, 69 mm (holotype, ANSP 47192. Courtesy of R. Robertson). 287, Hawaii, 86 mm (RH).
288-291, *C. maurus* (Broderip). 288, Anaa I., Tuamotu, 49 mm (lectotype, BMNH 197473). 289, Marquesas, 46.5 mm (RH). 290, locality unknown, 66.5 mm (holotype of *Murex steeriae* Reeve, UMZ, courtesy of UMZ). 291, Marquesas, 79.5 mm (RH).

FIGS 292-297. — Subgenus *C. (Triplex)*.
292, *C. dovi* Houart. Malindi, Kenya, 102 mm (holotype, IRSNB 26656/402).
293-294, *C. palmarosae* (Lamarck). 293, Sri Lanka, 89 mm (RH). 294, Philippines, 97 mm (MNHN).
295, *C. saulii* (Sowerby). Philippines, 88 mm (MNHN).
296-297, *C. torrefactus* (Sowerby). 296, Broome area, West Australia, 45.2 mm (AMS C106255). 297, Seychelles, 68 mm (MNHN).

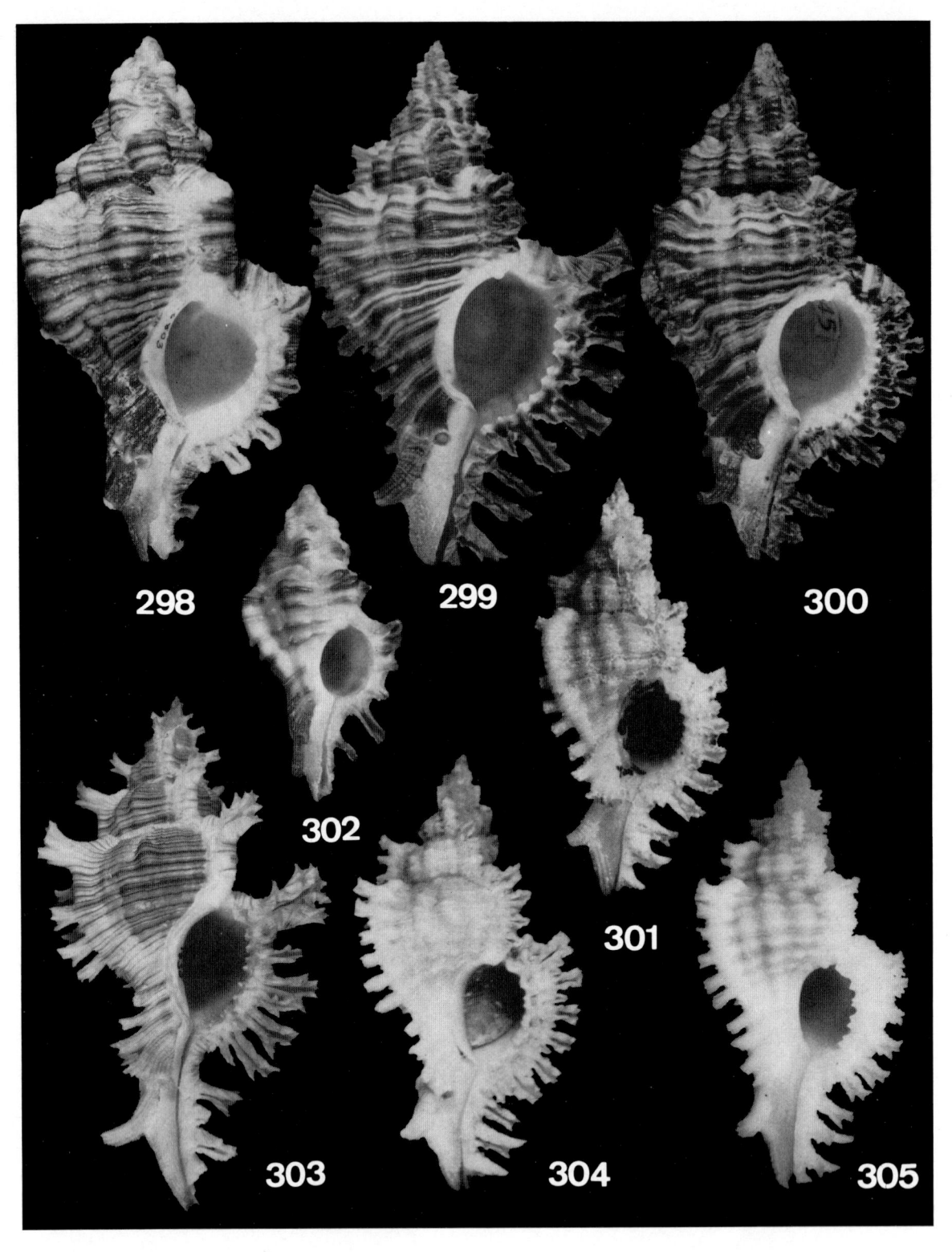

FIGS 298-305. — Subgenus *C. (Triplex)*.

298, *C. maurus* (Broderip). Gulf of California (erroneous), 82.3 mm (holotype of *Murex mexicanus* Stearns, USNM 46803. Courtesy of J. Rosewater).
299-301, *C. microphyllus* (Lamarck). 299, ?New Caledonia, 58.9 mm (lectotype of *Murex jousseaumei* Poirier, MNHN. Courtesy of P. Bouchet). 300, New Caledonia, 63.5 mm (lectotype of *Murex poirieri* Jousseaume, MNHN. Courtesy of P. Bouchet). 301, Tryon I., Queensland, Australia, 32 mm (AMS C106519).
303, *C. saulii* (Sowerby). Capul I., Philippines, 75 mm (holotype UMZ, courtesy of R.J. Symonds).
304-305, *C. strigatus* (Reeve). 304, Ryukyu I., 40.8 mm (lectotype of *Murex penchinati* Crosse, BMNH 1896.12.1.5). 305, Philippines, 32 mm (MNHN).

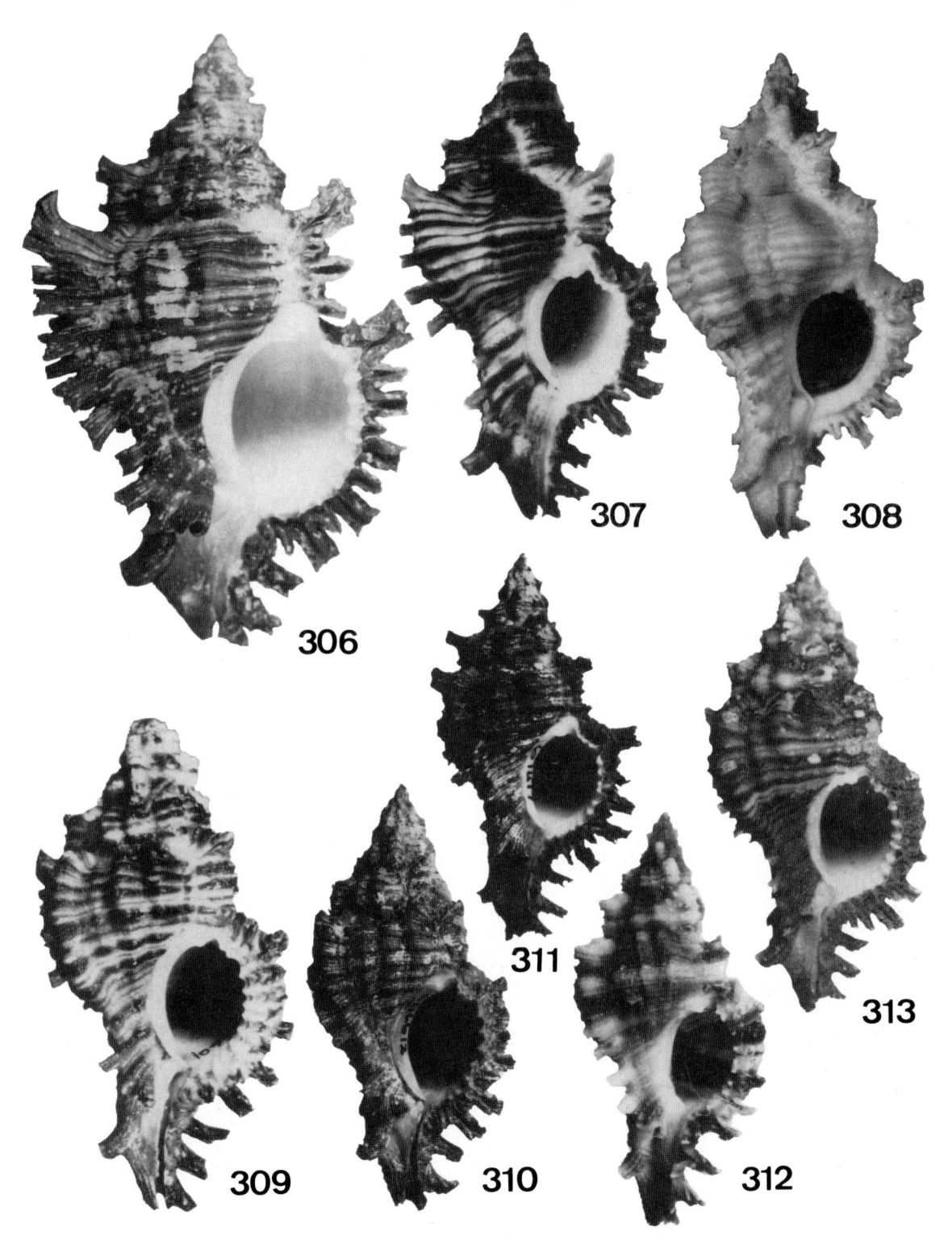

FIGS 306-312. — Subgenus *C. (Triplex)*.
306-308. *C. torrefactus* (Sowerby). 306, Broome, West Australia, 90 mm (RH). 307, Madagascar, 76 mm (RH). 308, Papua New Guinea, 86 mm, orange (RH).
309-313. *C. microphyllus* (Lamarck). 309, locality unknown, 61.2 mm (lectotype, MHNG 1099/22. Photo J. DAJOZ). 310, North Queensland, Australia, 49.8 mm (AMS C116612). 311, Queensland, Australia, 32 mm (AMS C106519). 312, New Caledonia, juvenile, 24 mm (MNHN). 313, Papua New Guinea, 60 mm (MNHN).

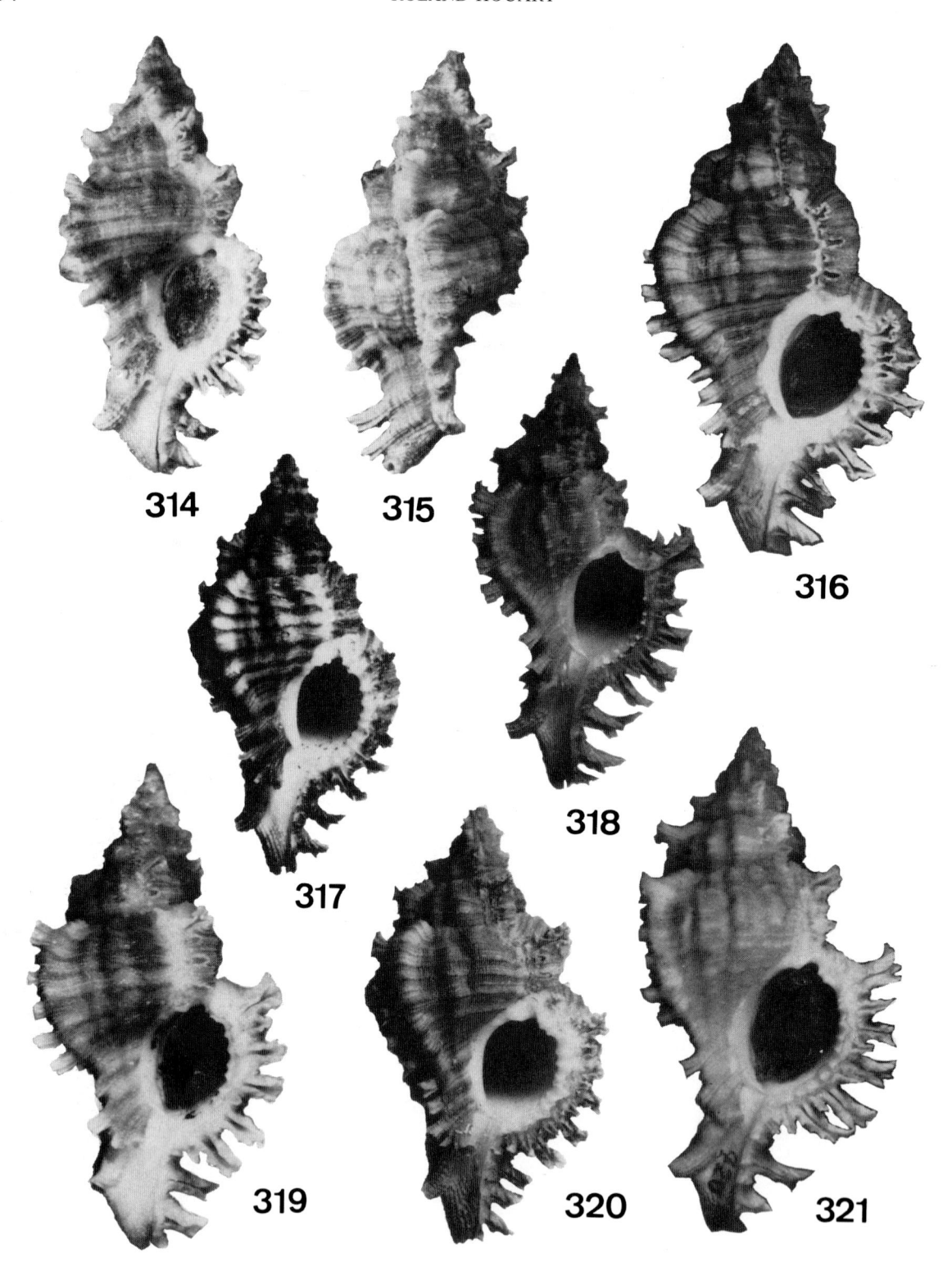

FIGS 314-321. — Subgenus *C. (Triplex)*.
314-321. *C. microphyllus* (Lamarck). 314-315, off Keppel I., Keppel Bay, Queensland, Australia, 51 mm (holotype of *Chicoreus akritos* Radwin & D'Attilio, SDNHM 53173). 316, Queensland, Australia, 79 mm (coll. O. RIPPINGALE). 317, Queensland, Australia, 48.5 mm (MNHN). 318, Philippines, 56 mm, orange (RH). 319, Queensland, Australia, 44 mm (AMS C106741). 320, Fiji, 40.5 mm (AMS C67477). 321, Queensland, Australia, 58 mm, orange (RH).

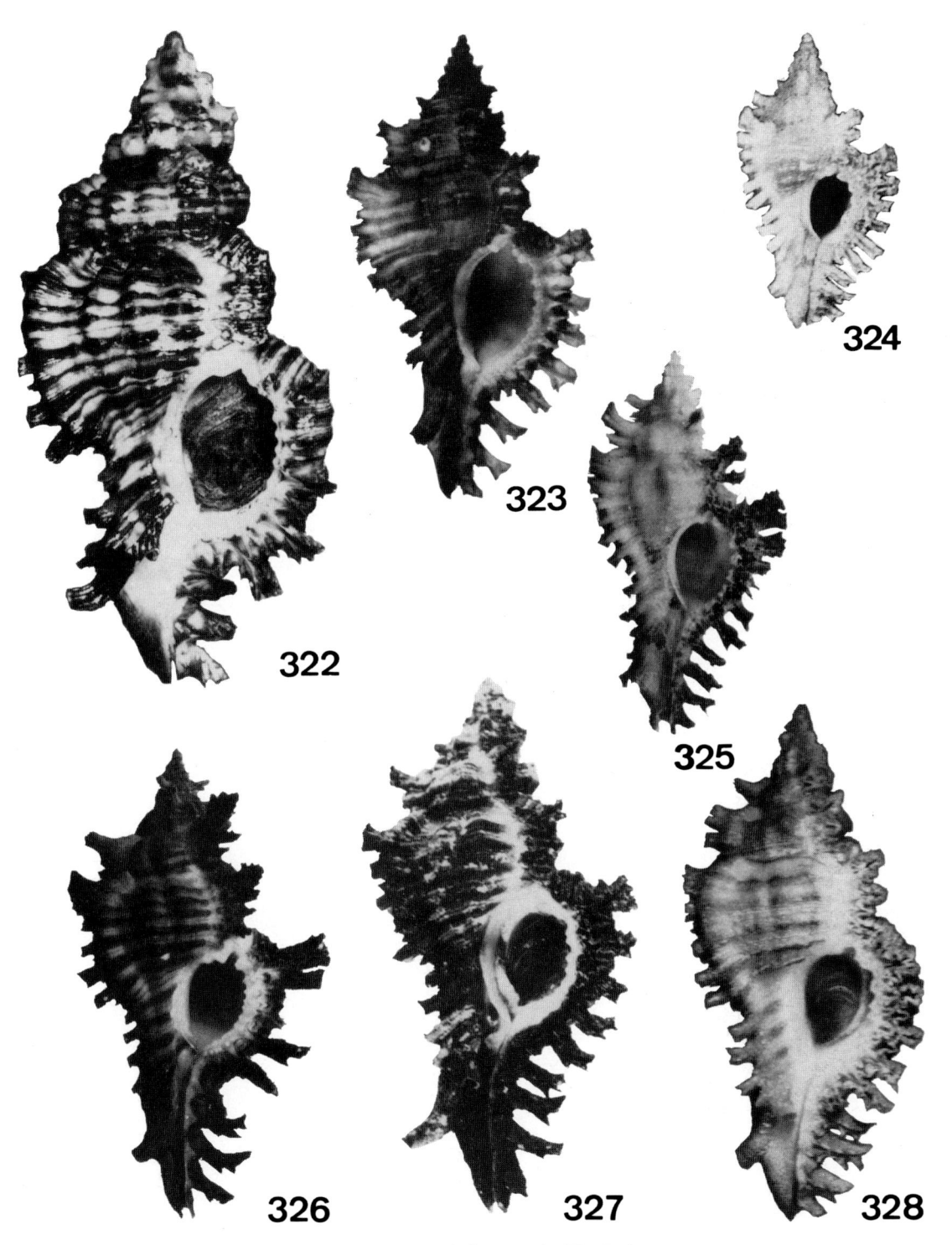

FIGS 322-328. — Subgenus *C. (Triplex)*.
322-323. *C. microphyllus* (Lamarck). 322, Queensland, Australia, 89 mm (AMS C106509). 323, Philippines, 40.5 mm (RH).
324-325. *C. strigatus* (Reeve). 324, locality unknown, 20.5 mm (holotype of *Murex multifrondosus* Sowerby, MNHN). 325, Philippines, 33 mm (RH).
326-328. *C. paini* Houart. 326, Honiara, Guadalcanal I., Solomon Is, 43 mm (holotype, IRSNB 26554/396). 327, Honiara, Guadalcanal I., Solomon Is, 53.5 mm (holotype of *Chicoreus kengaluae* Mühlhäusser & Alf, ZSM 1742. Courtesy of H. MÜHLHÄUSSER). 328, Solomon Is, 52 mm, orange (coll. R. Isaacs).

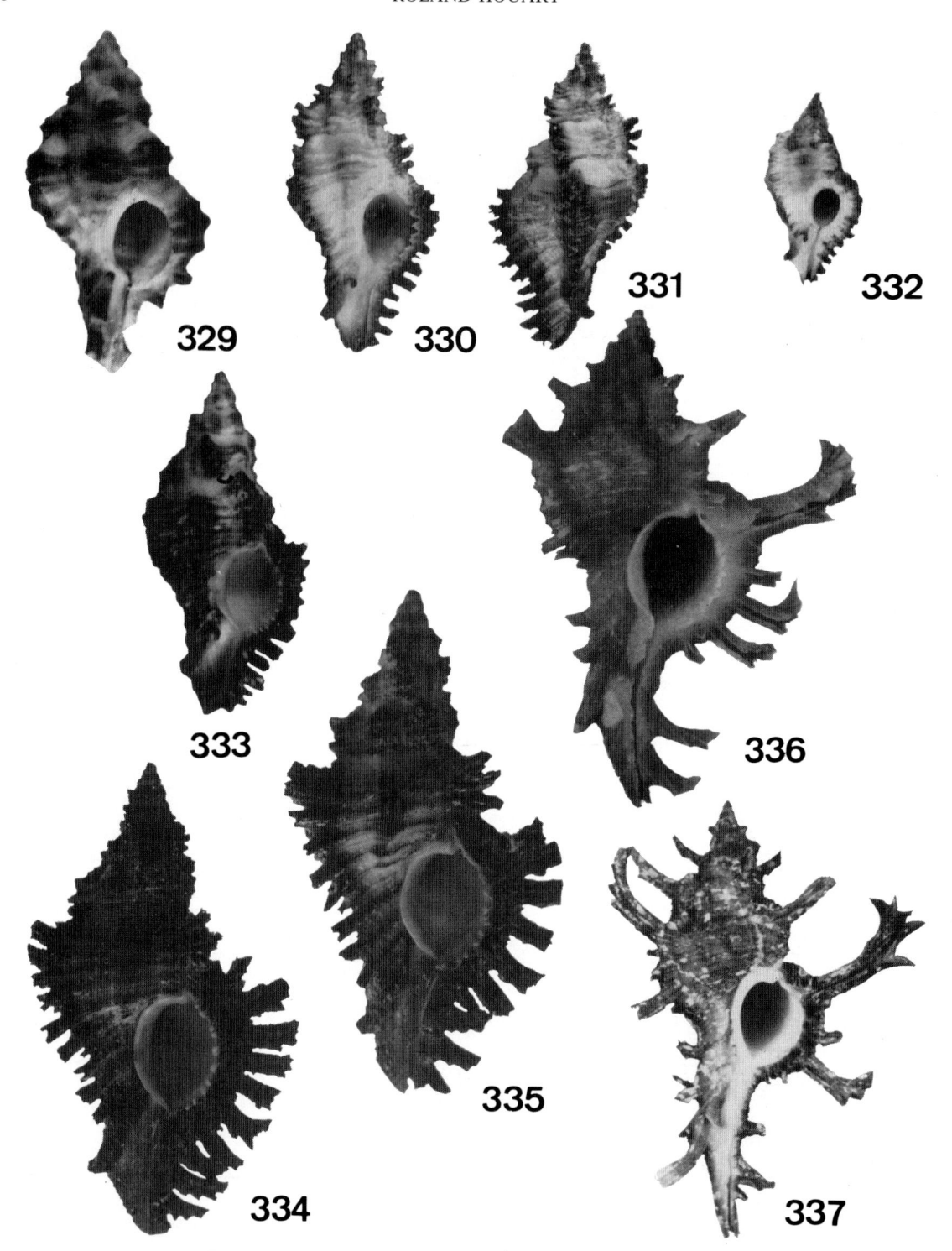

FIGS 329-337. — Subgenus *C. (Triplex)*.
329, *C. rubescens* (Broderip). Tuamotu, 30 mm (IRSNB IG 10591).
330-335, *C. trivialis* (A. Adams). 330-331, locality unknown, 37.5 mm (lectotype, BMNH 1980136). 332, locality unknown, 24.1 mm (paralectotype, BMNH 1980136). 333, Broome, West Australia, 42 mm (MNHN). 334, Cockatoo I., N.W. Australia, 49.5 mm (MNHN). 335, Cockatoo I., N.W. Australia, 54 mm (RH).
336-337, *C. axicornis* (Lamarck). 336, Mozambique, 39.5 mm (MNHN). 337, Celebes, 60.5 mm (MNHN).

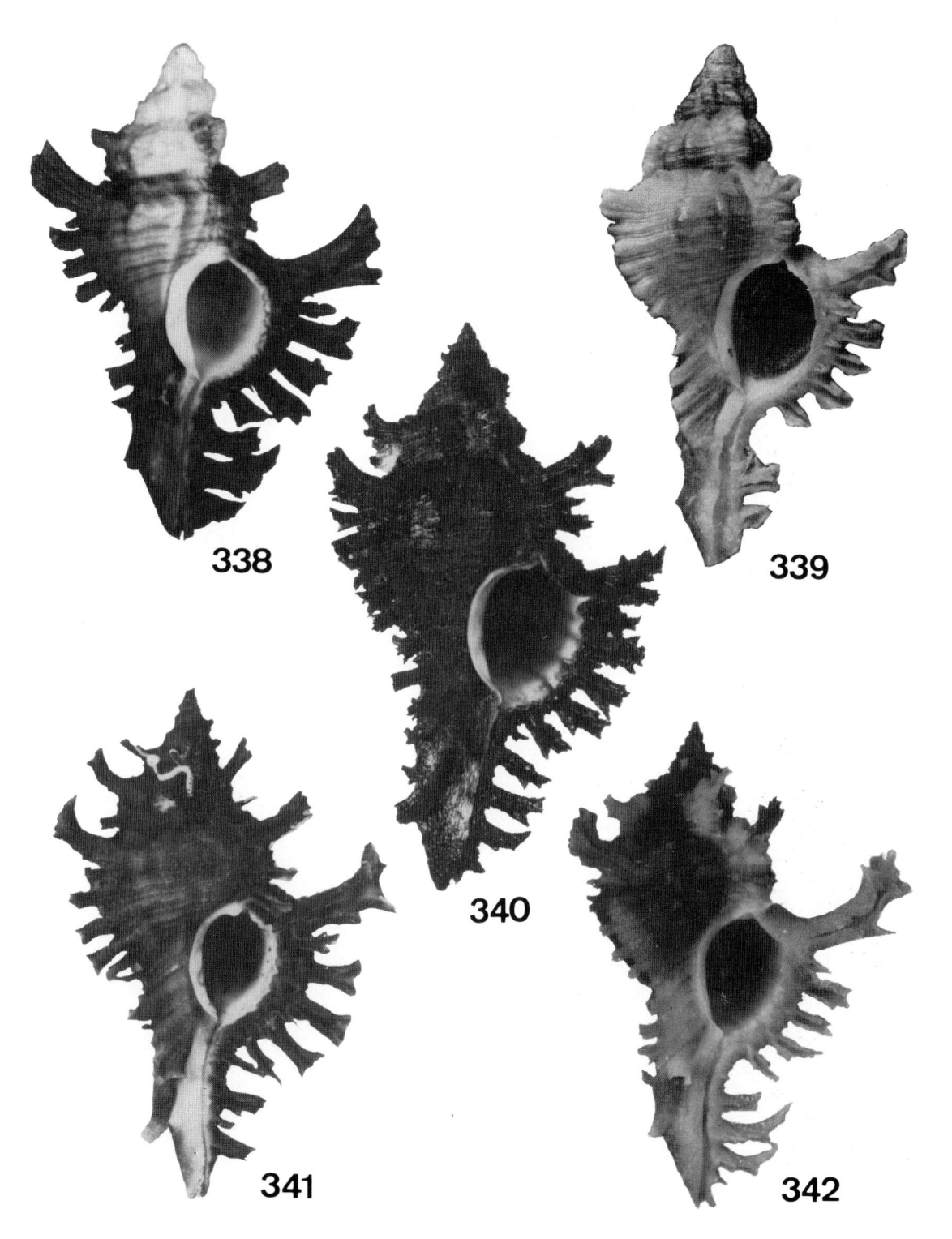

FIGS 338-342. — Subgenus *C. (Triplex)*.
338-342, *C. banksii* (Sowerby). 338, Philippines, 69 mm (MNHN). 339, locality unknown, 56.5 mm (Neotype, BMNH 1984076. Courtesy of E.H. VOKES). 340, Singapore, 65.5 mm (MNHN). 341, Philippines, 53.2 mm (coll. D. PELED). 342, Philippines, 49 mm (RH).

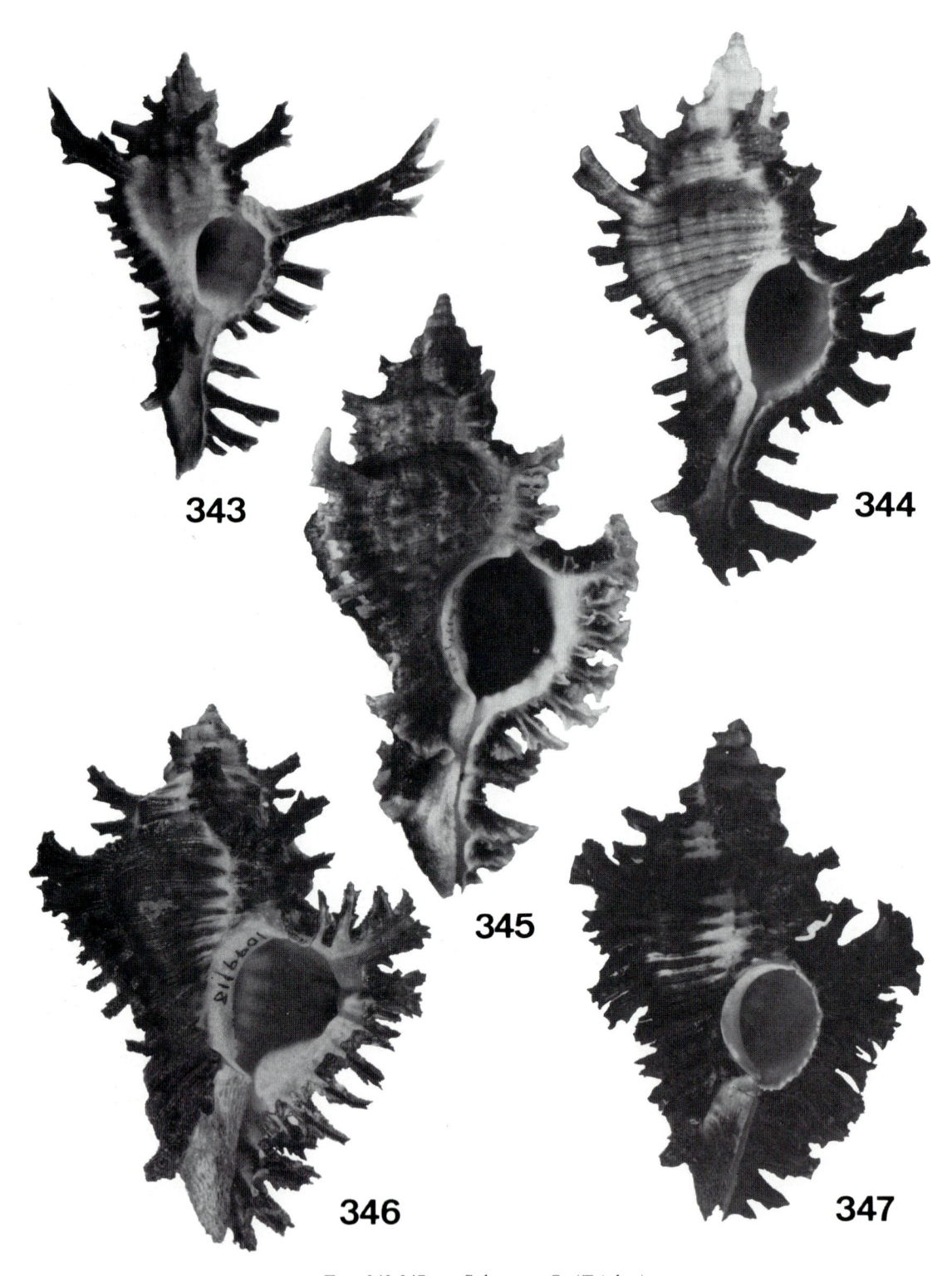

FIGS 343-347. — Subgenus *C. (Triplex)*.
343, *C. banksii* (Sowerby). North Queensland, Australia, 35.5 mm (MNHN).
344-345, *C. bourguignati* (Poirier). 344, Seychelles, 55 mm (MNHN). 345, Mozambique Channel, 78.5 mm (RH).
346-347, *C. brunneus* (Link). 346, Indian Ocean, 86.3 mm (lectotype of *Murex adustus* Lamarck, MHNG 1099/18. Photo J. DAJOZ). 347, Phuket, 80.5 mm (RH).

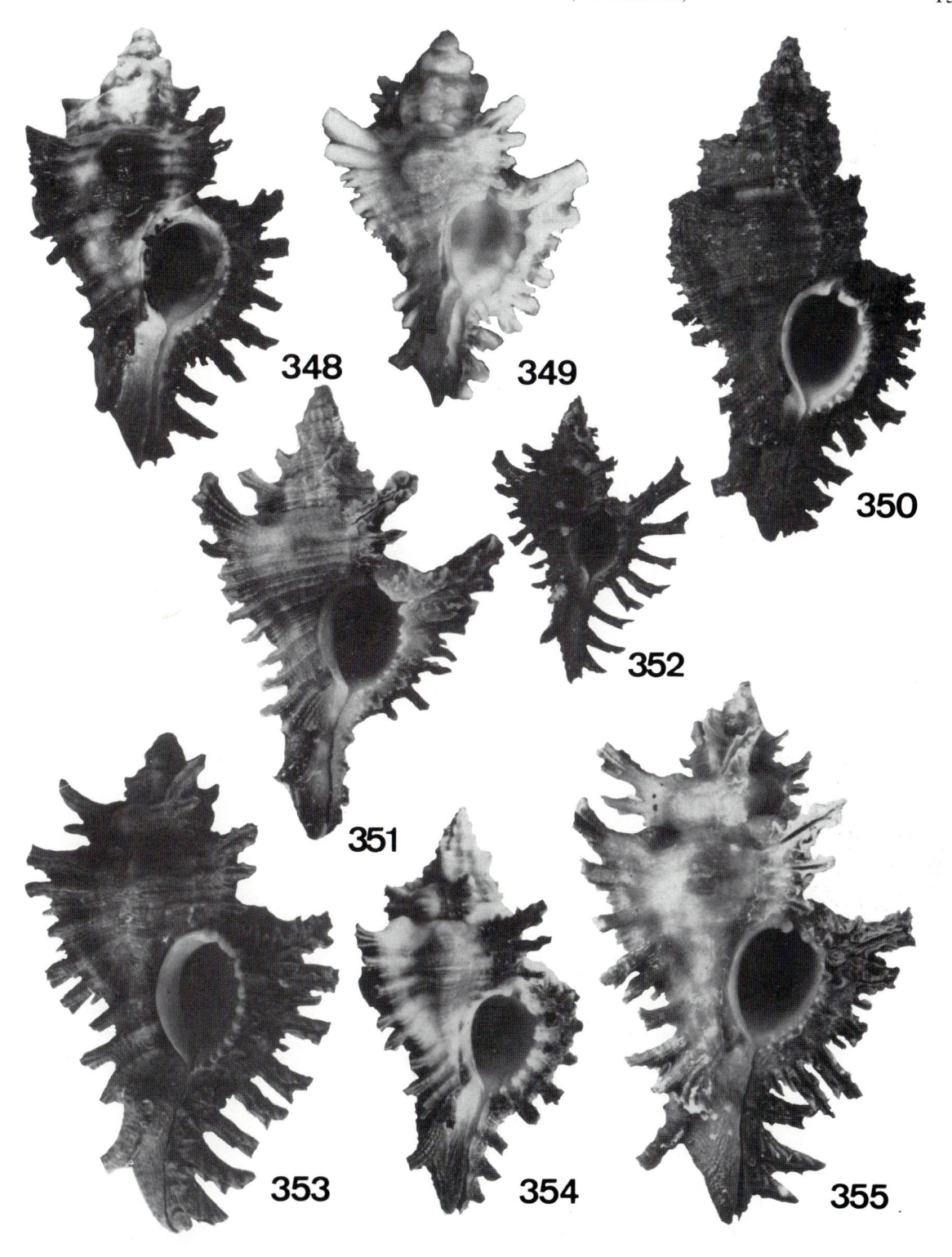

FIGS 348-355. — Subgenus *C. (Triplex)*.

348-355, *C. brunneus* (Link). 348, West Indies (erroneous), 53 mm (lectotype of *Murex despectus* A. Adams, BMNH 1980135). 349, Australia, 44 mm (lectotype of *Murex australiensis* A. Adams, BMNH 1980130). 350, New Caledonia, 52.5 mm (MNHN). 351, Chesterfields, 32 mm (MNHN). 352, New Caledonia, juvenile, 26.5 mm (MNHN). 353, New Caledonia, 49.5 mm (MNHN). 354, New Caledonia, 36 mm (MNHN). 355, Australia (no other data), 77.1 mm (RH).

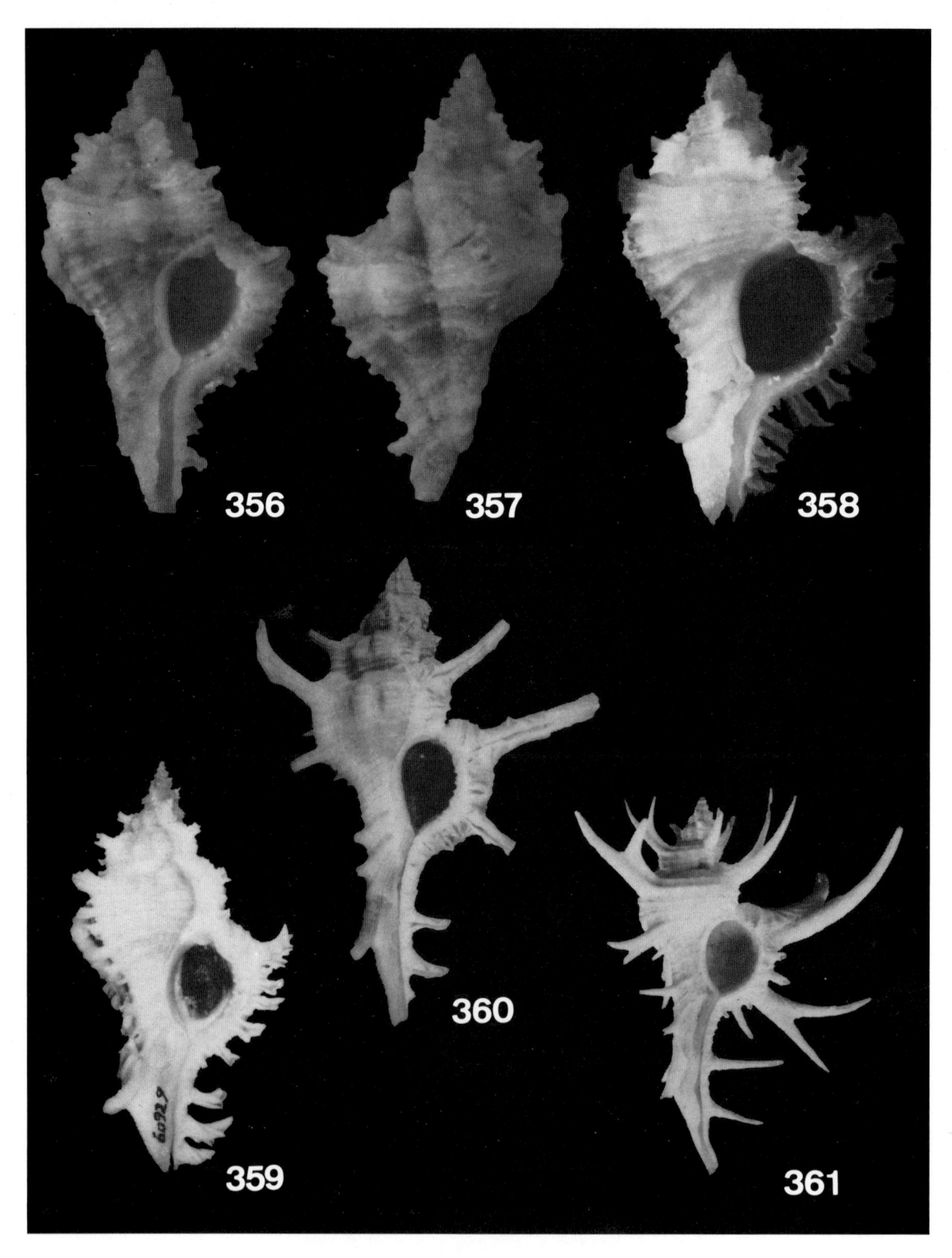

FIGS 356-361. — Subgenus *C. (Triplex)*.
356-357, *C. elisae* Bozzetti. Off Capo Ras Hafun, Somalia, 24.2 mm (paratype, RH).
358, *C. banksii* (Sowerby). Anna Plains, West Australia, 45.9 mm (RH).
359, *C. ryukyuensis* Shikama. Okinawa Is, 37.9 mm (holotype, NSMT 60929. Courtesy of T. OKUTANI).
360, *C. axicornis* (Lamarck). Southwest Taiwan, 77 mm (holotype of *Murex kawamurai* Shikama, NSMT 61245. Courtesy of T. OKUTANI).
361, *C. cervicornis* (Lamarck). Australia, 57 mm (holotype, MNHN).

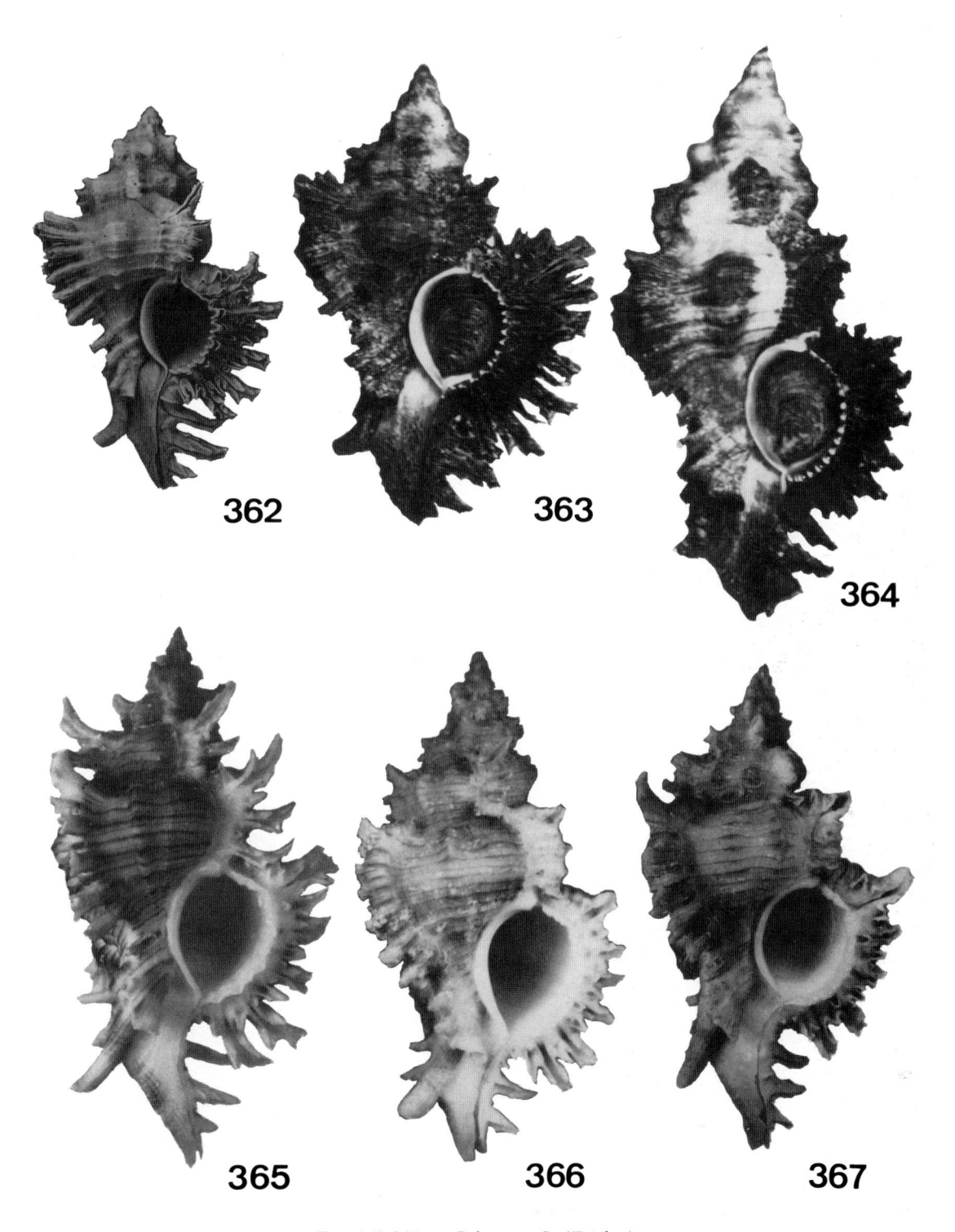

FIGS 362-367. — Subgenus *C. (Triplex)*.
362-364, *C. groschi* Vokes. 362, South-West Conducia Bay, Mozambique, 62.3 mm (holotype NM H192/T2134. Courtesy of E.H. VOKES). 363, Mozambique, 53 mm (coll. R. ISAACS). 364, Mozambique, 64.5 mm (coll. R. ISAACS).
365-367, *C. cnissodus* (Euthyme). 365, Taiwan, 86 mm (MNHN). 366, Sri Lanka, 79 mm (RH). 367, Taiwan, 87.1 mm (RH).

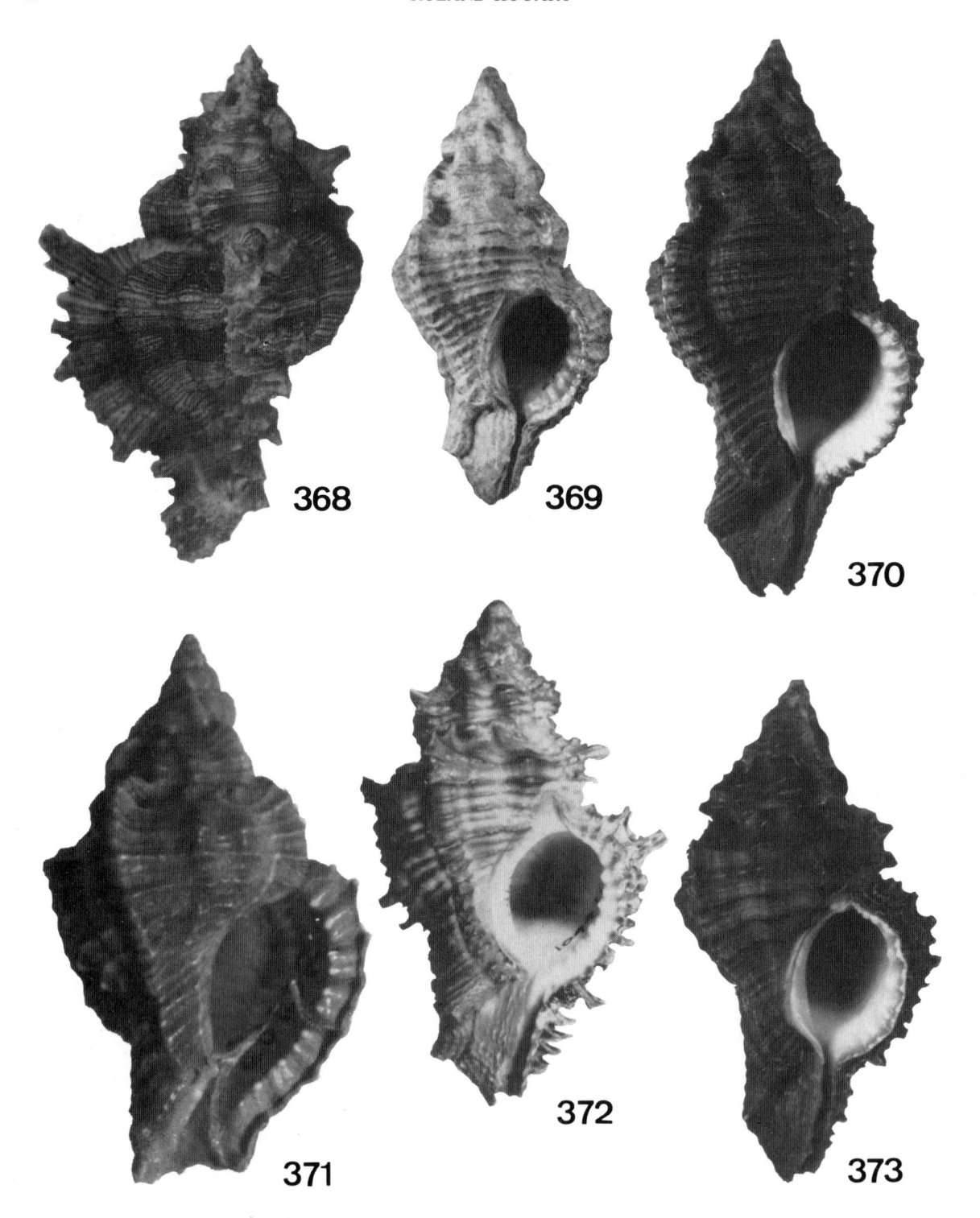

FIGS 368-373. — Subgenera *C. (Triplex)* and *C. (Rhizophorimurex)*.
368, *C. (Triplex) peledi* Vokes. Eilat, Red Sea, 58.8 mm (holotype, HUJ 10129. Courtesy of E.H. VOKES).
369-373, *C. (Rhizophorimurex) capucinus* (Lamarck). 369, off Mapoon, Queensland, Australia, 55 mm (holotype of *Murex permaestus* Hedley, AMS C14130). 370, Philippines, 60.1 mm (MNHN). 371, Australia, 35 mm (holotype of *Murex bituberculatus* Baker, CAS 20702. Courtesy of R. VASILE). 372, locality unknown, 70.4 mm (holotype of *Murex quadrifrons* Lamarck, MHNG 1099/45. Photo J. DAJOZ). 373, West Africa (erroneous), 54.9 mm (lectotype of *Murex lignarius* A. Adams, BMNH 1985229).

FIGS 374-379. — Subgenera *C. (Rhizophorimurex)* and *C. (Triplex)*.
374-375, *C. (Rhizophorimurex) capucinus* (Lamarck). 374, Philippines, 90 mm (MNHN). 375, North Queensland, Australia, 59 mm (AMS C114069).
376, *C. (Triplex) corrugatus corrugatus* (Sowerby). Eilat, Red Sea, 46 mm (coll. R. ISAACS).
377-378, *C. (Triplex) damicornis* (Hedley). 377, Queensland, Australia, 42.5 mm (MNHN). 378, Locality unknown, 54 mm (RH).
379, *C. (Triplex) denudatus* (Perry). Queensland, Australia, 45.5 mm (MNHN).

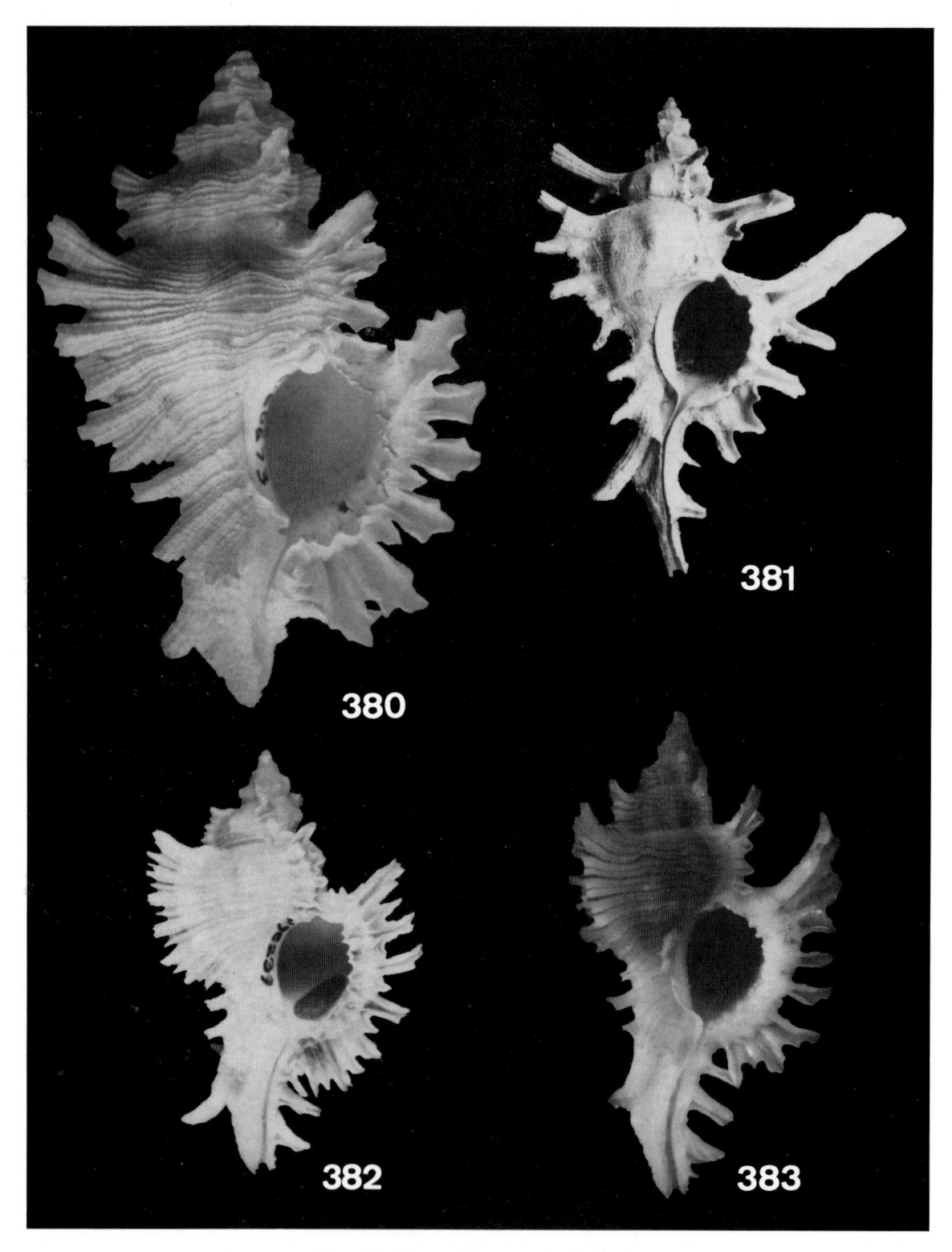

FIGS 380-383. — Subgenus *C. (Triplex)*.
380, *C. denudatus* (Perry). Sydney Harbour, 44.5 mm (holotype of *Torvamurex extraneus* Iredale, AMS C60673).
381, *C. damicornis* (Hedley). Shoalhaven Bight, New South Wales, Australia, 55.5 mm (holotype, AMS C16416).
382-383, *C. territus* (Reeve). 382, locality unknown, 52 mm (holotype, BMNH 198239). 383, New Caledonia, 22 mm (MNHN).

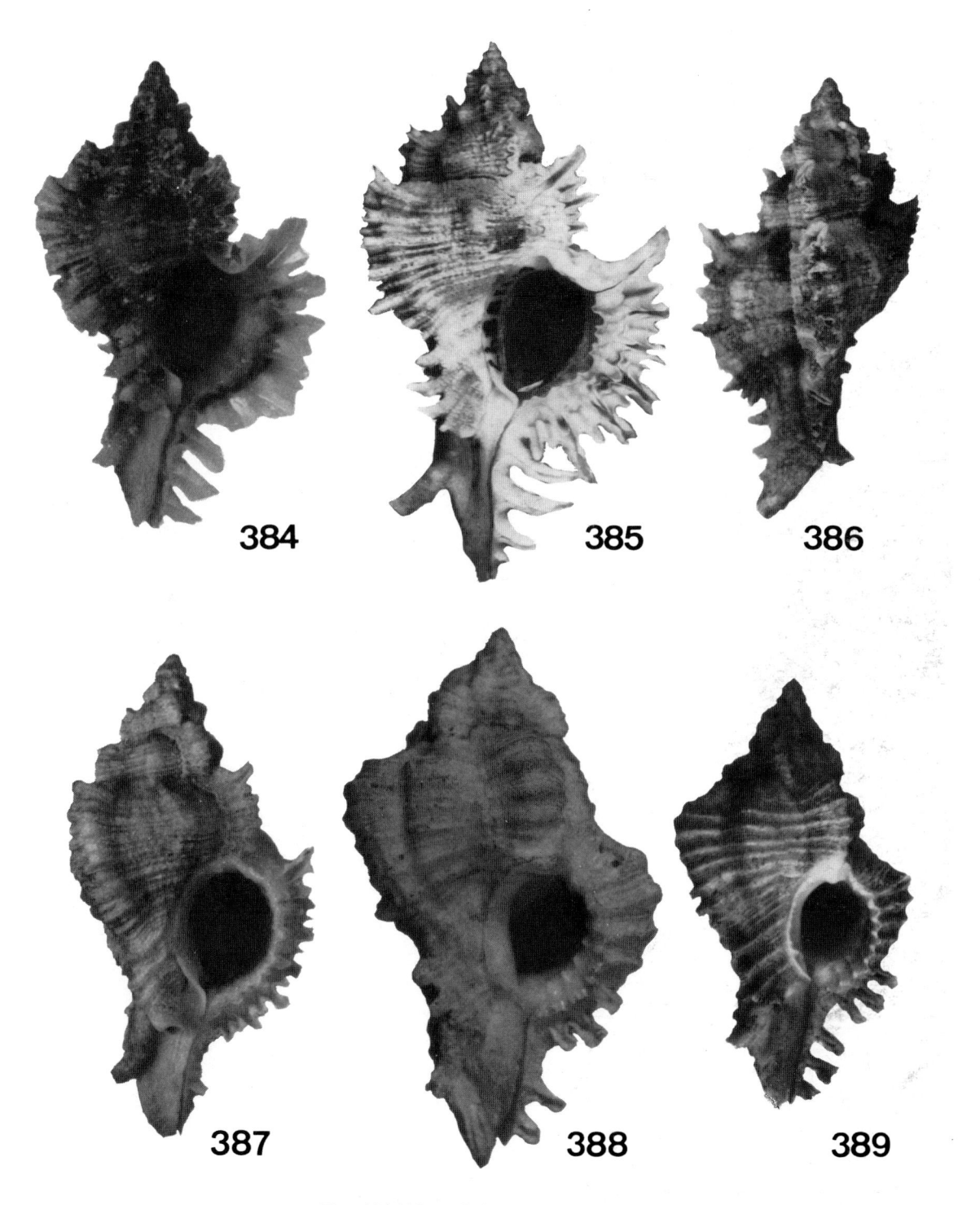

FIGS 384-389. — Subgenus *C. (Triplex)*.
384, *C. denudatus* (Perry). New South Wales, Australia, 27.5 mm, juvenile (RH).
385-387, *C. territus* (Reeve). 385, Queensland, Australia, 50 mm (AMS C113897). 386, Broome, West Australia (locality doubtful), 52 mm (RH). 387, Queensland, Australia, 43.2 mm (RH).
388-389, *C. thomasi* (Crosse). 388, Marquesas Is, 45.5 mm (lectotype, BMNH 1902.5.28.53). 389, Marquesas, 39.8 mm (IRSNB IG 10591).

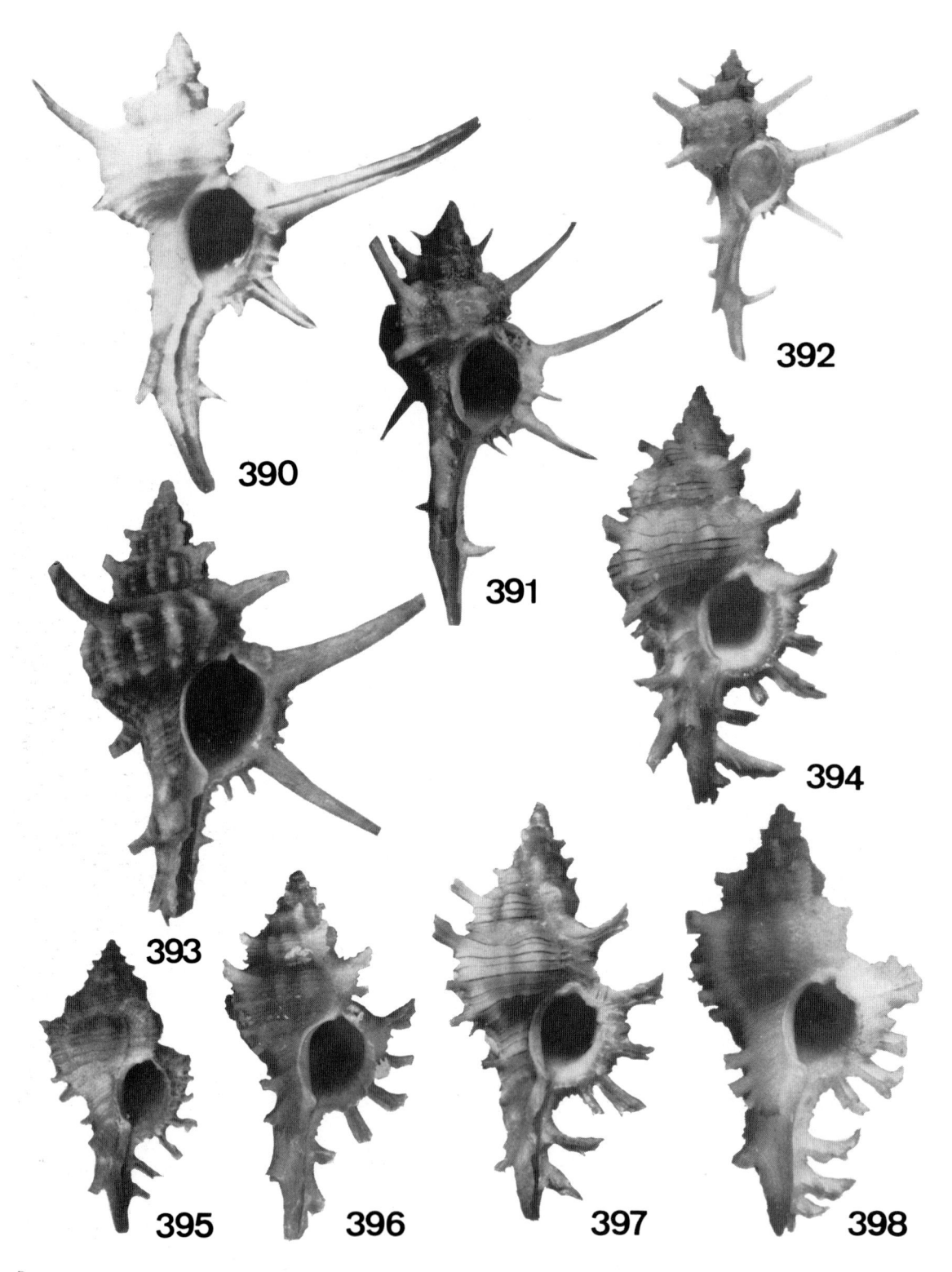

FIGS 390-398. — Subgenus *C. (Triplex)*.

390-393, *C. longicornis* (Dunker). 390, East Australia, 26°05′ S, 42 mm (lectotype of *Murex recticornis* Kobelt, LM. Courtesy of J. BOSCHEINEN). 391, East Australia, 26°05′ S, 35.2 mm (paralectotype of *Murex recticornis* Kobelt, ZMB). 392, off Cape Moreton, Queensland, Australia, 42 mm (holotype of *Poirieria kurranula* Garrard, MV F21118. Courtesy of S. BOYD). 393, Queensland, Australia, 39 mm (RH).
394, *C. aculeatus* (Lamarck). Wakayama, Japan, 52 mm (RH).
395, *C. rossiteri* (Crosse). New Caledonia, 26.5 mm (IRSNB IG 10591).
396, *C. crosnieri* Houart. South of Madagascar, 28 mm (paratype, MNHN).
397, *C. aculeatus* (Lamarck). Philippines, 38 mm (Neotype, MNHN).
398, *C. cloveri* Houart. Mauritius, 23 mm (holotype, MNHN).

FIGS 399-405. — Subgenus *C. (Triplex)*.
399-400, *C. boucheti* Houart. 399, New Caledonia, 35 mm (MNHN). 400,New Caledonia, 37.5 mm (MNHN).
401-405, *C. fosterorum* Houart. 401-402, Mzamba, Pondoland, Transkei, South Africa, 36.5 mm (holotype, NM 5343). 403-404, on Aliwal Shoal, south coast of Natal, South Africa, 41 mm (paratype, NM 5343/T 114). 405, on Aliwal Shoal, south coast of Natal, South Africa, 46.5 mm (paratype 86-083, coll. GLASS & FOSTER).

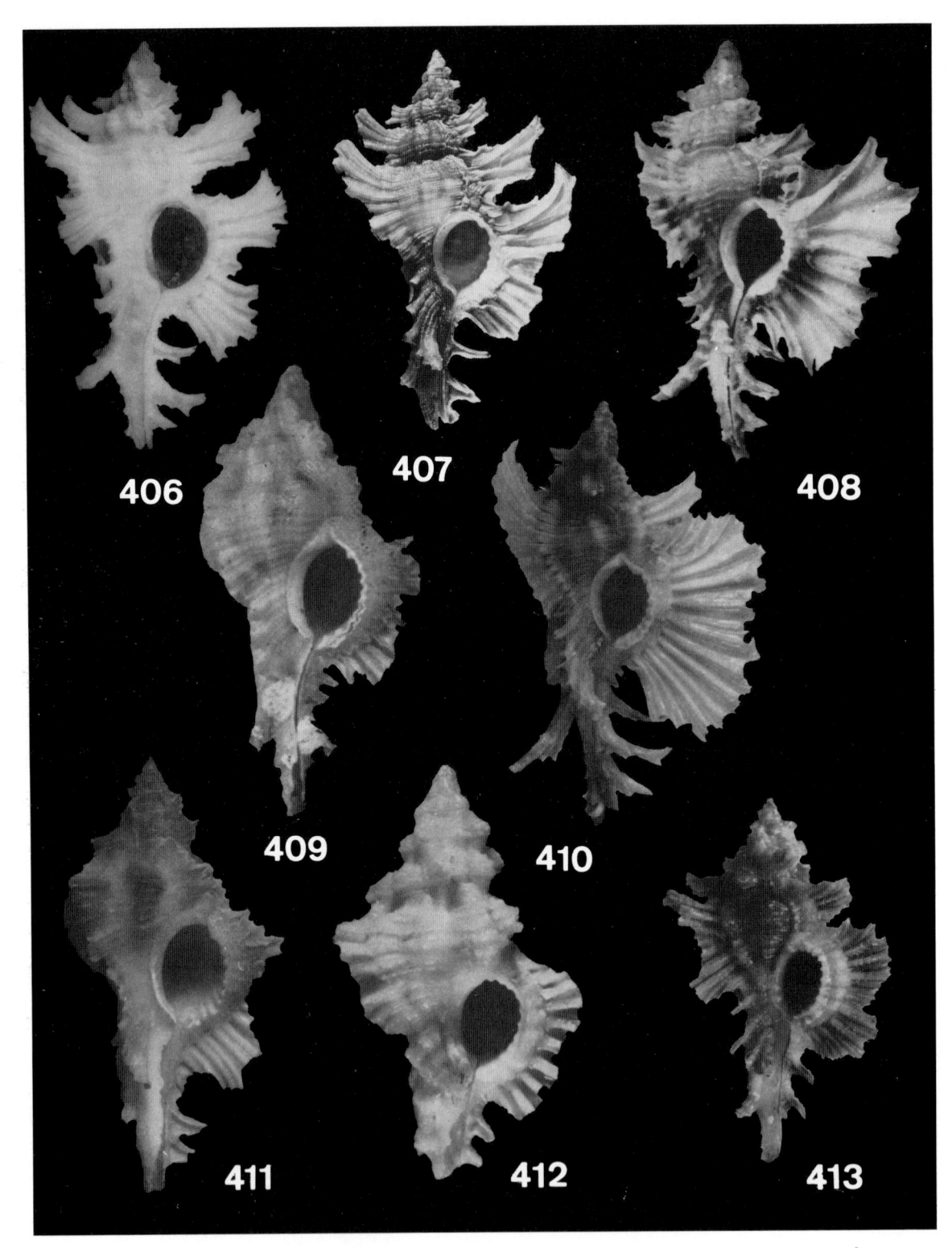

FIGS 406-408. — Subgenus *C. (Chicopinnatus)*.

406-410, *C. orchidiflorus* (Shikama). 406, Taiwan, 40.2 mm (holotype, NSMT 60927. Courtesy of T. OKUTANI). 407, northern-east Taiwan, 30 mm (holotype of *Chicoreus subtilis* Houart, IRSNB IG 25708). 408, off Bohol I., Philippines, 33.6 mm (holotype of *Pterynotus celinamarumai* Kosuge, IMT 80-55. Courtesy of S. KOSUGE). 409, New Caledonia, 24 mm (AMS C147586). 410, Philippines, 39.1 mm (RH).
411, *C. guillei* (Houart). Off Réunion, 29.5 mm (paratype, MNHN).
412-413, *C. laqueatus* (Sowerby). 412, Locality unknown, 30 mm (holotype, UMZ. Courtesy of A. D'ATTILIO). 413, Guam, 34 mm (RH).

FIGS 414-420. — Subgenus *C. (Siratus)*.
414-415, *C. alabaster* (Reeve). 414, Philippines, 141 mm (RH). 415, Philippines, 66.8 mm, juvenile (RH).
416-420, *C. pliciferoides* Kuroda. 416, Philippines, 38 mm, juvenile (RH). 417, Philippines, 38.2 mm, juvenile (RH), 418, Taiwan, 131.2 mm (MNHN). 419, Taiwan, 101.5 mm (RH). 420, northern New Caledonia, 68 mm (MNHN).

FIGS 421-427. — Genus *Chicomurex*.
421-423, *C. laciniatus* (Sowerby). 421, locality unknown, 53.2 mm (lectotype, BMNH 1974072. Courtesy of E.H. VOKES). 422, Philippines, 59.5 mm (MNHN). 423, North Queensland, Australia, 40.1 mm (AMS C116622).
424, *C. superbus* (Sowerby). Coral Sea, 24°02′ S, 159°38′ E, 83.5 mm (MNHN).
425-426, *C. venustulus* (Rehder & Wilson). 425, Papua New Guinea, 34.5 mm (RH). 426, New Caledonia, 47 mm (coll. GLASS & FOSTER).
427, *C. protoglobosus* n.sp. Off S.W. New Caledonia, 30.1 mm, juvenile (holotype, MNHN).

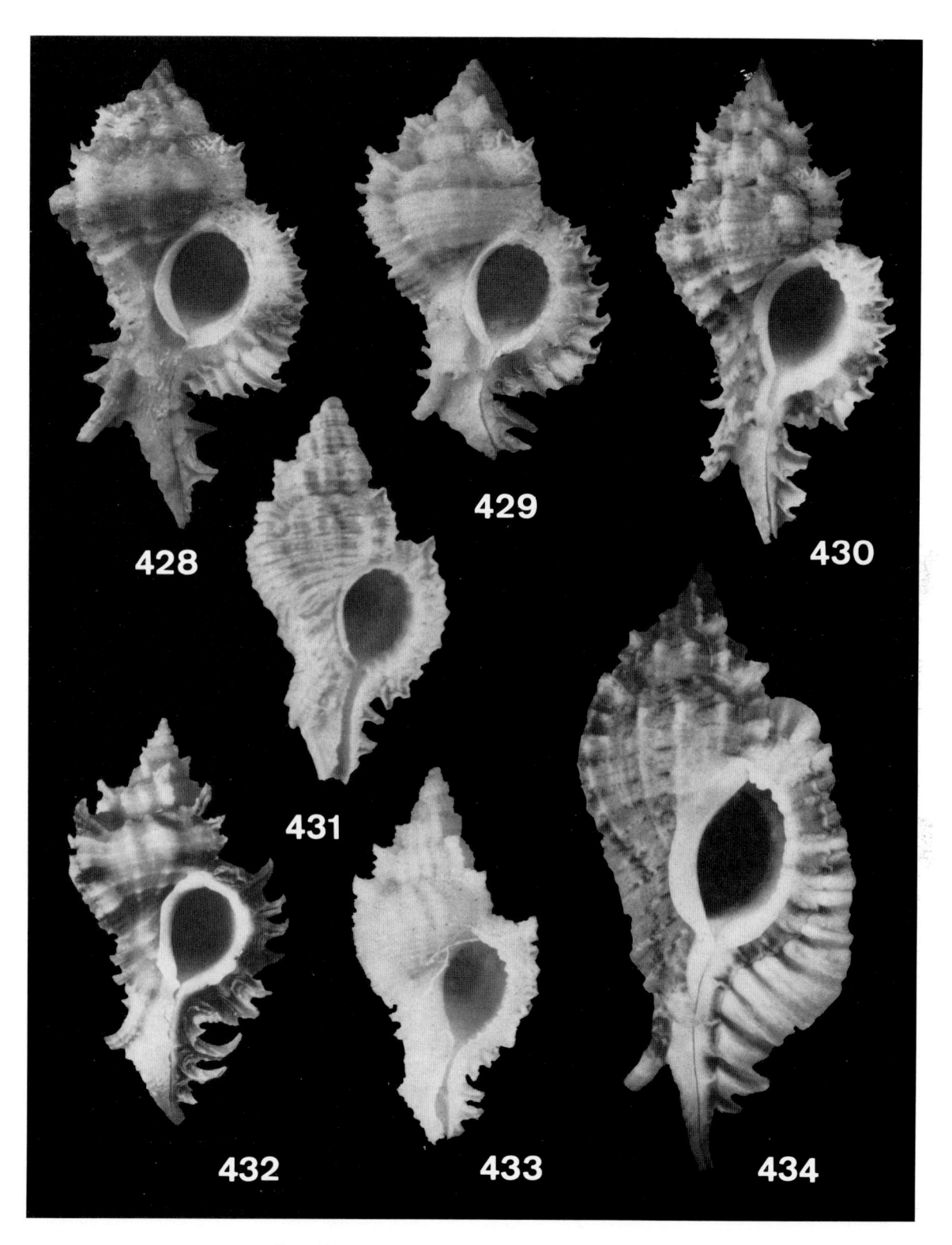

FIGS 428-434. — Genera *Chicomurex* and *Naquetia*.
428-430, *Chicomurex venustulus* (Rehder & Wilson). 428, Philippines, 51.2 mm (RH). 429, Papua New Guinea, 39.6 mm, subfossil (MNHN). 430, Taiwan, 70.1 mm (RH).
431-433, *C. turschi* (Houart). 431, South of New Caledonia, 13.1 mm, juvenile (MNHN). 432, Philippines, 37.5 mm (coll. J. COLOMB). 433, Ifaty Lagoon, Madagascar, 18.1 mm, juvenile (coll. J. DRIVAS).
434, *Naquetia barclayi* (Reeve). Philippines, 94.1 mm (RH).

FIGS 435-442. — Genus *Naquetia*.
435-436, *N. fosteri* D'Attilio & Hertz. Off Eilat, Gulf of Aqaba, Red Sea, 92.2 mm (holotype, SDNHM 91996. Courtesy of A. D'ATTILIO).
437-442, *N. cumingii* (A. Adams). 437-438, locality unknown, 37.2 mm (lectotype of *Murex trigonulus* Lamarck, 1822, MHNG 1099/55/1). 439, Philippines, 53 mm (MNHN). 440, Philippines, 18.5 mm, juvenile (coll. GLASS & FOSTER). 441, locality unknown, 39 mm (paralectotype of *Murex trigonulus* Lamarck, 1822, MHNG 1099/35/2). 442, Seychelles, 42 mm (MNHN).

FIGS 443-449. — Genus *Naquetia*.
443-445, *N. cumingii* (A. Adams). 443, Tulear, Madagascar, 40.1 mm (RH). 444, Dahlak, Red sea, 40 mm (RH). 445, Madat I., Ethiopia, 50.5 mm (coll. D. PELED).
446-447, *N. triqueter* (Born). 446, Papua New Guinea, 55.5 mm (MNHN).447, Philippines, 35.1 mm (holotype of *Murex roseotinctus* Sowerby, BMNH 1974100).
448-449, *N. vokesae* (Houart). 448, Mozambique, 56.2 mm (RH). 449, South east Nacala Bay, northern Mozambique, 66 mm (holotype, NM H213. Courtesy of E.H. VOKES, specimen whitened).

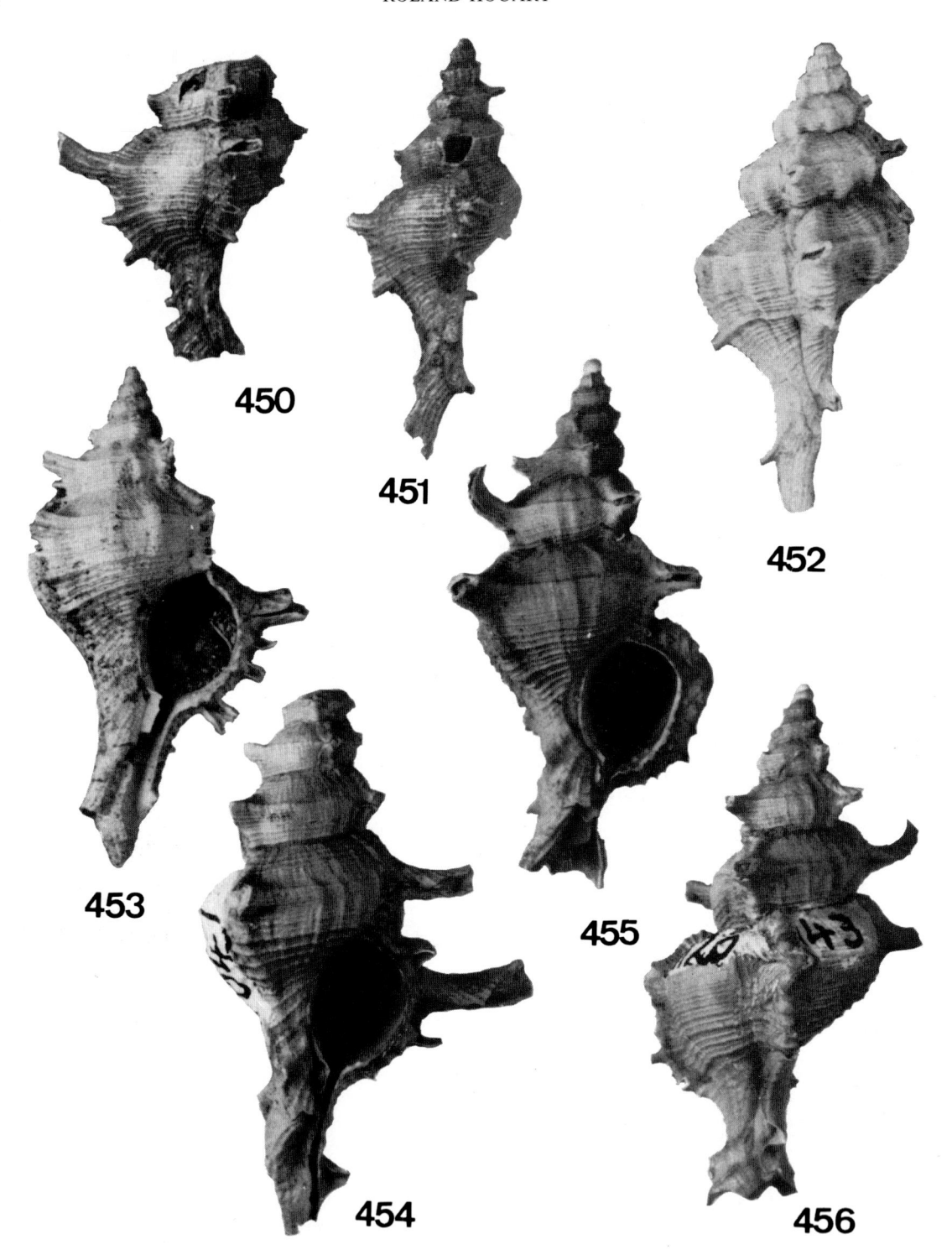

FIGS 450-456. — Subgenus *C. (Triplex)*.
450-451, *C. amblyceras* (Tate). 450, Victoria, Australia, Miocene, 17.5 mm (lectotype, SAM T428A). 451, Victoria, Australia, Miocene, 24.9 mm (paralectotype, SAM T428B).
452-453, *C. basicinctus* (Tate). 452, near Morgan, South Australia, Miocene, 35 mm (lectotype, SAM T417). 453, Clifden Bank, Muddy Creek, Victoria, Australia, Miocene, 32 mm (MV 112058-62).
454-456, *C.* cf. *amblyceras*. 454, Balcombe Bay, Fossil Beach, Mornington, Victoria, Australia, Miocene, 28.1 mm (MV 112039-43). 455-456. Same locality, 29.8 mm (MV 112039-43).

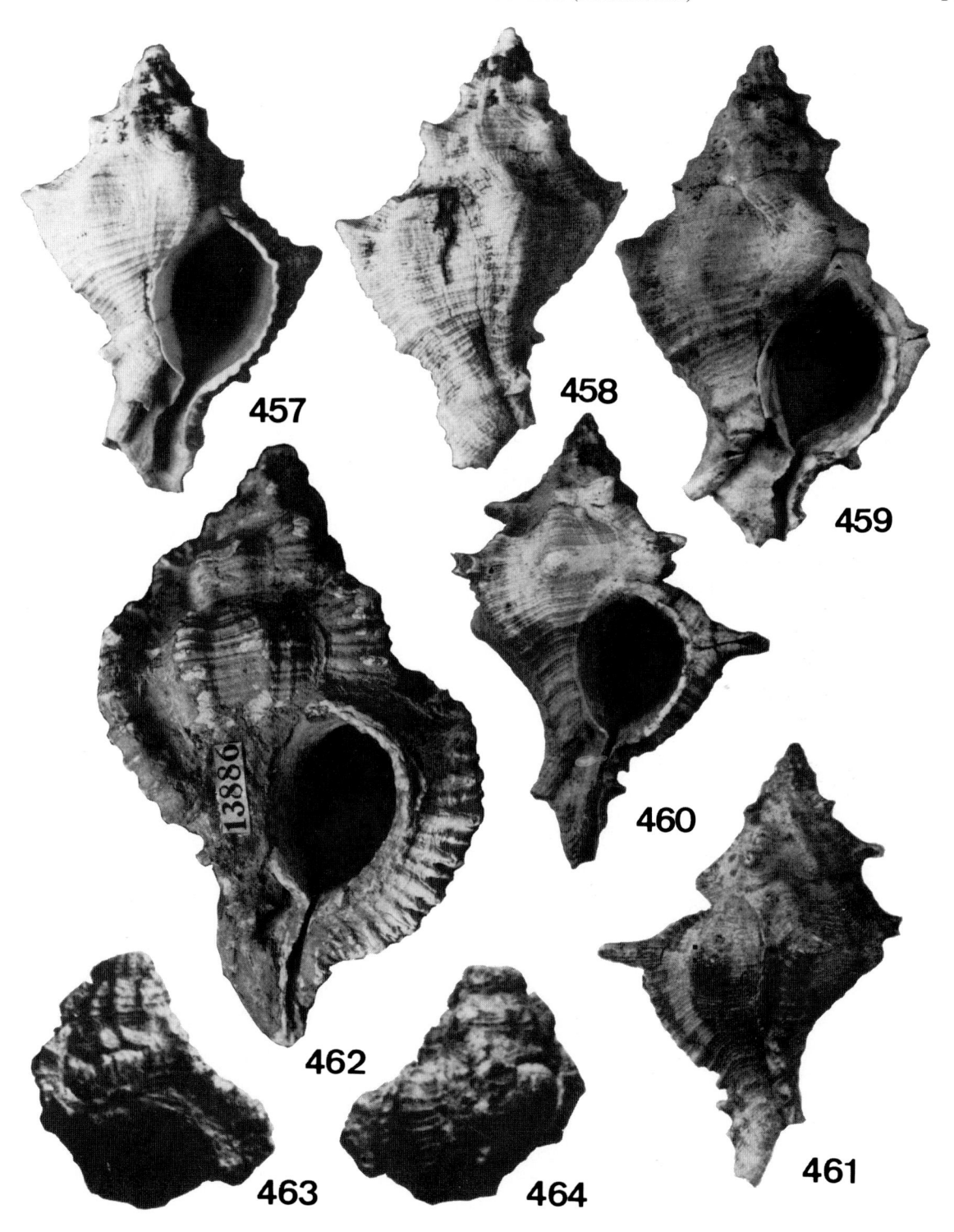

FIGS 457-464. — Subgenus *C. (Triplex)*.
457-459, *C. syngenes* Finlay. 457-458, Clifden, New Zealand, Miocene, 31.1 mm (holotype, AIM TM-164). 459, Calamity Point, Sandstone, Clifden, New Zealand, Miocene, 46 mm (NZGS 10344).
460-461, *C.* cf. *komiticus*. Pakaurangi Pt., Kaipara Harbour, 41.5 mm (NZGS 1161).
462, *C. timorensis* (Tesch). Timor, Miocene, 63 mm (holotype, MGM 13886. Courtesy of Chr. MAUGENEST).
463-464, *C. naricus* (Vredenburg). Nari, India, ?Oligocene, 38 mm (holotype, GSI 12571. Courtesy of S.C. SHAH).

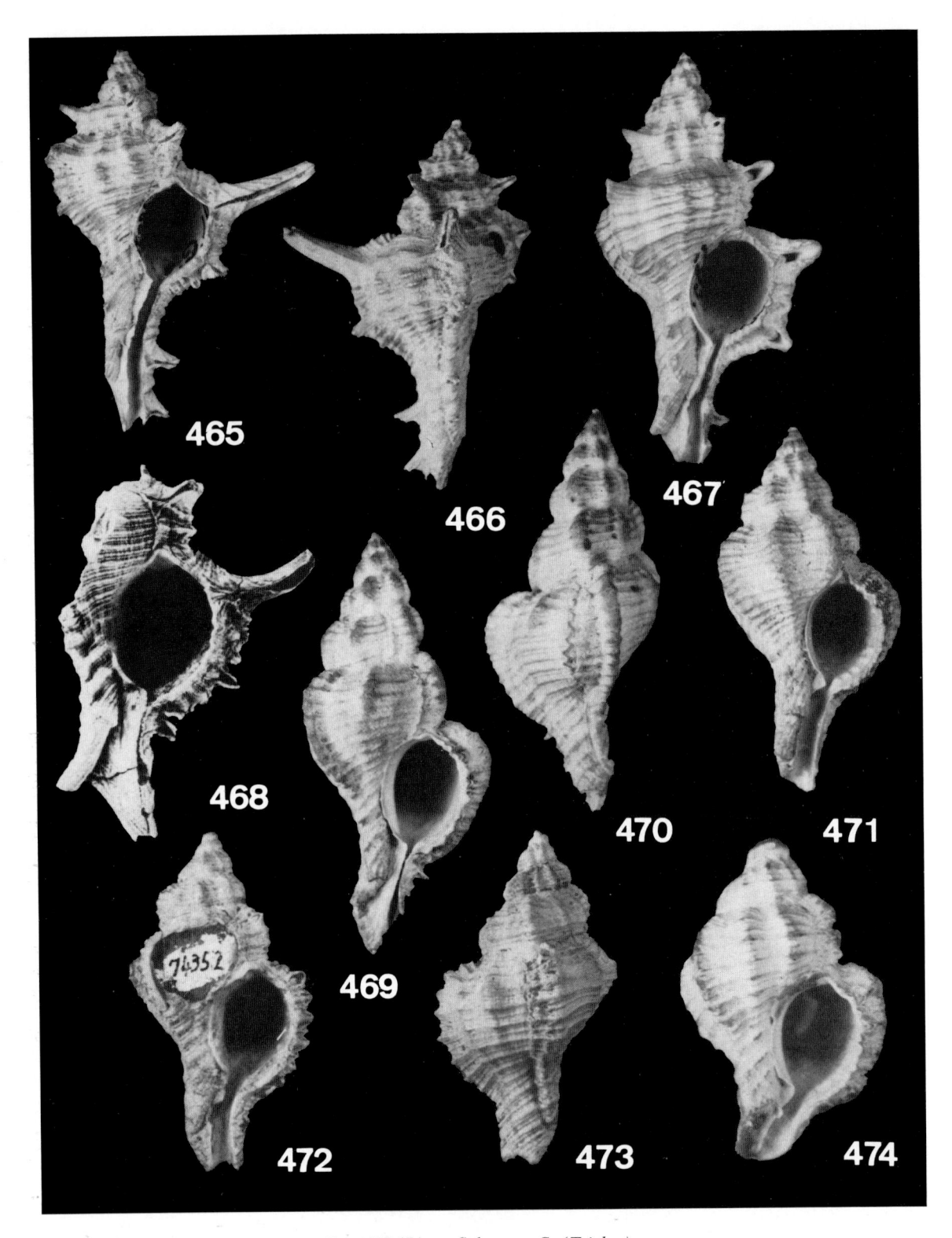

FIGS 465-474. — Subgenus *C. (Triplex)*.

465-467, *C. batavianus* (Martin). 465-466, Batavia, Java, Indonesia, Miocene, 33 mm (lectotype, RML 9689). 467, same locality, 42 mm (paralectotype, RML 9686).
468, *C. komiticus* (Suter). Kaipara Harbour, North Auckland, New Zealand, Lower Miocene, natural size (holotype OM C16.46. Courtesy of J.D. CAMPBELL).
469-471, *C. juttingae* (Beets). 469-470, Borneo, Miocene, 34 mm (holotype, RML 312.451). 471, same locality, 21.5 mm (paratype, RML 312.453).
472-473, *C. tateiwai* Hatai & Kotaka. Hamukyon-pukuton, Heiroku formation, Korea, Lower Miocene, 31.2 mm (holotype, IGPS 74352).
474, *C. karangensis* (Martin). Djilintung, Java, Indonesia, Miocene, 16 mm (lectotype, RML 9683).

FIGS 475-480. — Subgenus *C. (Triplex)*.
475-476, *C. lawsi* (Maxwell). Pakaurangi Point, Kaipara Harbour, New Zealand, Lower Miocene, 27 mm (holotype, NZGS TM 4912, from MAXWELL, 1971, with author's permission).
477-478, *C. rutteni* (Beets). Hill near Sekoerau, East Borneo, Pliocene, 44.5 mm (holotype, RML 113.166).
479-480, *C. lundeliusae* Ludbrook. Roe Plain, South Australia, Early Pleistocene, 83.7 mm (holotype, WAM 70.1137. Courtesy of F. WELLS).

REFERENCES

ANONYMOUS, 1976. — Family Muricidae. *La Conchiglia,* **8** (93-94) : 16-18.
ABBOTT, R. T. & DANCE, S. P., 1982. — *Compendium of Seashells.* New York, E. P. Dutton : i-x, 1-411.
ADAMS, A., 1853. — Descriptions of several new species of *Murex, Rissoina, Planaxis,* and *Eulima* from the Cumingian collection. *Proc. Zool. Soc. London* (1851), **19** : 267-72.
ADAMS, A., 1854. — Descriptions of new shells from the collection of H. Cuming Esq. *Proc. Zool. Soc. London* (1853), **21** : 69-74.
ARAKAWA, K. Y., 1964. — A study on the radulae of the Japanese Muricidae (2). *Venus,* **22** (4) : 355-364.
ASTARY, J. C., 1973. — Marquesan Muricidae. *Hawaiian Shell News,* **21** (5) : 7.
AZUMA, M., 1961. — Descriptions of six new species of Japanese marine Gastropoda. *Venus,* **21** (3) : 296-303.
BAKER, F. C., 1891. — Descriptions of new species of Muricidae, with remarks on the apices of certain forms. *Proc. Rochester Acad. Sci.,* **1** : 129-137.
BANDEL, K., 1984. — The radulae of Caribbean and other Mesogastropoda and Neogastropoda. *Zool. Verhandelingen,* **214** : 1-188, 22 pls.
BEETS, C., 1941. — Eine jungmiocäne molluskan-fauna von der Halbinsel Mangkalihat, Ost-Borneo. *Ver. Geol. Mijn. Nederland Kol. (Geol.),* **13** : 1-218.
BEETS, C., 1950. — Pliocene Mollusca from a coral limestone of a hill near Sekoerau, E. Borneo. *Leidse Geol. Mededelingen,* **15** : 241-264.
BORN, I., 1778. — *Index rerum naturalium Musei Caesarei Vindobonensis,* pt. 1, Testacea. Vienna, J. P. Krauss : XIII + 458 pp.
BORN, I., 1780. — *Testacea Musei Caesarei Vindobonensis,* Vienna, J. P. Krauss : XXXVI + 442 pp.
BOSCH, D. & BOSCH, E., 1982. — *Seashells of Oman,* England, Longman, G. L. : 1-206.
BOSCHEINEN, J., 1983. — Wiedergefundene muriciden-type. *Club Conchylia Informationen,* XV (5) : 16-20.
BOUCHET, P., 1987. — La protoconque des gastéropodes. Thesis, Paris : 1-181.
BOUCHET, P., 1989. — A review of Poecilogony in gastropods. *J. Moll. Stud.,* **55** : 67-78.
BOZZETTI, L., 1991. — Two new Muricidae from Somalia (Muricidae & Muricopsinae). *Conchiglia,* **22** (260) : 43-45.
BRODERIP, W. J. & SOWERBY, G. B., 1833. — Characters of new species of Mollusca and Conchifera collected by H. Cuming. *Proc. Zool. Soc. London,* **2** : 173-179, 194-202.
BROST, F. B. & COALE, R. D., 1971. — *A guide to shell collecting in the Kwajalein Atoll.* Tokyo, C. E. Tuttle Co. : i-xii : 1-157.
BURCH, J. Q., 1955. — A systematic outline of the Muricacea in the Eastern Pacific. *Minutes Conch. Club Southern Calif.,* **149** : 3-13.
CERNOHORSKY, W. O., 1967a. — The Muricidae of Fiji (Mollusca : Gastropoda) Part I — Subfamilies Muricinae and Tritonaliinae. *Veliger,* **10** (2) : 111-132.
CERNOHORSKY, W. O., 1967b. — *Marine Shells of the Pacific.* Sydney, Pacific Publications : 1-248.
CERNOHORSKY, W. O., 1971. — Contribution to the taxonomy of the Muricidae. *Veliger,* **14** (2) : 187-91.
CERNOHORSKY, W. O., 1972. — *Marine Shells of the Pacific II.* Sydney, Pacific Publications : 1-411.
CERNOHORSKY, W. O., 1978a. — The taxonomy of some Indo-Pacific Mollusca (6). *Rec. Auckland Inst. Mus.,* **15** : 67-86.
CERNOHORSKY, W. O., 1978b. — *Tropical Pacific Marine Shells.* Sydney, Pacific Publications : 1-352.
CERNOHORSKY, W. O., 1985. — The taxonomy of some Indo-Pacific Mollusca (12). *Rec. Auckland Inst. Mus.,* **22** : 47-67.
CLENCH, W. J. & PEREZ FARFANTE, I., 1945. — The genus *Murex* in the western Atlantic. *Johnsonia,* **1** (17) : 1-56.
COCHRANE, A., 1980. — *Typhinellas occulusus* (sic) from Northern Queensland. *Hawaiian Shell News,* **28** (10) : 7.
COUCOM, E., 1983. — Recent finds. *Keppel Bay Tidings,* **22** (1) : 1.
COUTURIER, M., 1907. — Étude sur les mollusques gastéropodes recueillis par L. G. Seurat dans les archipels de Tahiti, Paumotu et Gambier. *J. Conchyl.* **55** : 123-78.
COX, L. R., 1948. — Neogene Mollusca from the Dent Peninsula, British North Borneo. *Schweizer. Paleont. Abh.,* **66** (2) : 2-70.
CROSSE, H., 1861. — Description de deux *Murex* nouveaux. *J. Conchyl.,* **9** : 351-354.
CROSSE, H., 1872a. — Diagonoses molluscorum Novae Calendoniae. *J. Conchyl.,* **20** : 69-75.
CROSSE, H., 1872b. — Diagnoses molluscorum novorum. *J. Conchyl.,* **20** : 211-214.
CROSSE, H., 1873. — Descriptions d'espèces nouvelles. *J. Conchyl.,* **21** : 248-54.
D'ATTILIO, A., 1966. — Notes on the Japanese species of the family Muricidae. *Hawaiian Shell News,* **14** (7) : 4-5.

D'ATILLIO, A., 1967. — On the occurence of a labial tooth in the Muricacea. *Hawaiian Shell News,* **15** (7) : 6.

D'ATILLIO, A., 1981. — Note on the holotype of *Murex laqueatus* Sowerby, 1841 (Gastropoda : Muricidae) with a description of a specimen from Guam. *Festivus,* **13** (7) : 78-81.

D'ATILLIO, A., 1988. — On the systematic position of *Murex permaestus,* Hedley, 1915, a muricopsine species. *Festivus,* **20** (1) : 3-6.

D'ATILLIO, A. & HERTZ, C. M., 1987a. — Comparison of *Naquetia annandalei* (Preston, 1910) and *Naquetia barclayi* (Reeve, 1857). *Festivus,* **19** (6) : 56-60.

D'ATILLIO, A. & HERTZ, C. M., 1987b. — A new species of *Naquetia* (Muricidae) from the Gulf of Aqaba. *Veliger,* **30** (2) : 190-195.

D'ATILLIO, A. & MYERS, B. W., 1988. — A note on the distribution of *Siratus pliciferoides* (Kuroda, 1942). *Festivus,* **20** (2) : 11-14.

DE COUET, H. G. & MÜHLHÄUSSER, H., 1983. — Another view of *Murex aculeatus, M. artemis* and *M. rossiteri, Hawaiian Shell News,* **31** (3) : 3.

DODGE, H., 1957. — An historical review of the mollusks of Linnaeus. Pt. 5, *Murex. Bull. Amer. Mus. Nat. Hist.,* **113** (2) : 734-224.

DRIVAS, J. & JAY, M., 1988. — *Coquillages de la Réunion et de l'Ile Maurice.* Neuchâtel, Delachaux et Niestlé : 1-159.

DUNKER, W., 1864. — Funf neue mollusken. *Malak. Blätt.,* **11** : 99-102.

EARLE, J., 1980. — A look at Hawaii's rarest *Murex. Hawaiian Shell News,* **28** (10) : 1.

EISENBERG, J. M., 1981. — *A collector's guide to seashells of the world.* New York, Mc Graw-Hill Book Co. : 1-239.

EUTHYME (Le Frère), 1889. — Description de quelques espèces nouvelles de la faune marine exotique. *Bull. Soc. Malac. France,* **6** : 259-82.

FAIR, R. H., 1973. — A rare *Murex* from the Red Sea. *Hawaiian Shell News,* **21** (6) : 8.

FAIR, R. H., 1974a. — *Chicoreus elliscrossi* — a new name, *Hawaiian Shell News,* **22** (2) : 1 and 5.

FAIR, R. H., 1974b. — " Rediscovered ". *Hawaiian Shell News,* **22** (9) : 12.

FAIR, R. H., 1976. — *The Murex Book, an illustrated catalogue of Recent Muricidae (Muricinae, Muricopsinae, Ocenebrinae),* Honolulu : 1-138, 23 pls.

FINET, Y. & HOUART, R., 1989. — On the taxonomic status of *Murex trigonulus* Lamarck, 1816 and *Murex trigonulus* Lamarck, 1822 and related taxa (Gastropoda, Muricidae). *Apex,* **4** (1-2) : 1-8.

FINLAY, H. J., 1930. — New shells from New Zealand Tertiary beds — part 3, *Trans. N.Z. Inst.,* **61** : 49-86.

FISCHER, P., 1870. — Description d'espèces nouvelles. *J. Conchyl.,* **18** : 176-179.

FISCHER, P. H., 1939. — Sur la dent du labre de *Murex ramosus* L. *J. Conchyl.,* **83** : 39-41.

FRANÇOIS, P., 1891. — Mœurs d'un *Murex. Arch. Zool. Exp. Gen.,* **9** (2) : 240-42.

FREEMAN, D., 1986. — Unraveling the *Murex maurus* mystery, *Hawaiian Shell News,* **34** (12) : 9.

GARRARD, T. A., 1961. — Mollusca collected by M.V. " Challenger " off the east coast of Australia. *J. Malac. Soc. Aust.,* **5** : 3-38.

GMELIN, J. F., 1791. — *Caroli a Linné Systema naturae per regna tria, naturae.* Editio 13, vol. 1 (6), cl. 6. Vermes : pp. 3021-3910. Lipsiae.

GREGORIO, A. DE, 1885. — Studi su talune conchiglie Mediterranee, viventi e fossili con una rivista del gen. *Vulsella* e del gen. *Ficula* e con dei reffronti con specie di altre regioni e di altri bacini. *Boll. Soc. Malac. ital.,* **10** : 247-257.

HABE, T., 1968. — *Shells of the Western Pacific in color,* vol. 2 (revised English edition). Osaka, Hoikusha : 1-233.

HADAR, A., 1968. — Rare *Murex* collected in Red Sea. *Hawaiian Shell News,* **16** (6) : 5.

HARASEWYCH, J., 1973. — More on *Murex barclayi. Hawaiian Shell News,* **21** (4) : 4.

HATAI, K. & KOTAKA, T., 1952. — On some Lower Miocene shells. *Short Papers Inst. Geol. Paleont.,* Tohoku Univ., **4** : 70-86.

HEDLEY, C., 1903. — Scientific results of the trawling expedition of H.M.C.S. " Thetis " off the coast of New South Wales in February and March, 1898, pt. 6. *Mem. Aust. Mus.,* **4** (1) : 326-402.

HEDLEY, C., 1915. — Studies on Australian Mollusca, pt. 12. *Proc. Linn. Soc. New South Wales,* **39** : 695-755.

HIGA, L., 1978. — Recent finds. *Hawaiian Shell News,* **26** (8) : 5.

HINTON, A., 1972. — *Shells of New Guinea and the Central Indo-Pacific.* Milton, Jacaranda Press : i-xviii, 1-94.

HINTON, A., 1979. — *Guide to shells of Papua New Guinea.* Port Moresby, R. Brown & Associates : 1-68.

HOUART, R., 1977. — *Chicoreus subtilis,* espèce nouvelle de la famille des Muricidae. *Inf. Soc. Belge Malac.,* **5** (2) : 13-14.

HOUART, R., 1981a. — New Muricidae named after 1971. *La Conchiglia,* **13** (144-145) : 6-10.

HOUART, R., 1981b. — On the identity of 3 related muricids. *Hawaiian Shell News,* **29** (6) : 7.

HOUART, R., 1981c. — New Muricidae named after 1971. *La Conchiglia,* **13** (148-149) : 16-17.

HOUART, R., 1981d. — *Chicoreus (Chicomurex) turschi,* a new Muricidae from New Guinea. *Nautilus,* **95** (4) : 186-88.

HOUART, R.,1983a. — *C. artemis — rossiteri — nobilis. Hawaiian Shell News,* **31** (7) : 10.

HOUART, R., 1983b. — Three new tropical muricacean species (Gastropoda : Muricidae). *Venus,* **42** (1) : 26-33.

HOUART, R., 1984a. — New Muricidae named after 1971. *La Conchiglia,* **15** (178-179) : 11-13.

HOUART, R., 1984b. — Description of a new species of Muricidae (Mollusca : Gastropoda) from the Eastern African coast : *Chicoreus (Chicoreus) dovi* n. sp. *Venus,* **43** (1) : 55-59.

HOUART, R., 1985a. — Gros plan sur les *Naquetia* (Gastropoda : Muricidae). *Xenophora,* **29** : 8-14.

HOUART, R., 1985b. — Report on Muricidae (Gastropoda) recently dredged in the South-Western Indian Ocean. I. Description of eight new species. *Venus,* **44** (3) : 159-171.

HOUART, R., 1985c. — Report on Muricidae (Gastropoda) recently dredged in the South-Western Indian Ocean. II. List of species with remarks and illustrations. *Venus,* **44** (4) : 239-248.

HOUART, R., 1986a. — New muricids named after 1971. *La Conchiglia,* **18** (206-207) : 8-14.

HOUART, R., 1986b. — *Chicoreus (Naquetia) triqueter vokesae* subs. nov., a new name for a misidentified species (Gastropoda : Muricidae). *Apex,* **1** (3) : 95-96.

HOUART, R., 1986c (" 1985 "). — Résultats des campagnes Musorstom I & II. Philippines. Mollusca Gastropoda : Noteworthy Muricidae from the Pacific Ocean, with description of seven new species. *Mém. Mus. natn. Hist. nat.* (A), **133** : 427-455 (issued 31 March 1986).

HOUART, R., 1987 (" 1986 "). — Description of three new muricid gastropods from the South-Western Pacific Ocean with comments on new geographical data. *Bull. Mus. natn. nat.,* Paris (4), **8** (A, 4) : 757-767 (issued 5 June 1987).

HOUART, R., 1988. — Description of seven new species of Muricidae (Neogastropoda) from the south-western Pacific Ocean. *Venus,* **47** (3) : 185-196.

HOUART, R., 1989. — Description of two new species of the genus *Chicoreus* (Gastropoda : Muricidae) from southern Africa. *Veliger,* **32** (1) : 60-63.

HOUART, R., 1992. — Description of a new species of *Chicoreus* (*s.s.*) (Gastropoda : Muricidae) from Borneo. *Apex,* **7** (1) : 27-30.

HOUART, R. & PAIN, T., 1982. — On the designation of a neotype for *Chicoreus (Chicoreus) torrefactus* (Sowerby jr., 1841) and description of a new species : *Chicoreus (Chicoreus) kilburni* sp. nov. (Gastropoda : Muricidae, Muricinae). *Inf. Soc. Belge Malac.,* **10** (1-4) : 51-56.

HOUART, R. & PAIN, T., 1983a. — The *Chicoreus (Chicoreus) torrefactus* complex (I). *La Conchiglia,* **15** (168-169) : 16-19.

HOUART, R. & PAIN, T., 1983b. — The *Chicoreus (Chicoreus) torrefactus* complex (II). *La Conchiglia,* **15** (170-171) : 1-6.

IREDALE, T., 1915. — A commentary on Suter's " Manual of the New Zealand Mollusca ". *Trans. N.Z. Inst.,* **47** : 417-497.

IREDALE, T., 1936. — Australian molluscan notes 2. *Rec. Aust. Mus.,* **19** : 267-340.

JAZWINSKI, S., 1979. — A range extension for *Murex triqueter. Hawaiian Shell News,* **27** (5) : 11.

JOUSSEAUME, F., 1880. — Division méthodique de la famille des purpuridés. *Le Naturaliste,* **42** : 335-336.

JOUSSEAUME, F., 1881. — Diagnoses de mollusques nouveaux. *Le Naturaliste,* **44** : 349-350.

KAICHER, S. D., 1973, 1974, 1979, 1980. — *Card Catalogue of World-Wide Shells,* Muricidae packs I, II, IV, V. St. Petersburg, Florida, Privately published.

KAY, E. A., 1979. — *Hawaiian Marine Shells.* — Honolulu, Bishop Museum Press : i-xviii, 1-653.

KENNELLY, D. H., 1964. — *Marine Shells of Southern Africa.* S. Africa, Nelson : 1-92, 32 pls.

KIENER, L. C., 1842-43. — *Spécies général et iconographie des coquilles vivantes...* Vol. 7, Rocher (*Murex*) : 1-130, text (1842) ; pls. 1-47 (1843).

KILBURN, R. & RIPPEY, E., 1982. — *Sea Shells of Southern Africa.* South Africa, Johannesburg, Mac Millan : 1-249.

KIRA, T., 1965. — *Shells of the western Pacific in color,* Vol. 1 (English edn). Osaka, Hoikusha : 1-224.

KOBELT, W. & KÜSTER, H. C., 1843-78. — Die geschwantzen und bewehrten purpurschnecken (*Murex, Ranella, Tritonium, Trophon, Hindsia*). Systematisches Conchylien-Cabinet von Martini und Chemnitz. Nuremberg, Von Bauer & Raspe.

KOSUGE, S., 1980. — Descriptions of three new species of the family Muricidae (Gastropoda, Muricacea). *Bull. Inst. Malac. Tokyo,* **1** (4) : 53-58.

KOSUGE, S., 1985a. — Newly recorded or noteworthy molluscs from Okinawa Islands (Ryukyu Group) part 1. *Bull. Inst. Malac. Tokyo,* **2** (2) : 24-27.

KOSUGE, S., 1985b. — Emendations of two specific names. *Bull. Inst. Malac. Tokyo,* **2** (2) : 27.

KOSUGE, S., 1985c. — Noteworthy Mollusca from North-western Australia (1) (Preliminary report). *Bull. Inst. Malac. Tokyo,* **2** (3) : 58-59.

KURODA, T., 1942. — Two Japanese muricids whose names have been preoccupied. *Venus,* **12** (1-2) : 80-81.

KURODA, T., 1964. — A new muricid species from Japan. *Venus,* **23** : 129-130.

KURODA, T., HABE, T. & OYAMA, K., 1971. — *The Sea Shells of Sagami Bay.* Tokyo, Maruzen : i-xix, 1-741 (Japanese text), 1-489 (English text), 1-51 (index).

LAI, K. Y., 1987. — *Marine Gastropods of Taiwan* (2), Taipei, Taiwan Museum : 1-116.

LAMARCK, J. B. P. A. DE M. DE, 1816. — *Tableau encyclopédique et méthodique des trois règnes de la nature,* 23ᵉ partie, mollusques et polypes divers. Paris, H. Agasse : pls. 391-488.

LAMARCK, J. B. P. A. DE M. DE, 1822. — *Histoire naturelle des animaux sans vertèbres.* Vol. 7. Paris, Verdière : 1-232.

LAN, T. C., 1980. — *Rare shells of Taiwan in color.* Taiwan, T. C. Lan, Taipei : 1-143.

LAN, T. C., 1981. — Description of a new sub-species of Muricidae from the Philippines and Taiwan. *Bull. Malac. Soc. Rep. China,* **8** : 11-13.

LEEHMAN, E. G., 1973a. — Identity problems : *Murex (Latirus) barclayi. Hawaiian Shell News,* **21** (1) : 8.

LEEHMAN, E. G., 1973b. — *Murex* from Aqaba. *Hawaiian Shell News,* **21** (8) : 3.

LEEHMAN, E. G., 1974. — Another *Murex barclayi. Hawaiian Shell News,* **22** (5) : 6.

LEEHMAN, E. G., 1976a. — Mystery *Murex. Hawaiian Shell News,* **24** (1) : 9.

LEEHMAN, E. G., 1976b. — Four of a kind. *Hawaiian Shell News,* **24** (5) : 5.

LEEHMAN, E. G., 1977. — The *Murex laqueatus* story. *Hawaiian Shell News*, **25** (4) : 4.
LEEHMAN, E. G., 1978a. — A range extension of ultra-rare *Murex barclayi*. *Hawaiian Shell News*, **26** (2) : 9.
LEEHMAN, E. G., 1978b. — A muddle of *Murex*. *Hawaiian Shell News*, **26** (4) : 1.
LEEHMAN, E. G., 1978c. — A crowd of " little strangers ". *Hawaiian Shell News*, **26** (6) : 3.
LEEHMAN, E. G., 1979. — Range extension for Guam's " endemic " *Murex*. *Hawaiian Shell News*, **27** (3) : 6.
LEEHMAN, E. G., 1980a. — From rarity to availability. *Hawaiian Shell News*, **28** (3) : 14.
LEEHMAN, E. G., 1980b. — From Philippine bottom nets. *Hawaiian Shell News*, **28** (10) : 11.
LEEHMAN, E. G., 1981. — Big and beautiful. *Hawaiian Shell News*, **29** (7) : 9.
LINK, H. F., 1807. — *Beschreibung der Naturalien Sammlung der Universität zu Rostock*, Rostock : 1-160.
LINNE, C. VON, 1758. — *Systema naturae per regna tria naturae*. Editio decima, reformata. Laurentii Salvii, Stockholm, vol. 1, Regnum animale : 1-824.
LÖBBECKE, T. & KOBELT, W., 1879. — Diagnoses neuer Murices. *Jahr. Deutsch. Mal. Gesell.*, **6** : 78-79.
LÖBBECKE, T. & KOBELT, W., 1880. — Museum Loebbeckeanum, *Jahr. Deutsch. Malak. Gesell.*, **7** : 81-82.
LUDBROOK, N. H., 1978. — Quaternary mollusca of the western part of the Eucla Basin. *Geol. Surv. W. Australia Bull*, **125** : 1-286.
MAC DONALD, D. J., 1979. — Sixteen muricids from Kwajalein. *Hawaiian Shell News*, **27** (9) : 7-8.
MACPHERSON, J. H. & GABRIEL, C. J., 1962. — *Marine Molluscs of Victoria*. London, Cambridge Univ. Press : i-xv, 1-475.
MAKIYAMA, J., 1927. — Molluscan fauna of the lower part of the Kakegawa series in the province of Tôtômi, Japan. *Mem. Coll. Sci. Kyoto Univ.*, ser. B, **3** (1) : 1-147.
MARTIN, K., 1884. — Päläontologische Ergebnisse von Tiefbohrugen auf Java. *Samml. Geol. Reichsmus Leiden*; ser. **1** (3) : 43-240.
MARTIN, K., 1895. — Die fossielen von Java. *Geol. Reichsmus. Leiden Samml.*, new ser. **1** (2-5) : 1-132.
MARTINI, F. H., 1777. — *Neues systematisches Conchylien-Cabinet*. Nurnberg. Vol. **3** : i-vi, 1-434.
MAXWELL, P. A., 1971. — Notes on some Cenozoic Muricidae (Mollusca : Gastropoda) from New Zealand, with a review of the genus *Poirieria* Jousseaume, 1880. *N.Z. Journal of Geology and Geophysics*, **14** (4) : 757-781.
MIENIS, H. K., 1983. — *Naquetia annandalei* (Preston, 1910) (Muricidae). *Levantina*, **46** : 538.
MIENIS, H. K., 1984. — Range extension of *Chicoreus kilburni* with a note on the type locality of *Chicoreus maurus*. *La Conchiglia*, **16** (178-179) : 14.
MÖRCH, O. A. L., 1852. — *Catalogus Conchyliorum quae reliquit... Comes de Yoldi*. Klein, Hafniae, pt. 1 : 1-170.
MÜHLHÄUSSER, H. & ALF, A., 1983. — *Chicoreus kengaluae* n. sp. (Muricidae, Prosobranchia). *Spixiana*, **6** (2) : 101-104.
MÜHLHÄUSSER, H. & DE COUET, H. G., 1982. — Beitrage zur kenntnis der arten *Chicoreus aculeatus* (Lamarck, 1822) und *Chicoreus rossiteri* (Crosse, 1872). *Spixiana*, **5** (12) : 35-45.
NICOLAY, K., 1976. — Family Muricidae. — *La Conchiglia*, **8** (93-94) : 16-18.
OKUTANI, T., 1983. — *World seashells of rarity and beauty (Kawamura collection)*. Tokyo, National Science Museum : i-iii, 1-12, 48 pls.
OYAMA, K., 1950. — Studies of fossil molluscan biocoenosis, n° 1, Biocoenological studies on the mangrove swamps with descriptions of new species from the Yatuo Group. *Rept. Geol. Surv. Japan*, **132** : 1-14.
PERRY, G., 1810. — *Arcana, or the museum of natural history ; containing the most recent discovered objects*. London, George Smeeton : 251 unnumbered pages.
PERRY, G., 1811. — *Conchology, or the natural history of shells : containing a new arrangement of the genera and species, illustrated by coloured engravings executed from the natural specimens and including the latest discoveries*. London, Blumer & Co. : 1-4, 61 pls.
PILSBRY, H. A., 1921. — Marine mollusks of Hawaii. *Proc. Acad. Nat. Sci. Phila.*, **72** : 296-328.
POIRIER, J., 1883. — Révision des *Murex* du Muséum. *Nouv. Arch. du Mus. Hist. Nat.*, Paris, (2) **5** : 13-128.
PONDER, W. F. & VOKES, E. H., 1988. — Revision of the Indo-West Pacific fossil and Recent species of *Murex* s.s. and *Haustellum* (Mollusca : Gastropoda : Muricidae). *Rec. Australian Mus.*, Suppl. **8** : 1-160.
PRESTON, H. B., 1910. — Descriptions of five new species of marine shells from the Bay of Bengal. *Rec. Indian Mus.*, **5** : 117-21.
PURTYMUN, B., 1981. — On the reef. *Hawaiian Shell News*, **29** (8) : 6.
QUOY, J. R. C. & GAIMARD, J. P., 1833. — *Voyage de découvertes de « l'Astrolabe » exécuté par ordre du Roi, pendant les années 1826-1827-1828-1829, sous le commandement de M. J. Dumont d'Urville*. Paris, Zoologie, **2** : 321-674.
RADWIN, G. & D'ATILLIO, A., 1976. — *Murex shells of the World, an Illustrated Guide to the Muricidae*. Stanford, University Press : 1-284, 32 pls., 192 text figs.
REEVE, L., 1845-46. — *Conchologia Iconica, or illustrations of the shells of molluscous animals*. London, Reeve : Vol. 3, *Murex*, 36 pls.
REEVE, L., 1849. — *Conchologia Iconica, or illustrations of the shells of molluscous animals. Murex*. London, Reeve : suppl. 1.
REEVE, L., 1858. — Description of seven new shells from the collection of the Hon. Sir David Barclay of Pt. Louis, Mauritius. *Proc. Zool. Soc. London* (1857), **25** : 209-10.
REHDER, H. A. & WILSON, B. R., 1975. — New species of marine mollusks from Pitcairn Island and the Marquesas. *Smiths. Contr. Zool.*, **203** : 1-16.
RICHARDS, A., 1983. — (Untitled) *Hawaiian Shell News*, **31** (7) : 12.
RICHARDS, D., 1981. — *South African Shells*. Cape Town, C. Struik : 1-98.

RIPPINGALE, O. H., 1987. — *Murex Shells*. Published by the author, Margate Beach, Queensland : 1-37.

RÖDING, J. F., 1798. — *Museum Boltenianum*... Hamburg, J. C. Trapp : i-vii, 1-199.

SALVAT, B. & RIVES, C., 1975. — *Coquillages de Polynésie*. Papeete, Les éditions du Pacifique : 1-391.

SCHLÜTER, F., 1838. — *Kurzgefasstes verzeichniss meiner conchyliensammlung nebst Andeutung aller von mir bis jetzt bei halle gefundenen Land und Flussconchylien*, **7** : 1-40.

SCHUMACHER, H. C. F., 1817. — *Essai d'un nouveau système des habitations des vers testacés...* Copenhagen : i-iv, 1-287.

SHARABATI, D., 1984. — *Red Sea Shells*. London, KPI Limited : 1-128.

SHIKAMA, T., 1963. — *Selected shells of the world illustrated in colours*. Tokyo, Hokuryu-Kan. Vol. **1** : 1-154.

SHIKAMA, T., 1964. — *Selected shells of the world illustrated in colours*. Tokyo, Hokuryu-Kan. Vol. **2** : 1-212.

SHIKAMA, T., 1971. — On some noteworthy marine Gastropoda from southwestern Japan (III). — *Sci. Rept. Yokohama nat. Univ.*,(2), **18** : 27-35.

SHIKAMA, T., 1973. — Description of new marine Gastropoda from the East and South China seas. *Sci. Rept. Yokohama nat. Univ.*, (2), **20** : 1-8.

SHIKAMA, T., 1977. — Descriptions of new and noteworty [sic] Gastropoda from Western Pacific and Indian Oceans. *Sci. Rept. Yokohama nat. Univ.*, (2), **24** : 9-23.

SHIKAMA, T., 1978. — Description of new and noteworthy Gastropoda from western Pacific Ocean (I), *Sci. Rept. Yokosuka City Mus.*, **25** : 35-42.

SHORT, J. W. & POTTER, D. G., 1987. — *Shells of Queensland and the Great Barrier Reef, Marine Gastropods*. Bathurst, Robert Brown & Associates : i-vi, 1-135.

SHUTO, T., 1983. — Larval development and geographical distribution of the Indo-West Pacific *Murex*. *Bull. Marine Sci.*, **33** (3) : 536-44.

SMITH, M., 1953. — *An illustrated catalog of the Recent species of rock shells*. Windermere, Tropical Laboratory : i-x, 1-34.

SMYTHE, K., 1982. — *Seashells of the Arabian Gulf*. London, G. Allen & Unwin : 1-123.

SOWERBY, G. B. (1st), 1825. — *A catalogue of the shells contained in the collection of the late Earl of Tankerville... together with an appendix containing descriptions of many new species*. London : 1-92 + 1-34 (appendix).

SOWERBY, G. B. (2nd), 1834-41. — The Conchological illustrations, *Murex*. London, Sowerby : pls. 58-67 ; 1841 : pls. 187-99 and catalogue : 1-9.

SOWERBY, G. B. (2nd), 1841b. — Descriptions of some new species of *Murex*, principally from the collection of H. Cuming Esq. *Proc. Zool. Soc.* London (1840) : 137-147.

SOWERBY, G. B., 1860. — Descriptions of new shells in the collection of H. Cuming. *Proc. Zool. Soc. London* (1859), **27** : 428-429.

SOWERBY, G. B., 1879. — *Thesaurus conchyliorum*. Vol. 4, pts. 33-34. London, Sowerby : 1-55, pls. 380-402.

SOWERBY, G. B. (3rd), 1889. — Descriptions of fourteen new species of shells from China, Japan and the Andaman Islands... *Proc. Zool. Soc. London* (1888), **56** : 565-70, pl. 28.

SPEDEN, I. G. & KEYES, I. W., 1981. — Illustrations of New Zealand fossils. *New Zealand Department of Scientific and Industrial Research, Infor. Series*, **150** : 1-109.

SPRINGSTEEN, F. J., 1982. — The Identity of *Chicoreus crocatus* (Reeve, 1845). *Carfel Philippine Shell News,* **4** (6) : 9-10.

SPRINGSTEEN, F. J. & LEOBRERA, F. M., 1986. — *Shells of the Philippines*. Manila, Carfel Seashell Museum : 1-377.

SPRY, J. F., 1961. — *The Sea Shells of Dar es Salaam*. Tanzania Society, Dar es Salaam : 1-33.

STEARNS, R. E. C., 1893. — Report on the mollusk fauna of the Galapagos Islands with descriptions of new species. *Proc. U.S. Nat. Mus.*, **16** (942) : 353-450.

STOTT, D. E., 1983. — Prize *M. axicornis*, *Hawaiian Shell News*, **31** (2) : 4.

SUGGATE, R. P. & al., 1978. — *The geology of New Zealand*.2 vols. : Wellington, Government Printer : 1-820.

SUTER, H., 1917. — Descriptions of new Tertiary Mollusca occuring in New Zealand, accompanied by a few notes on necessary changes in nomenclature. Part I. *N.Z. Geol. Surv. Paleont. Bull.* **5** : 1-93.

TAPPARONE-CANEFRI, C. M., 1875. — Viaggio dei Signori O. Antinori, O. Beccari ed Ad. Issel nel Mar Rosso,... durante glianni 1870-71, studio monografico sopra I. Muricidi... *Ann. Mus. Civ. Stor. Nat. Genova*, **7** : 560-640.

TATE, R., 1888. — The gastropods of the older Tertiary of Australia. Part I. *Trans. Roy. Soc. Aust.* **10** : 91-176.

TESCH, P., 1915. — Jungtertiäre und Quartäre mollusken von Timor, Teil 1 ; Lieferung 5 (9), Stuttgart, *of Paläontologie von Timor* : 1-70.

TRYON, G. W., 1880. — *Manual of Conchology*, Vol. 2, Muricidae, Purpuridae. Privately published : 1-289.

VAN REGTEREN ALTENA, C. O., 1950. — The marine Mollusca of the Kendeng Beds (East Java). Gastropoda. Part 5 (families Muricidae-Volemidae inclusive). *Leidse Geol. Mededelingen* **15** : 205-240.

VAUGHT, K. C., 1989. — *Chicoreus banksii* (Sowerby, 1841) (Gastropoda : Muricidae). *La Conchiglia*, **21** (233-236) : 3-7.

VOKES, E. H., 1964. — Supraspecific groups in the subfamilies Muricinae and Tritonaliinae (Gastropoda : Muricidae). *Malacologia* **2** : 1-41.

VOKES, E. H., 1965. — Cenozoic Muricidae of the western Atlantic region. Pt. 2. *Chicoreus* s.s. and *Chicoreus (Siratus)*. *Tulane Stud. Geol. Paleont.*, **3** (4) : 181-204.

VOKES, E. H., 1967. — Cenozoic Muricidae of the western Atlantic region. Pt. 3. *Chicoreus (Phyllonotus)*. *Tulane Stud. Geol. Paleont.*, **5** (3) : 133-66.

VOKES, E. H.? 1968. — On the identity of *Murex trigonulus* of authors. *Jour. of Conchology*, **26** (5) : 300-304.

VOKES, E. H., 1970. — Some Comments on Cernohorsky's " Muricidae of Fiji " (The Veliger, 1967). *Veliger,* **13** (2) : 182-187.
VOKES, E. H., 1971. — Catalogue of the genus *Murex* Linné (Mollusca : Gastropoda). Muricinae, Ocenebrinae. *Bull. Am. Paleont.,* **61** (268) : 1-141.
VOKES, E. H., 1973. — *Murex varius* Sowerby, and the systematic validity of the genus *Hexaplex* (Gastropoda : Muricidae). *Of Sea and Shore,* Spring 1973 :15-16.
VOKES, E. H., 1973. — More on *Murex barclayi. Hawaiian Shell News,* **21** (4) : 5.
VOKES, E. H., 1974. — On the identity of *Murex triqueter* Born (Gastropoda : Muricidae). *Veliger,* **16** (3) : 258-263.
VOKES, E. H., 1978. — Muricidae from the eastern coast of Africa. *Ann. Natal Mus.,* **23** (2) : 375-418.
VREDENBURG, E., 1925. — Descriptions of Mollusca from the Post-Eocene Tertiary formation of North-western India : Cephalopoda, Opisthobranchiata, Siphonostomata. *Mem. Geol. Survey India,* **50** (1) : i-xii, 1-351.
WELLS, F. E. & BRYCE, C. W., 1986. — *Seashells of Western Australia.* Perth, Western Australian Museum : 1-207.
WILSON, B. R. & GILLETT, K., 1971. — *Australian Shells.* Sydney, A. H. & A. W. Reed : 1-168.
WOOD, W., 1828. — *Supplement to the Index testaceologicus, or a catalogue of shells, British and foreign...* London, privately printed : i-vi + 59 pp.
WRIGHT, B., 1878. — *Murex huttoniae,* sp. nov. *Ann. Soc. Malac. Belge,* **13** : 85-86, pl. 9.
ZHANG, F. S., 1965. — Studies on the species of Muricidae off the China coasts. I. *Murex, Pterynotus* and *Chicoreus. Studia Marina Sinica,* **8** : 11-24.

INDEX

Species names printed in *italics* are the valid species with the here adopted combination. Other references, including objective or subjective synonyms, preoccupied names, *nomen nudum, nomen dubium*, are printed in boldfaces. Species names preceded by an asterisk are fossils.

abortiva, Triplex 13
Acanthina 34
aculeata Aranea 96
aculeatus, Chicoreus 78, 94, 100
aculeatus, Chicoreus (Chicoreus) 95, 100
aculeatus, Chicoreus (Triplex) 20, 36, 48, 94, **95**, 96, 99, 101, 102, 103, 104, 166
aculeatus, Murex 78, 94, 96, 100, 101, 102
aculeatus, Murex (Chicoreus) 94
aculeatus, Murex (Chicoreus) axicornis 95, 100
aculeatus Muricites 96
adustus, Chicoreus 72, 73
adustus, Murex 72, 158
affinis, Murex 55, 56, 57, 58, 149
akritos, Chicoreus 59, 154
alabaster, Chicoreus (Siratus) 23, **109**, 110, 169
alabaster, Murex 109
alabaster, Murex (Siratus) 109
alabaster, Siratus 109
altenai, Chicoreus (Triplex) **134**, 135, 138
*altenai, Murex 134
amanuensis, Murex (Chicoreus) triqueter var. 128
amblyceras, Chicoreus (Triplex) 24, **135**, 136, 137, 174
*amblyceras, Murex 135
anguliferus, Murex 44, 45
annadalei Naquetia 126
annandalei, Naquetia 126, 127, 128, 130, 131
annandalei, Chicoreus (Naquetia) 126, 130
annandalei, Murex 127, 130
annandalei, Pteronotus 127, 128, 130
aranea, Murex 40
Arca 34
argyna, Murex 52
artemis, Chicoreus 95, 96, 102
artemis, Chicoreus (Chicoreus) 95, 100
artemis, Murex 95, 100
artemis, Murex (Chicoreus) 95, 102
asianus, Chicoreus 36, 37
asianus, Chicoreus (Chicoreus) **36**, 37, 38, 39
asianus, Murex 37
asianus, Murex (Chicoreus) 37
australiensis, Murex 72, 73, 86, 159
australis, Murex 84, 85, 86
austramosus, Chicoreus 38
austramosus, Chicoreus (Chicoreus) 15, **38**, 39, 147
axicornis, Chicoreus 67, 69
axicornis, Chicoreus (Chicoreus) 67, 69
axicornis, Chicoreus (Triplex) 13, 17, 43, 44, 47, **67**, 68, 136, 156, 160
axicornis, Euphyllon 67, 69
axicornis, Murex 41, 67, 68, 69
axicornis, Murex (Chicoreus) 67, 69
banksi, Chicoreus 69
banksii, Chicoreus 69, 70, 107

banksii, Chicoreus (Chicoreus) 69, 70
banksii, Chicoreus (Triplex) 13, 18, 28, 47, 67, 68, **69**, 70, 71, 72, 73, 77, 95, 157, 158, 160
banksii, Murex 69, 70, 107
banksii, Murex (Chicoreus) 69
barclayi, Chicoreus (Naquetia) 126, 127, 130
barclayi, Murex 126, 127, 129
barclayi, Murex (Latirus) 126
barclayi, Murex (Naquetia) 126
barclayi, Naquetia 21, **126**, 128, 131, 171
barclayi, Pterynotus 126
barcleyi, Murex (Naquetia) 126
basicinctus, Chicoreus (Triplex) 24, 135, **136**, 174
*basicinctus, Murex 136
batavianus, Chicoreus (Triplex) **136**, 137, 176
*batavianus, Murex 136
benedictinus, Chicoreus 56, 58, 79
benedictinus, Chicoreus (Chicoreus) 56, 82
benedictinus, Murex 55, 56, 57, 58, 79, 80
bitubercularis, Murex 106, 107, 109, 162
*borni Murex 140
boucheti, Chicoreus 88, 89, 92
boucheti, Chicoreus (Triplex) 20, 25, 28, 48, **88**, 93, 137, 144, 167
bourguignati, Chicoreus (Triplex) 18, 47, 69, **70**, 72, 76, 145, 158
bourguignati, Murex 70
brevifrons, Chicoreus (Triplex) 37, 38, 44
brevifrons, Murex 38
brunnea, Chicoreus 64, 73
brunnea Purpura 64, 72, 73
brunneus, Chicoreus 73
brunneus, Chicoreus (Chicoreus) 73
brunneus, Chicoreus (Chicoreus) brunneus 73
brunneus, Chicoreus (Triplex) 13, 18, 25, 28, 44, 47, 61, 66, **72**, 74, 77, 109, 140, 158, 159
brunneus, Murex (Chicoreus) 72
bundharmai, Chicoreus (Chicoreus) 15, **39**, 40, 149
cancellata, Purpura 131
capuchinus, Murex 35
capucina, Naquetia 107
capucina, Purpura 13
capucinus, Chicoreus 106, 107
capucinus, Chicoreus (Chicoreus) 107
capucinus, Chicoreus (Naquetia) 106
capucinus, Chicoreus (Rhizophorimurex) 13, 19, 25, 28, 69, **106**, 108, 109, 137, 162, 163
capucinus, Murex 106, 107, 108
capucinus, Murex (Naquetia) 107
capucinus, Naquetia 106
carneola, Purpura 13, 56
carneolus, Chicoreus 56
castaneus, Murex 106, 107, 109
celinamarumai, Chicoreus (Chicoreus) 114

celinamarumai, Pterynotus 114, 115, 168
cerinamarumai, Chicoreus 114
cerinamarumai, Pterynotus 114
cervicornis, Chicoreus 89
cervicornis, Chicoreus (Chicoreus) 29
cervicornis, Chicoreus (Triplex) 20, 48, **89**, 91, 137, 160
cervicornis, Euphyllon 89
cervicornis, Murex 89
cervicornis, Murex (Euphyllon) 89
cervicornis, Murex (Murex) 89
Chicomurex 1, 2, 25, 115, 116, 120, 121, 125, 140
Chicopinnatus 1, 2, 35, 36, 112
Chicoreus 1, 2, 13, 25, 34, 35, 36, 46, 58, 73, 92, 106, 109, 112, 113, 115, 125, 134, 136, 138, 139, 140
cloveri, Chicoreus 96
cloveri, Chicoreus (Triplex) 20, 35, 36, 48, **96**, 97, 99, 166
cnissodus, Chicoreus 78
cnissodus, Chicoreus (Chicoreus) 78
cnissodus, Chicoreus (Triplex) 19, 47, **78**, 79, 95, 161
cnissodus, Murex 78
cnissodus, Murex (Chicoreus) 78
colpos, Murex 51, 52
consuela, Chicoreus (Siratus) 127
consuela Murex 128, 131
cornucervi, Chicoreus 41
cornucervi, Chicoreus (Chicoreus) 1, **40**, 41, 148
cornucervi, Euphyllon 41
cornucervi, Murex 41
cornucervi, Murex (Chicoreus) 41
cornucervi, Murex (Euphyllon) 41
cornucervi, Purpura 34, 40
cornucervi, Triplex 69
cornudama, Purpura 13
corrugatus, Chicoreus 81
corrugatus, Chicoreus (Chicoreus) 81
corrugatus, Chicoreus (Triplex) corrugatus 19, 47, **81**, 82, 163
corrugatus, Murex 81
crocatus, Chicoreus 69, 70
crocatus, Chicoreus (Chicoreus) 69
crocatus, Murex 69, 70
crocatus, Murex (Chicoreus) 69
crosnieri, Chicoreus 98
crosnieri, Chicoreus (Triplex) 21, 36, 48, **98**, 99, 105, 144, 166
cumingii Chicoreus (Naquetia) 128
cumingii, Murex 128, 129, 132
cumingii, Naquetia 21, 22, 127, **128**, 129, 131, 133, 172, 173
cyacantha, Murex 45, 46
damicornis, Chicoreus 83, 84
damicornis, Chicoreus (Chicoreus) 83
damicornis, Chicoreus (Triplex) 19, 47, **83** 135, 136, 163, 164
damicornis, Euphyllon 83
damicornis, Murex 83
damicornis, Murex (Chicoreus) 83
damicornis, Torvamurex 83
*dennanti, Chicoreus (Triplex) 23, 135, **136**
*dennanti, Murex 136
denudata, Triplex 34, 81, 84, 85
denudatus, Chicoreus 81, 84, 86
denudatus, Chicoreus (Chicoreus) 84
denudatus, Chicoreus (Triplex) 14, 19, 47, 82, 83, **84**, 85, 135, 138, 163, 164, 165
denudatus, Murex (Chicoreus) 86
denudatus, Torvamurex 84
despectus, Murex 72, 73, 74, 159
dovi, Chicoreus 48
dovi, Chicoreus (Triplex) 16, 35, 47, **48**, 49, 145, 151
*dujardini, Murex 34
elisae, Chicoreus 75
elisae, Chicoreus (Triplex) 47, **75**, 160
elliscrossei, Chicomurex 116
elliscrossi, Chicomurex **116**, 117, 118, 119, 120, 121, 122, 146
elliscrossi, Chicoreus 116
elliscrossi, Chicoreus (Chicomurex) 116
elliscrossi, Murex 116
elliscrossi, Siratus 116
elongata, Purpura 13
elongatus, Murex 37, 38
erithrostomus, Murex 72, 73
erythraeus, Murex 45
ethiopius, Chicoreus 82
ethiopius, Chicoreus corrugatus 82
ethiopius, Chicoreus (Triplex) corrugatus 47, 56, 58, 81, **82**, 83
Euphyllon 34, 84
extraneus, Chicoreus 84
extraneus, Torvamurex 84, 85, 164
falsinii, Chicoreus 83, 84
ferrugo, Murex 44, 46
fiatus, Murex (Pirtus) 34
filialis, Chicoreus 117, 118, 119
filiaris, Chicoreus 117, 118
flavicundus, Chicoreus (Chicoreus) brunneus 73
flavicunda, Triplex 72, 73, 74
flexuosa, Triplex 125, 131
foliatus, Triplex 34, 52
fortispinna, Murex 43
fosteri, Naquetia 21, 127, 128, **130**, 131, 172
fosterorum, Chicoreus 99, 100
fosterorum, Chicoreus (Triplex) 20, 36, 48, **99**, 167
frondosa, Triplex 84, 85
Frondosaria 34
frondosus, Murex 43
fusiformis, Purpura 43
gloriosus, Chicoreus 124
groschi, Chicoreus 70, 76
groschi, Chicoreus (Chicoreus) 76
groschi, Chicoreus (Triplex) 18, 47, 65, 70, 75, **76**, 77, 161
guillei, Chicoreus (Chicopinnatus) 23, **112**, 113, 145, 168
guillei Pterynotus 112
guillei, Pterynotus (Pterynotus) 112
hirasei, Siratus 110, 111
huttoniae, Chicoreus 59, 73
huttoniae, Chicoreus (Chicoreus) 59, 73
huttoniae, Chicoreus brunneus var. 73
huttoniae, Murex 59, 72, 73, 74
immunitus, Torvamurex 84, 85, 86
incarnata, Purpura 43
inflata, Frondosaria 34
inflatus, Murex 43, 44
insularum Chicoreus 50
insularum, Chicoreus (Chicoreus) 50, 51
insularum, Chicoreus (Triplex) 47, 49, **50**, 51, 52, 150
insularum, Murex 51
insularum, Murex torrefactus 50
jickelii, Chicoreus (Naquetia) 128
jickelii, Murex 128, 129
jickelii Naquetia 128
jickelii, Pterynotus (Naquetia) 128
jousseaumei, Murex 59, 60, 152
*juttingae, Chicoreus (Triplex) 24, **137**, 176
*juttingae, Murex (Chicoreus) 137
*karangensis, Chicoreus (Triplex) **137**, 176
*karangensis, Murex 137
kawamurai, Chicoreus 67
kawamurai, Murex 67, 68
*kendengensis, Chicoreus 140
*kendengensis, Chicomurex **140**, 141
kengaluae, Chicoreus 60, 61, 62, 155
kilburni, Chicoreus 52, 56, 57, 58
kilburni, Chicoreus (Chicoreus) 55, 56
*komiticus, Chicoreus (Triplex) **137**, 175, 176
*komiticus, Murex zelandicus var. 137
kurranula, Poirieria 91, 166

laciniatus, Chicomurex 23, 30, **117**, 118, 121, 124, 146, 170
laciniatus, Chicoreus 116, 118, 119
laciniatus, Chicoreus (Chicomurex) 118
laciniatus, Chicoreus (Chicoreus) 118
laciniatus, Murex 116, 117, 118, 128, 129
laciniatus, Murex (Phyllonotus) 118
laciniatus, Naquetia 118
laciniatus, Naquetia cf. 116, 118
laciniatus, Phyllonotus 118, 128
laciniatus, Siratus 118
lactuca, Purpura 13
laqueata, Marchia 113
laqueata, Murex (Marchia) 113
laqueata, Naquetia 113
laqueatus, Chicoreus (Chicopinnatus) **113**, 168
laqueatus, Murex 113
laqueatus, Pterynotus 113
*lawsi, Chicoreus (Triplex) **137**, 177
*lawsi, Murex 137, 138
lignarius, Murex 106, 107, 108, 109, 162
litos, Chicoreus 42, 148
litos, Chicoreus (Chicoreus) 15, **42**, 43, 67, 147
longicornis, Chicoreus (Chicoreus) 91
longicornis, Chicoreus (Triplex) 14, 20, 48, 89, **91**, 92, 166
longicornis, Euphyllon 91
longicornis, Murex 91, 92
longicornis, Murex (Chicoreus) 89, 91
longicornis, Murex (Murex) 91
*lophoessus, Chicomurex 24, **140**
*lophoessus, Murex 140, 141
*lundeliusae, Chicoreus (Chicoreus) 138
*lundeliusae, Chicoreus (Triplex) **138**, 177
maurus, Chicoreus 51, 55, 86, 87
maurus, Chicoreus (Chicoreus) 51, 55
maurus, Chicoreus (Triplex) 15, 47, 49, 50, **51**, 52, 56, 150, 152
maurus, Murex 51, 55
maurus, Murex (Chicoreus) 51, 55
mexicanus, Murex 50, 52, 152
mexicanus, Murex (Chicoreus) palma-rosae 51
microphyllus, Chicoreus 56, 59, 64
microphyllus, Chicoreus (Chicoreus) 56, 59
microphyllus, Chicoreus (Triplex) 13, 14, 16, 17, 26, 35, 47, 48, 49, 58, **59**, 60, 61, 62, 64, 73, 134, 136, 139, 140, 152, 153, 154, 155
microphyllus, Murex 56, 59, 64
microphyllus, Murex (Chicoreus) 59, 64
monodon, Murex 34, 40
multifrondosus, Chicoreus 64
multifrondosus, Murex 63, 64, 65, 155
Murex 24, 92, 138
Naquetia 1, 2, 25, 115, 125
*naricus, Chicoreus (Triplex), **138**, 175
*naricus, Murex (Haustellum) 138
nobilis, Chicoreus 100
nobilis, Chicoreus (Chicoreus) 100
nobilis, Chicoreus (Triplex) 21, 25, 29, 36, 48, 94, 95, 97, 99, **100**, 101, 102, 103, 143
nubilis, Murex 86
oligacanthus, Murex 72, 73, 74
orchidifloris, Chicoreus 114
orchidifloris, Murex 114
orchidiflorus, Chicoreus (Chicopinnatus) 23, 31, 113, **114**, 145, 168
orchidiflorus, Chicoreus 114
orchidiflorus, Chicoreus (Chicoreus) 114
orchidiflorus, Pterynotus 35, 114
orchidiformis, Pterynotus 114
orientalis, Chicoreus 36, 38
paini, Chicoreus 60
paini, Chicoreus (Triplex) 17, 27, 47, **60**, 61, 62, 155

palmarosae, Chicoreus 53
palmarosae, Chicoreus (Chicoreus) 53
palmarosae, Chicoreus (Triplex) 14, 16, 44, 47, 49, **52**, 53, 54, 55, 151
palmarosae, Murex 34, 52, 54
palmarosae, Murex (Chicoreus) 53
palmiferus, Chicoreus 56, 81
palmiferus, Murex 56, 84
palmiferus, Murex (Chicoreus) 84
paucifrondosus, Chicoreus 92
paucifrondosus, Chicoreus (Triplex) 20, 38, **92**, 93, 144
peledi, Chicoreus 58, 79, 80
peledi, Chicoreus (Chicoreus) 79
peledi, Chicoreus (Triplex) 47, 56, **79**, 143, 162
penchinati, Chicoreus 59, 64, 140
penchinati, Chicoreus (Chicoreus) 64
penchinati, Murex 59, 63, 64, 65, 152
penchinati, Murex (Chicoreus) 64
permaestus, Chicoreus 107
permaestus, Murex 106, 107, 109, 162
permaestus, Pterynotus (Naquetia) 107
philippensis, Typhis 136
Phyllonotus 25, 123, 124
Pirtus 34
pliciferoides, Chicoreus 110
pliciferoides, Chicoreus (Siratus) 23, 29, **110**, 111, 169
pliciferoides, Murex 110
pliciferoides, Murex (Siratus) 110
pliciferoides, Siratus 110
pliciferus, Murex 110
poirieri, Murex 59, 152
pomum, Phyllonotus 25
ponderosus, Murex 45, 46
ponderosus, Murex (Chicoreus) anguliferus 45
Potamididae 96
problematica (Chicomurex) 119
problematicum, Chicoreus (Chicomurex) superbus 119
problematicum, Phyllonotus superbus 119
problematicus Chicomurex 116, **119**, 120, 121, 122, 146
problematicus, Chicoreus superbus 119
problematicus, Chicoreus (Chicomurex) superbus 119
propinquus, Murex 110, 111
propinquus, Murex (Siratus) pliciferoides 110
protoglobosus, Chicomurex 1, 23, 31, **120**, 121, 170
Pterynotus 1, 11, 36, 113
quadrifrons, Chicoreus (Chicoreus) 106
quadrifrons, Murex 106, 107, 108, 109, 162
quadrifrons, Murex (Phyllonotus) 106
ramosus, Chicoreus 34, 43
ramosus, Chicoreus (Chicoreus) 15, 39, 40, **43**, 44, 46, 138, 147, 148
ramosus, Murex 11, 34, 43, 44
ramosus, Murex (Chicoreus) 43
recticornis, Chicoreus 91
recticornis, Murex 91, 166
recticornis, Murex (Chicoreus) 91
Rhizophorimurex 1, 2, 35, 106
rochebruni, Murex 55, 56, 57, 58, 149
rosana, Purpura 13
rosaria, Triplex 52
rosarium Purpura 52
rosarius, Chicoreus 53
rosarius, Chicoreus (Chicoreus) 53
rosarius Chicoreus (Triplex) 53
roseotinctus, Murex 131, 173
rossiteri, Chicoreus 95, 101, 102
rossiteri, Chicoreus (Chicoreus) 100, 102
rossiteri, Chicoreus (Triplex) 21, 25, 29, 36, 48, 94, 95, 99, **102**, 103, 104, 105, 144, 166
rossiteri, Murex 102
rossiteri, Murex (Chicoreus) 102

rubescens, Chicoreus 62
rubescens, Chicoreus (Chicoreus) 59
rubescens, Chicoreus (Triplex) 19, 47, **62**, 63, 75, 143, 156
rubescens, Murex 59, 62
rubescens, Murex (Chicoreus) 62
rubicunda, Triplex 72, 73
rubicundus, Chicoreus 72
rubicundus, Murex (Chicoreus) 72
rubiginosus, Chicoreus 56, 58
rubiginosus, Chicoreus (Chicoreus) 56
rubiginosus, Murex 55, 56, 57, 58, 149
rudis, Purpura 44
*rutteni, Chicoreus (Triplex) **138**, 177
*rutteni, Murex (Phyllonotus) 138
ryosukei, Chicoreus 69, 77
ryosukei, Chicoreus (Triplex) 21, 35, 47, 69, 71, **77**, 78, 104, 105
ryukyuensis, Chicoreus (Triplex) 36, 48, 77, 99, 100, **104**, 105, 143, 160
saltatrix, Chicoreus 102, 104
saltatrix, Chicoreus (Chicoreus) 102
sauliae, Murex (Chicoreus) maurus 53, 54
saulii, Chicoreus 54, 56
saulii, Chicoreus (Chicoreus) 54
saulii, Chicoreus (Triplex) 16, 47, 49, 51, **54**, 151, 152
saulii, Murex 54, 56
saulii, Murex (Chicoreus) 54
scabra, Purpura 72, 73
scabrosus, Murex 117, 118
senegalensis, Murex 35
sinensis, Murex 36, 37, 38
sirat, Purpura 35
Siratus 1, 2, 35, 46, 109, 138
steeriae (Chicoreus) 51
steeriae, Chicoreus (Chicoreus) 51
steeriae, Murex 51, 52, 150
strigatus, Chicoreus (Chicoreus) 64
strigatus, Chicoreus (Triplex) 17, 26, 47, 59, 62, **63**, 64, 65, 73, 140, 152, 155
strigatus, Murex 63, 64, 65
subtilis, Chicoreus 114, 115, 168
subpalmatus, Chicoreus 92
subpalmatus, Chicoreus (Triplex) 20, 29, 48, **92**, 93, 94, 144
superbus, Chicomurex 22, 31, 116, 117, 119, **121**, 122, 124, 170
superbus, Chicoreus 121, 124
superbus, Chicoreus (Chicomurex) 121
superbus, Murex 115, 116, 118, 119, 121
superbus, Murex (Phyllonotus) 121
superbus, Phyllonotus 116, 118, 119, 121
superbus, Siratus 119, 121
supersus, Murex (Chicoreus) 121
*syngenes, Chicoreus 139
*syngenes, Chicoreus (Triplex) 137, **139**, 175
*tateiwai, Chicoreus 139
*tateiwai, Chicoreus (Triplex) **139**, 140, 176
territus, Chicoreus 86
territus, Chicoreus (Chicoreus) 86
territus, Chicoreus (Triplex) 19, 47, 84, **86**, 87, 135, 164, 165
territus, Murex 86
thomasi, Chicoreus 86
thomasi, Chicoreus (Chicoreus) 86

thomasi, Chicoreus (Triplex) 19, 47, **86**, 88, 139, 165
thomasi, Murex 86
*timorensis, Chicoreus (Triplex) 138, **139**, 175
*timorensis, Murex 139
tirondus, Murex 13, 82
torrefactus, Chicoreus 51, 55, 56
torrefactus, Chicoreus (Chicoreus) 56
torrefactus, Chicoreus (Triplex) 13, 16, 25, 27, 35, 47, 48, 49, 51, 52, 54, **55**, 56, 57, 58, 59, 73, 80, 81, 108, 109, 149, 151, 153
torrefactus, Murex 51, 55, 57, 58
torrefactus, Murex (Chicoreus) 55
Torvamurex 34
*totomiensis, Chicoreus (Triplex) **139**
*totomiensis, Murex 139
trigonula, Naquetia 128
trigonulus Chicoreus (Naquetia) 128
trigonulus, Murex 128, 129, 131, 132, 172
trigonulus, Murex (Pteronotus) 127, 129, 130
trigonulus, Naquetia 128
trigonulus, Pterynotus (Naquetia) 128
Triplex 1, 2, 25, 34, 46, 86, 125, 134, 137
triqueter, Chicoreus (Naquetia) 131, 133
triqueter, Murex 125, 128, 129, 131, 132, 133
triqueter, Murex (Naquetia) 131
triqueter, Murex (Pteronotus) 128, 131
triqueter, Naquetia 21, 31, 128, **131** 132, 133, 173
triqueter, Pterynotus 128, 131
triqueter, Pterynotus (Naquetia) 128, 131
triquetor, Pterynotus 131, 133
triquetra, Chicoreus (Naquetia) 131
triquetra, Naquetia 131, 133
trivialis, Chicoreus 65
trivialis, Chicoreus (Chicoreus) 64, 65, 76
trivialis, Chicoreus (Triplex) 13, 14, 47, 61, 62, **65**, 66, 75, 77, 156
trivialis, Murex 64, 65, 76
tubulatus, Murex 13
turschi, Chicomurex 22, 30, 120, **123**, 124, 145, 171
turschi, Chicoreus (Chicomurex) 123
Tympanotos 96
variegata, Murex 13
variegata, Purpura 131
venustulus, Chicomurex 22, 30, 116, 117, 119, 121, 122, **124**, 125, 146, 170, 171
venustulus, Chicoreus 124
venustulus, Chicoreus (Chicomurex) 124
venustulus, Siratus 124
versicolor, Murex 72, 73, 74
vicdani, Siratus 110, 111
virgineus, Chicoreus 45
virgineus, Chicoreus (Chicoreus) 15, 27, **44**, 45, 46, 139, 148
virgineus, Murex (?Siratus) 45
virgineus, Purpura 44
virgineus, ?Siratus 45
vokesae, Chicoreus (Naquetia) triquiter 132, 133
vokesae, Naquetia 22, 131, **133**, 134, 173
zululandensis, Chicoreus 105
zululandensis, Chicoreus (Triplex) 20, 36, 48, 99, **105**, 106, 143

ACHEVÉ D'IMPRIMER
EN OCTOBRE 1992
SUR LES PRESSES
DE
L'IMPRIMERIE F. PAILLART
À ABBEVILLE

Date de distribution : 20 octobre 1992.
Dépôt légal légal : octobre 1992.
N° : d'impression : 8096.